Smart Tools for Smart Applications

Smart Tools for Smart Applications: New Insights into Inorganic Magnetic Systems and Materials

Editors

Francesca Garello
Roberto Nisticò
Federico Cesano

MDPI • Basel • Beijing • Wuhan • Barcelona • Belgrade • Manchester • Tokyo • Cluj • Tianjin

Editors
Francesca Garello
University of Torino
Italy

Roberto Nisticò
Polytechnic of Torino
Italy

Federico Cesano
University of Torino
Italy

Editorial Office
MDPI
St. Alban-Anlage 66
4052 Basel, Switzerland

This is a reprint of articles from the Special Issue published online in the open access journal *Inorganics* (ISSN 2304-6740) (available at: https://www.mdpi.com/journal/inorganics/special_issues/Inorganic_Magnetic_Systems).

For citation purposes, cite each article independently as indicated on the article page online and as indicated below:

LastName, A.A.; LastName, B.B.; LastName, C.C. Article Title. *Journal Name* **Year**, *Volume Number*, Page Range.

ISBN 978-3-0365-0234-2 (Hbk)
ISBN 978-3-0365-0235-9 (PDF)

Cover image courtesy of Federico Cesano.

© 2021 by the authors. Articles in this book are Open Access and distributed under the Creative Commons Attribution (CC BY) license, which allows users to download, copy and build upon published articles, as long as the author and publisher are properly credited, which ensures maximum dissemination and a wider impact of our publications.

The book as a whole is distributed by MDPI under the terms and conditions of the Creative Commons license CC BY-NC-ND.

Contents

About the Editors vii

Francesca Garello, Roberto Nisticò and Federico Cesano
Smart Tools for Smart Applications: New Insights into Inorganic Magnetic Systems and Materials
Reprinted from: *Inorganics* **2020**, *8*, 56, doi:10.3390/inorganics8100056 1

Roberto Nisticò, Federico Cesano and Francesca Garello
Magnetic Materials and Systems: Domain Structure Visualization and Other Characterization Techniques for the Application in the Materials Science and Biomedicine
Reprinted from: *Inorganics* **2020**, *8*, 6, doi:10.3390/inorganics8010006 7

Panagiota S. Perlepe, Diamantoula Maniaki, Evangelos Pilichos, Eugenia Katsoulakou and Spyros P. Perlepes
Smart Ligands for Efficient 3d-, 4d- and 5d-Metal Single-Molecule Magnets and Single-Ion Magnets
Reprinted from: *Inorganics* **2020**, *8*, 39, doi:10.3390/inorganics8060039 67

Irene Fernández-Barahona, María Muñoz-Hernando, Jesus Ruiz-Cabello, Fernando Herranz and Juan Pellico
Iron Oxide Nanoparticles: An Alternative for Positive Contrast in Magnetic Resonance Imaging
Reprinted from: *Inorganics* **2020**, *8*, 28, doi:10.3390/inorganics8040028 113

Anastasiia A. Kozlova, Sergey V. German, Vsevolod S. Atkin, Victor V. Zyev, Maxwell A. Astle, Daniil N. Bratashov, Yulia I. Svenskaya and Dmitry A. Gorin
Magnetic Composite Submicron Carriers with Structure-Dependent MRI Contrast
Reprinted from: *Inorganics* **2020**, *8*, 11, doi:10.3390/inorganics8020011 135

Fabio Carniato and Giorgio Gatti
^{1}H NMR Relaxometric Analysis of Paramagnetic Gd_2O_3:Yb Nanoparticles Functionalized with Citrate Groups
Reprinted from: *Inorganics* **2019**, *7*, 34, doi:10.3390/inorganics7030034 147

Marcos E. Peralta, Santiago Ocampo, Israel G. Funes, Florencia Onaga Medina, María E. Parolo and Luciano Carlos
Nanomaterials with Tailored Magnetic Properties as Adsorbents of Organic Pollutants from Wastewaters
Reprinted from: *Inorganics* **2020**, *8*, 24, doi:10.3390/inorganics8040024 157

Lisandra de Castro Alves, Susana Yáñez-Vilar, Yolanda Piñeiro-Redondo and José Rivas
Efficient Separation of Heavy Metals by Magnetic Nanostructured Beads
Reprinted from: *Inorganics* **2020**, *8*, 40, doi:10.3390/inorganics8060040 185

About the Editors

Francesca Garello obtained her Ph.D. in Pharmaceutical and Biomolecular Sciences in 2015 at the Department of Molecular Biotechnology and Health Sciences of the University of Turin, Turin, Italy. She is currently working in the group of Professor E. Terreno at the molecular and preclinical imaging center in Turin as a member of the research team in the development and testing of innovative molecular imaging probes. Her interest is mainly focused on the visualization and monitoring of inflammatory processes using 1H and 19F magnetic resonance and optical and photoacoustic Imaging. Most of her research activities deal with the active targeting and tracking of the immune system cells in vivo, the visualization of the inflamed endothelium, and cell surveillance after transplantation using newly synthesized nano- and microsystems.

Roberto Nisticò obtained his Ph.D. in Chemical and Materials Sciences at the University of Torino (Department of Chemistry, Italy). His research is focused on several aspects at the interface between nanotechnology and materials science, always looking for novel and appealing solutions for a sustainable future. His principal fields of interest are magnetic and/or metallic nanomaterials, functional/porous coatings, plasma treatments, biomaterials (for biomedical applications), valorization of natural resources, (bio)polymers and carbons, nanomaterials for photocatalysis, and AOPs.

Federico Cesano received his Degree in Chemistry in 1999 at the University of Torino. After spending two years at the Italian National Research Council (2000–2002), he completed his Ph.D. in Material Science and Technology in 2005. Since 2006, he has been working at the Chemistry Dept. of the University of Turin. He is co-author of more than 70 papers and several book chapters published in the main journals of chemistry and materials science. His main research interests are 1D, 2D, and 3D nanostructured materials (including oxides, carbon nanomaterials, transition metal dichalcogenides, polymers), either alone or combined to form hybrid structures and composites.

Editorial

Smart Tools for Smart Applications: New Insights into Inorganic Magnetic Systems and Materials

Francesca Garello [1], Roberto Nisticò [2,†] and Federico Cesano [3,*]

[1] Department of Molecular Biotechnology and Health Sciences, Molecular and Preclinical Imaging Centers, University of Torino, Via Nizza 52, 10126 Torino, Italy; francesca.garello@unito.it
[2] Department of Applied Science and Technology DISAT, Polytechnic of Torino, C.so Duca degli Abruzzi 24, 10129 Torino, Italy; roberto.nistico0404@gmail.com
[3] Department of Chemistry, University of Torino, Via P. Giuria, 7, 10125 Torino, Italy
* Correspondence: federico.cesano@unito.it; Tel.: +39-011-6707548
† Current address: Independent Researcher, Via Borgomasino 39, 10149 Torino, Italy.

Received: 22 September 2020; Accepted: 29 September 2020; Published: 10 October 2020

Abstract: This Special Issue, consisting of four reviews and three research articles, presents some of the recent advances and future perspectives in the field of magnetic materials and systems, which are designed to meet some of our current challenges.

Keywords: magnetic materials; magnetic particle and nanoparticles; single-molecule magnets; molecular magnetism; magnetic separation; magnetic resonance imaging; MRI contrast agents; magnetic domain visualization; paramagnetic properties; magnetically-guided drug delivery systems

---

In the recent years, the research in the field of magnetic materials and systems has been very active as documented by the increasing number of contributions (Figure 1). Micro/nanosystems with magnetic properties have been extensively investigated in many fields, ranging from physics and chemistry to mathematics and medicine. The research is consequently very broad and multidisciplinary, from basic studies to more applicative contributions (Figure 2).

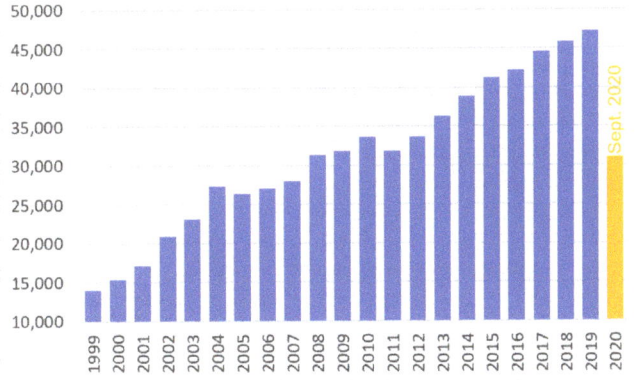

Figure 1. Number of documents published in the last 10 years (source: Scopus).

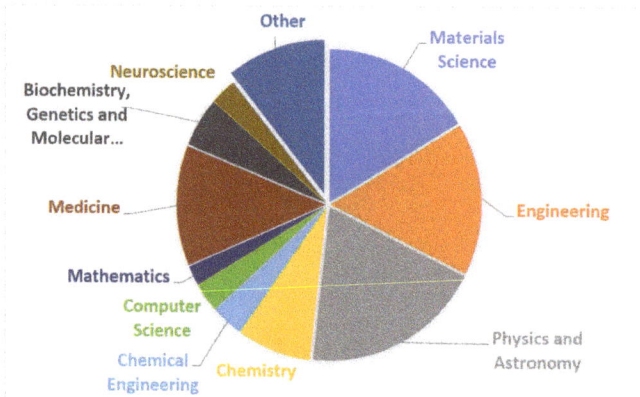

Figure 2. Subject areas of contributions dedicated to magnetic materials and systems (source: Scopus).

The research in these areas has recently shown that if the magnetic compounds are opportunely functionalized and modified with moieties and specific functional groups, a plethora of challenging multidisciplinary applications is available, including the development of magnetically-controlled particles, stimuli-responsive materials, magnetically-guided chemical/drug-delivery systems, sensors, spintronics, separation and purification of contaminated groundwater and soils, ferrofluids and magnetorheological fluids, contrast agents for MRI, and internal sources of heat for the thermo-ablation of cancer. Magnetic compounds have been found to be highly selective and effective in all these application fields, from the molecular to the microscale level. Furthermore, the research on new magnetic systems is very active as documented by recent achievements. Such systems—for example, two-dimensional magnetic materials [1], ferrofluid droplets exhibiting reversible paramagnetic-to-ferromagnetic transformation [2], and oxide heterostructures containing cation defects able to tune magnetism [3]—can be considered materials at the frontiers, which will receive growing attention in the coming years. This Special Issue aims at underlining the latest advances in the field of magnetic compounds, nanosystems, and materials, covering a large variety of topics related to novel synthesis and functionalization methods, properties, applications, and use of magnetic systems in chemistry, materials science, diagnostics, and medical therapy.

The present Special Issue, composed of four reviews and three research articles, showcases some of the latest achievements and future perspectives in the field of the magnetic materials and systems designed to meet some of our present challenges.

Nisticò et al. [4] reviewed the subjects of the domain structure visualization and other characterization techniques to be applied in materials science and biomedicine. In the review, the current understanding of the usage, advances, advantages, and disadvantages of many techniques currently available to investigate magnetic systems are presented with the aim to help the reader in the choice of the most suitable methodology. Due to the multidisciplinary approach characteristic of these studies, in most cases, these very specific characterization techniques are, for a fact, little known (or fully unknown) to most of the users. In the present review, the characterization techniques were classified into three sections and properly discussed with examples from the literature. Section I is dedicated to the definitions of magnetism and magnetization (hysteresis) techniques. Section II is dedicated to the morphological aspects, thus illustrating all the different visualization methods of magnetic domains. Finally, Section III is dedicated to the principal physicochemical characterization methods, with a final section particularly devoted to biomedical applications, including the exploitation of magnetism in imaging for cell tracking/visualization of pathological alterations in living systems (mainly by magnetic resonance imaging, MRI).

Among all fields of magnetism, single-molecule magnets (SMMs) and single-ion magnets (SIM) belong to an extremely interesting and innovative branch of modern magnetism. Perlepe et al. [5] reviewed a few inorganic and organic ligands in the chemistry of 3D-, 4D-, and 5D-metal SMMs and SIMs, through selected examples. Azide ion, cyanido group, tris(trimethylsilyl)methanide, cyclopentanienido group, soft (based on the Hard-Soft Acid-Base model) ligands, metallacrowns combined with click chemistry, deprotonated aliphatic diols, and the family of 2-pyridyl ketoximes including some of its elaborate derivatives are the selected ligands to be discussed with particular emphasis on the rationale behind the selection of the ligands. As underlined by the authors, the contribution is not an exhaustive and comprehensive review of the field, but rather takes a simple approach to the topic without containing large amounts of structural and magnetic information, synthetic discussions and chemical equations. A reader with a good general chemical background will find this material very accessible. Finally, current interests, actual limitations in the field, and perspectives are highlighted.

Fernández-Barahona et al. [6] reviewed the use of iron oxide nanoparticles (IONPs) as positive contrast agents for MRI. The authors highlighted the increasing interest in the development of innovative positive MRI contrast agents, due to the toxicity and retention issues associated with routinely administered Gd-based contrast agents [7]. After an overview of the mechanism of T_1 (longitudinal or spin lattice relaxation time)-based MRI contrast and a critical survey on the most remarkable Gd- and Mn-based nanosystems, the authors discussed the main physicochemical properties that IONPs must possess to act as T_1 agents, i.e., ultrasmall core size with moderate crystallinity (usually maghemite (γ-Fe_2O_3)) and high colloidal stability with hydrodynamic sizes ranging from 5 to 20 nm. The synthetic procedures useful to achieve these properties are then clearly summarized and are thus easily accessible to the readers. Finally, the authors reported the main in vivo applications of T_1-IONPs, not only for MRI but also for multimodal imaging, highlighting that even if longitudinal relaxivity values of IONPs are still far from those of some Gd nanoparticles, there is great potential in the development of these systems, given the status of the area as an emerging research field. Of course, biocompatibility, pharmacokinetics, and delivery pathways must be studied in advance to guarantee their clinical translation.

In this context, Kozlova et al. [8] reported the possibility of modulating the T_1 or T_2 (transversal or spin–spin relaxation time) contrast generated by submicron carriers containing Fe_3O_4 particles, according to their core-shell structure. The authors synthesized three different magnetic submicron core–shells, displaying a single layer of magnetite in the shell and various amounts of Fe_3O_4 particles in the core. They found that all three systems act as dual T_1/T_2 contrast agents. Remarkably, the highest T_1 and T_2 contrast in gradient echo mode can be observed from the core–shell suspension with magnetite nanoparticles contained only in the shell [9]. The addition of magnetite nanoparticles in the core, in fact, seems to impair the contrast properties due to an increase in packing density of magnetite nanoparticles and in the number of interactions between them. However, in the T_1 spin-echo mode, surprisingly the tendency is the inverse, with the greatest T_1 signal enhancement displayed by submicron carriers with one layer of magnetite and four loadings of Fe_3O_4 particles in the core. The authors thus practically proved that different combinations of MRI acquisition modalities and submicron magnetite carrier structures enabled magnetic systems suitable for both T_1 and T_2 MRI that can be also controlled and delivered to the site of interest by an external magnetic field.

Carniato and Gatti [10] contributed to the Special Issue with an interesting research article dealing with Gd_2O_3 nanoparticles doped with various amounts of Yb^{3+}. These mixed oxide nanoparticles were already proposed as a potential dual computed tomography (CT) and positive MRI contrast agent [11]. Carniato and Gatti proposed a cheap and fast co-precipitation synthesis procedure along with functionalization of the particle surface with citrate molecules, in order to confer high hydrophilicity, improve stability, and increase the interaction of the metal ions exposed on the surface with the water molecules. The relaxometric study carried out on the developed nanosystem displayed high relaxivity values at a high magnetic field (with a maximum close to 60 MHz) with respect to the clinically used Gd^{3+}-chelates and comparable to those of similar nanosytems. These features, together with the chemical stability of the nanoparticles in biological fluid and in the presence of a chelating agent, make these nanoparticles suitable for dual MRI-CT diagnostic analyses.

Peralta et al. [12] reviewed the most promising magnet-responsive nanomaterials used in groundwater and wastewater remediation processes. In particular, the authors proposed an overview of the main relevant synthetic methods, surface properties, and clean-up adsorption applications associated with magnetic core–shell nanoparticles and nanocomposites. The discussion is organized into five main sections. Section I is dedicated to silica-based materials, with a specific focus on the incorporation mechanisms of magnetic species (i.e., metallic iron and iron oxides) into silica structures (acting as functional coatings) to produce core-shell systems with freely available functionalities at the surface (namely, silanols and further modifications), as well as on magnetic nanocomposites made of magnetic nanoparticles dispersed in mesoporous silica matrices and hollow particles. Section II is dedicated to clay-based materials, with a specific focus on the incorporation of magnetic nanoparticles within the clays' porous system. Section III is dedicated to carbon-based materials with a particular emphasis on magnetic carbon hybrid nanocomposites. Section IV is dedicated to polymer-based materials, where polymers are chemically anchored or physically adsorbed at the surface of magnetic nanoparticles to form core–shell systems. Lastly, Section V is dedicated to the production of waste-derived magnetic systems produced by means of incorporation processes involving the functionalization of magnetic species (e.g., iron oxides) with waste-derived substances isolated from agricultural residues and biowaste, paving the way for the concept of "waste for cleaning waste", in line with the guide-principles of the circular economy.

In this context, the study reported by de Castro Alves et al. [13] is focused on the production and testing of magnetic alginate activated carbon beads for the removal of heavy metals (i.e., Cd(II), Hg(II), and Ni(II)) from aqueous environments. The study investigated the effect in terms of sorption capacity over different experimental conditions (pH, recycling, and reusability) for mono-metallic systems, as well as the competitive interactions in ternary systems (thus simulating the composition of a real wastewater derived from industrial and mining effluents). Results established a higher affinity of the tested material for Cd(II) ions in both mono-metal and ternary systems, whereas recycling experiments demonstrated that magnetic beads are re-usable for at least five consecutive adsorption/desorption cycles.

We truly hope that the contributions published within this Special Issue can help readers to increase their knowledge in the field of magnetic systems, providing inspiration for novel relevant publications. In this regard, we thank the authors for their valuable contributions; the referees for their insightful and appropriate comments, of paramount importance to enhance the scientific standard of this Special Issue; and the editorial staff, for their constant and unparalleled support.

Author Contributions: The editorial was written through contributions of all authors. All authors have read and agreed to the published version of the manuscript.

Funding: This work was supported by MIUR (Ministero dell'Istruzione, dell'Università e della Ricerca), INSTM Consorzio and NIS (Nanostructured Interfaces and Surfaces) Inter-Departmental Centre of University of Torino.

Acknowledgments: We thank all authors, reviewers, and editors who assisted in the present Special Issue. We thank Min Su, Silivia Luo, Edward Zhang, and all the editorial staffs who assisted us.

Conflicts of Interest: The authors declare no conflict of interest.

References

1. Mak, K.F.; Shan, J.; Ralph, D.C. Probing and controlling magnetic states in 2D layered magnetic materials. *Nat. Rev. Phys.* **2019**, *1*, 646–661. [CrossRef]
2. Liu, X.; Kent, N.; Ceballos, A.; Streubel, R.; Jiang, Y.; Chai, Y.; Kim, P.Y.; Forth, J.; Hellman, F.; Shi, S.; et al. Reconfigurable ferromagnetic liquid droplets. *Science* **2019**, *365*, 264–267. [CrossRef] [PubMed]
3. Park, D.S.; Rata, A.D.; Maznichenko, I.V.; Ostanin, S.; Gan, Y.L.; Agrestini, S.; Rees, G.J.; Walker, M.; Li, J.; Herrero-Martin, J.; et al. The emergence of magnetic ordering at complex oxide interfaces tuned by defects. *Nat. Commun.* **2020**, *11*, 3650. [CrossRef] [PubMed]

4. Nisticò, R.; Cesano, F.; Garello, F. Magnetic materials and systems: Domain structure visualization and other characterization techniques for the application in the materials science and biomedicine. *Inorganics* **2020**, *8*, 6. [CrossRef]
5. Perlepe, P.S.; Maniaki, D.; Pilichos, E.; Katsoulakou, E.; Perlepes, S.P. Smart Ligands for Efficient 3d-, 4d- and 5d-Metal Single-Molecule Magnets and Single-Ion Magnets. *Inorganics* **2020**, *8*, 39. [CrossRef]
6. Fernández-Barahona, I.; Muñoz-Hernando, M.; Ruiz-Cabello, J.; Herranz, F.; Pellico, J. Iron Oxide Nanoparticles: An Alternative for Positive Contrast in Magnetic Resonance Imaging. *Inorganics* **2020**, *8*, 28. [CrossRef]
7. Minaeva, O.; Hua, N.; Franz, E.S.; Lupoli, N.; Mian, A.Z.; Farris, C.W.; Hildebrandt, A.M.; Kiernan, P.T.; Evers, L.E.; Griffin, A.D.; et al. Nonhomogeneous Gadolinium Retention in the Cerebral Cortex after Intravenous Administration of Gadolinium-based Contrast Agent in Rats and Humans. *Radiology* **2020**, *294*, 377–385. [CrossRef] [PubMed]
8. Kozlova, A.A.; German, S.V.; Atkin, V.S.; Zyev, V.V.; Astle, M.A.; Bratashov, D.N.; Svenskaya, Y.I.; Gorin, D.A. Magnetic Composite Submicron Carriers with Structure-Dependent MRI Contrast. *Inorganics* **2020**, *8*, 11. [CrossRef]
9. German, S.V.; Bratashov, D.N.; Navolokin, N.A.; Kozlova, A.A.; Lomova, M.V.; Novoselova, M.V.; Burilova, E.A.; Zyev, V.V.; Khlebtsov, B.N.; Bucharskaya, A.B.; et al. In vitro and in vivo MRI visualization of nanocomposite biodegradable microcapsules with tunable contrast. *PCCP* **2016**, *18*, 32238–32246. [CrossRef] [PubMed]
10. Carniato, F.; Gatti, G. 1H NMR Relaxometric Analysis of Paramagnetic Gd_2O_3:Yb Nanoparticles Functionalized with Citrate Groups. *Inorganics* **2019**, *7*, 34. [CrossRef]
11. Liu, Z.; Pu, F.; Liu, J.; Jiang, L.; Yuan, Q.; Li, Z.; Ren, J.; Qu, X. PEGylated hybrid ytterbia nanoparticles as high-performance diagnostic probes for in vivo magnetic resonance and X-ray computed tomography imaging with low systemic toxicity. *Nanoscale* **2013**, *5*, 4252–4261. [CrossRef] [PubMed]
12. Peralta, M.E.; Ocampo, S.; Funes, I.G.; Onaga Medina, F.; Parolo, M.E.; Carlos, L. Nanomaterials with Tailored Magnetic Properties as Adsorbents of Organic Pollutants from Wastewaters. *Inorganics* **2020**, *8*, 24. [CrossRef]
13. De Castro Alves, L.; Yáñez-Vilar, S.; Piñeiro-Redondo, Y.; Rivas, J. Efficient Separation of Heavy Metals by Magnetic Nanostructured Beads. *Inorganics* **2020**, *8*, 40. [CrossRef]

© 2020 by the authors. Licensee MDPI, Basel, Switzerland. This article is an open access article distributed under the terms and conditions of the Creative Commons Attribution (CC BY) license (http://creativecommons.org/licenses/by/4.0/).

Review

Magnetic Materials and Systems: Domain Structure Visualization and Other Characterization Techniques for the Application in the Materials Science and Biomedicine

Roberto Nisticò [1,†], Federico Cesano [2] and Francesca Garello [3,*]

1. Department of Applied Science and Technology DISAT, Polytechnic of Torino, C.so Duca degli Abruzzi 24, 10129 Torino, Italy; roberto.nistico0404@gmail.com
2. Department of Chemistry and NIS Centre, University of Torino, Via P. Giuria 7, 10125 Torino, Italy; federico.cesano@unito.it
3. Molecular and Preclinical Imaging Centers, Department of Molecular Biotechnology and Health Sciences, University of Torino, 10126 Torino, Italy
* Correspondence: francesca.garello@unito.it; Tel.: +39-011-670-6452
† Current address: Independent Researcher, Via Borgomasino 39, 10149 Torino, Italy.

Received: 23 September 2019; Accepted: 31 December 2019; Published: 17 January 2020

Abstract: Magnetic structures have attracted a great interest due to their multiple applications, from physics to biomedicine. Several techniques are currently employed to investigate magnetic characteristics and other physicochemical properties of magnetic structures. The major objective of this review is to summarize the current knowledge on the usage, advances, advantages, and disadvantages of a large number of techniques that are currently available to characterize magnetic systems. The present review, aiming at helping in the choice of the most suitable method as appropriate, is divided into three sections dedicated to characterization techniques. Firstly, the magnetism and magnetization (hysteresis) techniques are introduced. Secondly, the visualization methods of the domain structures by means of different probes are illustrated. Lastly, the characterization of magnetic nanosystems in view of possible biomedical applications is discussed, including the exploitation of magnetism in imaging for cell tracking/visualization of pathological alterations in living systems (mainly by magnetic resonance imaging, MRI).

Keywords: magnetic materials; nanostructured materials; magnetic nanoparticles; magnetometry; magnetic hysteresis; magnetic domain visualization; magnetic resonance imaging; magnetic fluid hyperthermia; magnetic particle toxicity

---

1. Introduction

Since the early beginning of our society, magnetism catalyzed the attention of scientists worldwide due to its intrinsic capability to naturally attract/move inanimate matter [1,2]. However, it is with the discoveries of Pauli's exclusion principle and Heisenberg's quantum theory that the "Modern Theory of Magnetism" was finally coined in the 1920s, unveiling the strict correlation existing between magnetism and the number/motion of electrons [3]. From here, the scientific community reached several steps forward toward the production of more and more advanced magnetic (nano)materials and (nano)systems that found applications in many useful scientific/technological fields, such as in (bio)medicine [4,5], drug-delivery [6–8], imaging [9–11], spintronics and electronics [12], data storage [13], robotics [14,15], environmental remediation processes [14–19], (nano)engineering [20–22], and miniaturized devices [23].

Due to the growing interests around the exploitation of magnetic (nano)materials, a detailed comprehension of this phenomenon is becoming more and more important, if not crucial. Many characterization techniques are used daily to qualitatively and/or quantitatively determine the magnetic response in materials [1,24]. However, being very specific, these techniques could be unfamiliar to a wide audience. The analysis of the state-of-art pointed out that the scientific literature is very rich in reviews focused on the production/testing of magnetic materials in various fields [25–28], assuming as elementary the comprehension of the adopted characterization techniques. On the basis of the authors' experience, a superficial (and simplistic) interpretation of these data could leave to misleading (and in some cases incorrect) analysis [1]. In this context, it is worthy of note that there are many previous publications related to these subjects, including reviews and books. Some of them [24,29–36] can be still considered as "classical" as they are constantly used in many laboratories around the world.

Therefore, aim of this review is to provide (in a simple, but precise way) a technical summary of the main relevant characterization techniques mandatory for determining magnetism-related phenomena in (nanoscopic) materials and systems and some of the most recent advances in the field, new methods and approaches. Obviously, the number of techniques exploitable for this purpose can be extremely various and it is almost impossible to provide an enough-detailed analysis of all the possible variants and approaches (for a much detailed comprehension of each technique, readers should refer to dedicated papers and the afore mentioned literature). Thus, for the sake of clarity, authors have decided to focus the discussion on some relevant methods illustrated in the literature, in correlation also with their peculiar expertise. Hence, the following paragraphs were organized introducing three main topics: A brief introduction dedicated to the determination of the magnetization (hysteresis) curves (fundamental for recognizing not only the level of magnetism in materials/particles, but also the types of magnetism, *vide infra*) and their interpretation, the visualization and description of magnetism at mesoscale (including the correlation between nanomagnetism and morphology), and the exploitation of magnetism in imaging for cell tracking or visualization of pathological alterations in living systems (mainly by magnetic resonance imaging, MRI). Concerning this last topic, a particular attention will be devoted to the characteristics that magnetic systems shall possess to be safely and successfully employed in living organisms (both in vitro and in vivo).

The final goal of this review is to draw guidelines beneficial for the correct comprehension of the magnetism-related literature, even for not insiders, as well as to point out how a magnetic system should be designed and characterized in order to be suitable for in vivo applications. The multidisciplinary approach here presented is the result of different viewpoints, in particular the merging of the physical and morphological peculiar characteristics of magnetic nanosystems applied to the biomedical field. To facilitate the document's readability, specific case studies were taken as reference examples, key points and criticalities highlighted. With this work, the authors' hope is to have unequivocally disclosed any possible complex aspects in the field, thus facilitating the proliferation of interesting (and optimistically outstanding) future studies.

2. Magnetism and Magnetization (Hysteresis) Curves

On the basis of the "Modern Theory of Magnetism", the appearance of strong magnetic phenomena in materials and molecular structures is due to the presence of chemical elements with a particular electronic configuration, namely: Iron (Fe), nickel (Ni), cobalt (Co), manganese (Mn), chromium (Cr), and some rare earth metals [37]. Independently from the types of magnetism, the most common method for evaluating the magnetic response in materials is the determination of the magnetization (hysteresis) curves by means of a magnetometer [38]. Even if there are several configurations of magnetometers, the most common one is the vibrating sample mode (VSM). In a VSM magnetometer, the test specimen is subjected to a magnetization-demagnetization loop process by varying the external magnetic field applied. The material's magnetization, intended as the vector field which indicates the density of magnetic moments (i.e., vector relating the alignment on the material by applying an external magnetic field with respect to the field vector), is measured indirectly as electric current variation/formation of the inductive coils surrounding the sample-holder (it should be remembered that both electric and magnetic fields are strictly correlated between each other since being orthogonal). According to the "IEEE Magnetic Society" [39], magnetization is expressed in two different forms: as total (volume) magnetization (**M**, an expression of the magnetic moments per unit of volume, units of measurement A/m for the International System of units (SI) and emu/cm^3 for the Centimeter-Gram-Second system of units (CGS)) and as total (mass) magnetization. Another useful property is the magnetic induction (**B**, magnetic flux density in the sample, units of measurements T for SI and G for CGS) [40]. These units are correlated between each other [41] according to the Equation (1) (derived from the Maxwell's equation):

$$\mathbf{B} = \mu_0 (\mathbf{H} + \mathbf{M}) \tag{1}$$

where **H** is the applied magnetic field (external magnetic stimulus, units of measurement A/m for SI and Oe for CGS) and μ_0 is the vacuum permeability (a constant value equal to $4\pi \times 10^{-7}$ H/m, as defined in SI). Sometimes, it is better to remind that the CGS system is preferred respect to the SI, thus for any clarification concerning the units of measurements of magnetic properties, please refers also to [1]. At this point, it can be useful for the entire discussion, introducing some important physical quantities with their definitions. In details, the saturation magnetization (M_s) is the maximum magnetic moment induced by an external magnetic field applied, the intrinsic coercivity (H_{ci}) is the reverse field required to bring the magnetization **M** to zero, and the magnetic remanence (M_r) is the residual magnetization at zero external magnetic field ($H = 0$) [40].

The main differences between these two configurations (namely, **M** vs. **H** and **B** vs. **H**) are related to both shapes and physical quantities obtained [39]. As shown in Figure 1, B_r is the residual induction at $H = 0$, whereas H_c is the coercivity (or the reverse field necessary to bring the B to zero).

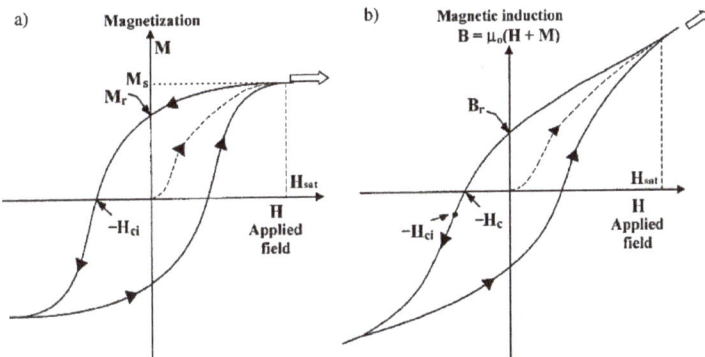

Figure 1. Ferromagnetic material hysteresis curves expressed as **M** vs. **H** (**a**) and **B** vs. **H** (**b**) curves. Legend: M_s is the saturation magnetization, M_r is the remnant magnetization at $H = 0$, H_{ci} is the intrinsic coercivity, B_r is the remnant induction (or remanence) at $H = 0$, and H_c is the coercivity. For hard magnets: $H_{ci} \neq H_c$; for soft-magnets: $H_{ci} \approx H_c$. Reprinted with permission from [40], published by Elsevier, 2003.

The more enlarged the hysteresis loop (such as in the case of hard-magnets, *vide infra*), the higher the discrepancies between the two coercivity values (H_c and H_{ci}). This suggests that when the hysteresis loop becomes very narrow (or negligible), the two representations of the magnetization profiles tend to be similar, thus justifying in some ways the wrong interchanges mostly found in the literature. For the sake of clearness, from here only **M** vs. **H** curves were considered.

Figure 2 reports the possible different profiles of magnetization (hysteresis) curves depending on the form of magnetism. Being more precise, there are five relevant forms of magnetism, and among these the more intense (and, consequently, macroscopically-detectable by human eyes without specific techniques) are only two: ferromagnetism and ferrimagnetism. Superparamagnetism is a thermal/size-induced particular response of the previous two forms of magnetism, while diamagnetism, paramagnetism, and antiferromagnetism are weaker forms of magnetism.

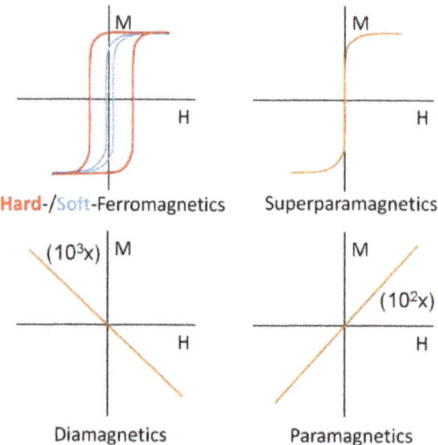

Figure 2. Magnetization (hysteresis) curves associated with the different classes of magnetic materials (i.e., Hard/Soft-ferromagnetics, superparamagnetics, diamagnetics, and paramagnetics).

Ferromagnetism consists in a spontaneous magnetization/alignment of the matter (even without applying an external magnetic stimulus) of the order of ca. 10^6 A/m. Ferromagnetism is generated by the self-alignment of the unpaired (same-spin) electrons forming the material. Since this phenomenon is energetically favored only at short-range, it is reflected by the formation of randomly aligned magnetic domains. In fact, at the macro level, the energetically favored anti-alignment organization of adjacent poles is still the more predominant one. Vice versa, in presence of an external magnetic field applied, domains aligned themselves according to the external magnetic field directions [42].

Interestingly, Fe, Ni, and Co (3d metals) are the only three pure elements with ferromagnetic properties at room temperature (RT). Ferromagnetic materials are characterized by having a well-defined M_s, and high H_{ci} and M_r.

Additionally ferromagnetic materials can be classified as hard (permanent magnet, with high H_{ci}) and soft (easily (de)magnetized, with low H_{ci}) [1,35,43,44]. Lastly, the high values of H_{ci} and M_r are an expression of the capability of ferromagnetic materials of retaining a memory of their magnetic history. Moreover, ferromagnets are sensible to temperature. In fact, by increasing the temperature above the Curie point (a critical temperature value typical of each magnetic material), ferromagnetic materials start behaving as paramagnetic materials (*vide infra*), with formation of random domains [45]. This reversible phenomenon is due to disordered motions of electrons caused by an overall increment of entropy in the system. From the magnetization curve in Figure 2, the formation of the hysteresis phenomenon is attributable to a certain magnetic anisotropy due to structural parameters (such as: Crystal structure, shape/dimensions of grains/particles, stress/tension, interaction with (anti)ferromagnetic materials), and this is particularly strengthened in the case of hard-magnets, which show high H_{ci} (see [1,35] and references therein). Figure 3 reports the schematic view of a ferromagnetic system in absence and in the presence of an external magnetic field applied.

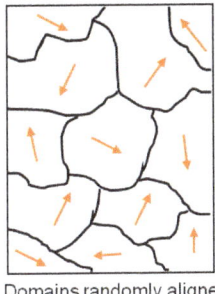 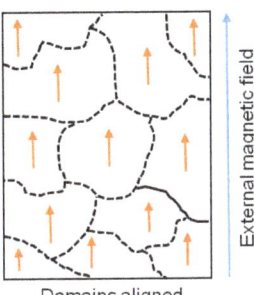

Domains randomly aligned   Domains aligned

Figure 3. Schematic representation of ferromagnetic domains in absence (**left**) and in presence (**right**) of an external magnetic field applied. In the latter case, boundaries are dashed since when domains are aligned (right panel), the sample reaches the saturation point and there are not any domains walls.

Interestingly, Fitta and co-workers [46] reported the layer-by-layer deposition of a bilayer system composed of hard $Ni_{3.38}[Fe(CN)_6]_2 \cdot nH_2O$ (indicated as NiFe) and soft $Ni_{3.1}[Cr(CN)_6]_2 \cdot nH_2O$ (indicated as NiCr) ferromagnetic compound. Figure 4 reports the magnetic hysteresis loops at 2 K for bilayer sample against their orientations respect to the direction of external magnetic field (namely, 0°, 45°, and 90°). As reported in the figure, when sample is parallel oriented (0°), a two-phase hysteresis was observed: (i) A drop in magnetization by decreasing the magnetic field (at small value) due to the presence of the NiCr layer (soft-magnet), and (ii) a pronounced hysteresis loop due to the presence of the NiFe layer (hard-magnet). By varying the orientation toward perpendicular (90°), the magnetization process is much more gradual. This variation of the magnetization curve with respect to the film orientation is attributable to the anisotropic properties of the NiCr layer: Parallel orientation indicates easy magnetization direction, whereas perpendicular one indicates a hard magnetization direction.

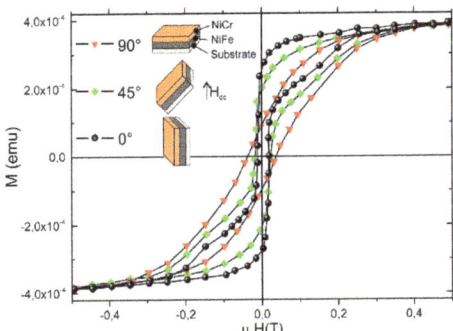

Figure 4. Hysteresis loops obtained at $T = 2$ K for sample oriented at 0°, 45°, and 90° in respect of the direction of external magnetic field. Reprinted with permission from [46], published by John Wiley & Sons, 2017.

Ferrimagnetism is a very particular form of magnetism. This phenomenon occurs in materials organized into two interpenetrating structures located at the different sublattice, showing an anti-alignment of spins with not-equal magnetic moments, thus resulting in an overall valuable magnetic moment below the Curie temperature [47]. Analogously to ferromagnetic materials, ferrimagnetic systems exhibit spontaneous magnetization, hysteresis loop, and a Curie point [1,35]. Typical ferrimagnetic systems are magnetite (Fe_3O_4, a mixed ferrous-ferric oxide with the spinel crystal structure AB_2O_4) and other ferrites (Fe-containing mixed oxides with the spinel crystal structure MFe_2O_4, with M being a general metal). For the sake of clarity, Figure 5a depicts the spinel crystal structure of MnV_2O_4, where the two substructures are made by the V-octahedral and Mn-tetrahedral sites [48]. Interestingly, by making a comparison between Figures 5b and 5c, it is possible to appreciate the effect induced by the temperature variation on the phase crystal structure (i.e., cubic-to-tetragonal transition during cooling) and consequently on the organization of the magnetic structure (from collinear to non-coplanar ferrimagnetic order). Additionally, also rare earth alloys (eventually in combination with magnetic transition metals) are widely adopted as permanent magnets [49–51]. Among the lanthanides, the most used ones are the neodymium (Nd)-based magnets [52] and the samarium (Sm)-based ones [53].

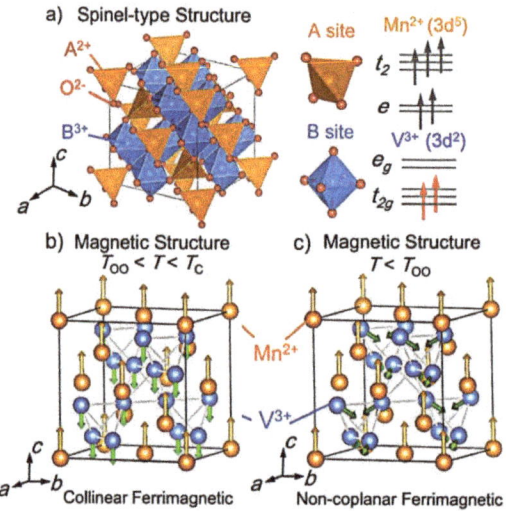

Figure 5. Effect by temperature variations on the phase crystal structure and magnetic structure organization. Panel (**a**) spinel-type oxide MnV_2O_4 crystal structure A-site are surrounded by oxygen tetrahedron and B-site ions by octahedron. The V^{3+} ion has orbital degeneracy in the t_{2g} orbital. Panel (**b**) The collinear ferrimagnetic structure. Panel (**c**) Magnetic structure in the non-coplanar ferrimagnetic tetragonal phase (c < a). Reprinted with permission from [48], published by Elsevier, 2018.

On the contrary, paramagnetism is the tendency of materials of displaying a net magnetic moment if exposed to a magnetic field. As paramagnetism comes from the partially-filled orbitals, either forming bands or being localized, only certain systems lead to paramagnetic properties, in principle metals and insulators having localized moments. Macroscopically, when paramagnetic materials are subjected to an external magnetic field, they tend to align following the magnetic field direction, giving both magnetization and susceptibility slightly positive (as depicted in Figure 2) [54]. However, it must be noticed that paramagnetism might be associated also to other forms of magnetism. For example, Preller and co-workers [55] reported the production of FePt nanomaterials. As summarized in Figure 6, Fe_3Pt nanomaterial shows clearly a soft magnetic behavior (Figure 6a), whereas $FePt_3$ predominantly exhibits a paramagnetic character associated with traces of ferromagnetism due to the presence of a minor ferromagnetic phase (Figure 6b).

As reported in a previous study [1], superparamagnetism is a particular form of paramagnetism below the Curie point in both ferro/ferri-magnetic materials when organized in the shape of small "single-domain" nanoparticles (thus it is a thermal and size dependent particular response of ferro/ferri-magnetic materials) [24]. In detail, since single-domain nanoparticles (with dimensions below 20 nm size) can randomly change their directions of the magnetization with temperature/time fluctuations, the average magnetization value of such systems in absence of an external magnetic field is close to zero [10]. In presence of an external magnetic field, nanoparticles align following the magnetic field direction as for paramagnetic materials. However, due to their ferro/ferri-magnetic origin, such nanomaterials show very high magnetic susceptibility (larger than common paramagnetic materials, i.e., "super"paramagnetism) and absence of hysteresis (H_{ci}-values close to zero, and M_r-values very low) [38].

Conversely to paramagnetism, diamagnetism is the capability of repel/oppose to an applied external magnetic field, due to the absence of unpaired electrons, giving weak negative magnetization and susceptibility [56]. This phenomenon is widely exploited for the levitation/floating of bodies, such as in the case of Maglev train (i.e., superconductors are perfect diamagnets).

Finally, analogously to ferrimagnetism, antiferromagnetism consists in magnetic phenomenon generated by the presence of two interpenetrating structures characterized by having an equal antialignment of electrons' spins, and consequently an overall zero magnetization at the macroscale. This phenomenon is evident at low temperature, whereas at temperature higher than the Néel point it is registered the antiferromagnetism-to-paramagnetism transition (in analogy to Curie transitions of ferromagnetism) [1,24].

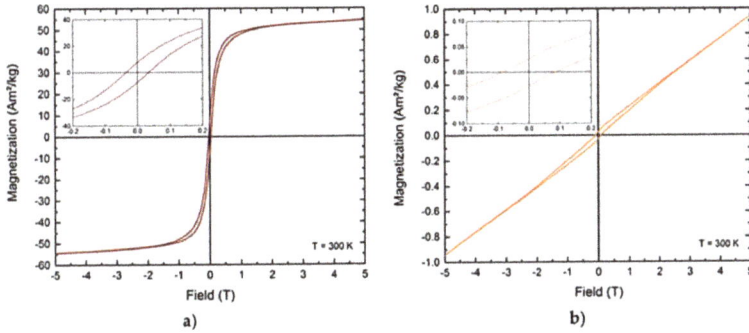

Figure 6. Magnetization curves of FePt nanomaterials. Panel (**a**) ferromagnetic Fe$_3$Pt nanocrystals. Panel (**b**) predominantly paramagnetic FePt$_3$ nanocrystals. Reprinted from [55] (published by MDPI, 2018) licensed under CC BY 4.0.

Figure 7 reports a summary of the main features of the principal different forms of magnetism [57]. On the basis of the state-of-art literature, several case studies reported the exploitation of magnetization (hysteresis) curves for the determination of the magnetic properties of materials, making this technique an extremely powerful method for investigating magnetic systems.

Paulo et al. [58] studied the formation of metallic (Ni, Fe, Co) ferromagnetic nanotubes via chemical electrodepositing using porous polycarbonate membranes as hard-templating system. Magnetic curves were used for evaluating not only the magnetic response, but also to determine the effect of parallel/perpendicular fields with respect to saturation, remanence and coercivity. Another interesting study is the one reported by Venkata Ramana and co-workers [59], focused on the evaluation of Fe-doping on BTCZO, acronym of (Ba$_{0.85}$Ca$_{0.15}$)(Ti$_{0.9}$Zr$_{0.1}$)O$_3$. Evidences confirmed that undoped BTCZO is diamagnetic, whereas the Fe doping caused a gradual evolution toward ferromagnetism. Peixoto et al. [60] produced CoFe$_2$O$_4$ nanoparticles embedded inside a TEOS-derived SiO$_2$ matrix via sol–gel (for sol–gel processes, refer to [61,62]) and evaluated the magnetic properties by fixing the temperature at either 5 K or 100–200 K confirming a superparamagnetic regime. In this context, it is important to point out that several cases studies reported magnetization curves detected at different temperatures (going from cryogenic conditions to temperature values above/below the Curie point), and this is fundamental for evaluating specific magnetic transition in the (nano)material/system analyzed. Wang et al. [63] reported a comprehensive study focused on the magnetic response evaluation in perovskite crystal systems. In recent studies, magnetic composite nanomaterials made by either magnetite/chitosan [64,65] or magnetite/humic-like substances [18,66] isolated from municipal biowaste [67] were thermally-converted via pyrolysis at 550–800 °C.

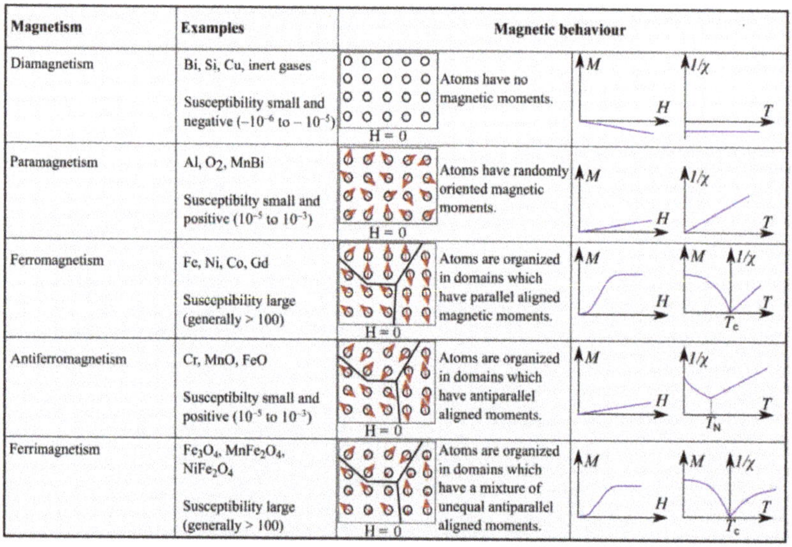

Magnetism	Examples	Magnetic behaviour		
Diamagnetism	Bi, Si, Cu, inert gases Susceptibility small and negative (-10^{-6} to -10^{-5})	Atoms have no magnetic moments. $H = 0$	M vs H	$1/\chi$ vs T
Paramagnetism	Al, O_2, MnBi Susceptibilty small and positive (10^{-5} to 10^{-3})	Atoms have randomly oriented magnetic moments. $H = 0$	M vs H	$1/\chi$ vs T
Ferromagnetism	Fe, Ni, Co, Gd Susceptibility large (generally > 100)	Atoms are organized in domains which have parallel aligned magnetic moments. $H = 0$	M vs H	M, $1/\chi$ vs T; T_c
Antiferromagnetism	Cr, MnO, FeO Susceptibilty small and positive (10^{-5} to 10^{-3})	Atoms are organized in domains which have antiparallel aligned moments. $H = 0$	M vs H	$1/\chi$ vs T; T_N
Ferrimagnetism	Fe_3O_4, $MnFe_2O_4$, $NiFe_2O_4$ Susceptibility large (generally > 100)	Atoms are organized in domains which have a mixture of unequal antiparallel aligned moments. $H = 0$	M vs H	M, $1/\chi$ vs T; T_c

Figure 7. Summary of the main relevant forms of magnetism and their features. Reprinted with permission from [57], published by Woodhead Publishing and Elsevier, 2016.

The analysis of the magnetization curves resulted to be fundamental to indirectly understand the chemical modifications induced by the pyrolysis treatment monitoring the magnetic properties in the final material. In detail, it should be remembered that magnetite is converted into wustite (FeO) upon heating (at 575 °C) under inert atmosphere and in the presence of a carbon source (e.g., chitosan or the humic-like substances). Since instable, FeO disproportionates, giving metallic Fe and magnetite phases. The analysis of the magnetization curves pointed out that pristine magnetic nanocomposites (containing magnetite) have saturation values lower than bare magnetite. Pyrolysis at temperatures below 575 °C affects only the magnetization profiles with an increment of saturation, coercivity and remanence due to the consumption of the organic layer toward the formation of carbon. Conversely, pyrolysis at temperature higher than 575 °C caused a remarkable increment of the saturation level (even higher than the reference bare magnetite), suggesting the growth of a different magnetic phase, namely Fe (which is ferromagnetic). Moving toward the organic-based and molecule-based magnets, Miller [68] resumes the main findings obtained for the family of [TCNE]$^{\bullet-}$-based magnets. Namely, by modifying the chemical structure of this family, different forms of magnetism were registered, such as: ferromagnetism for [Fe^{III}(C$_5$Me$_5$)$_2$]$^{\bullet+}$[TCNE]$^{\bullet-}$ and V[TCNE]$_2$, weak ferromagnetism for Li$^+$[TCNE]$^{\bullet-}$, ferrimagnetism for Mn^{II}(TCNE)$_{3/2}$(I$_3$)$_{1/2}$, and antiferromagnetism for M^{II}(TCNE)[C$_4$(CN)$_8$]$_{1/2}$ (with M = Mn, Fe). In all cases the analysis of the magnetization (hysteresis) curves resulted fundamental not only for recognizing the form of magnetism, but also for the evaluation of the magnetic properties. Shirakawa et al. [69] measured the magnetic properties of an organic radical ferromagnet, β-p-NPNN, below and above the T_c. Interestingly, in their study Shum and co-workers [70] reported evidences of pressure-induced metamagnetic-like transition from antiferromagnetism to ferromagnetism for [Ru$_2$(O$_2$CMe)$_4$]$_3$[Cr(CN)$_6$] system. Lastly, also carbon allotropes (mostly 2D organic systems [71–73] and carbon nanotubes [74,75]) show unique optoelectronic and magnetic properties. In most cases, for practical applications the doping of these materials with heteroatoms (in particular to extend their magnetic ordering at long-scale) is mandatory. In this context, Tucek and co-workers [76] synthesized flourographene (thus introducing F atoms in the carbon texture) with hydroxyl groups showing antiferromagnetic ordering. However, in the literature it is widely questioned the possible role of metal impurities (Fe, Ni, Co) derived from the synthesis of such carbon allotropes (and derivatives) in their magnetic response [77].

At the end of this section, authors suggest the following specific literature for further deepening the theory and influence of parameters affecting the magnetization (hysteresis) curves [32,34,78,79].

Moreover, magnetization processes (and equipments) can be classified into quasi-static (DC) or dynamic (AC) processes [80,81]. In detail, DC magnetometry consist in magnetizing the samples and registering the magnetic moment by applying a constant magnetic field (one above all the VSM set up). On the contrary, in the AC magnetometry, an AC magnetic field is superimposed on a DC field, thus causing a time-dependent magnetic moment, which allows the magnetization without moving the sample. In the AC magnetometry, at low frequencies the **M** vs. **H** curve is very similar to the DC one and the slope of the **M** vs. **H** profile is the susceptibility. The main advantage of this technique is the high sensitivity, thus making detectable even very small magnetic variations. At high frequencies, instead, the magnetic moment registered differs from the DC one due to dynamic effects. For these reasons, the AC magnetic susceptibility is defined by two parameters: The magnitude and the phase shift, which are sensitive to thermodynamics phase changes or temperature-induced magnetic transitions (e.g., the superparamagnetism and superconductivity as well as the determination of the Curie/Néel point). Interestingly, by plotting the magnetization against the temperature or the susceptibility against the temperature, one can determine the magnetic phase transition (i.e., Curie/Néel point). For completeness, authors suggest the following references dedicated [82–86].

Furthermore, another very promising equipment that deserve the attention of experts is the SQUID (acronym of Superconductive QUantum Interference Device), which is an extremely sensible magnetometer widely exploited for the determination of very weak magnetic field (below 10^{-18} T). SQUIDs can be classified as either direct current (DC) or radio frequency (RF), and can be integrated into chips. In general SQUIDs are made by Nb or Pb alloys operating at temperature closed to the absolute zero (i.e., the system deserves being cooled down with liquid He or N_2). The analysis of the literature revealed that SQUIDs are widely used for biology studies (due to the very weak intensity of the magnetic fields registered) as well as detectors in magnetic field imaging (MFI) or in relaxometry (*vide infra*), however a complete and detailed discussion on this very specific technique is out from the scope of this review, which instead is focused on the technique easily findable in a magnetism-dedicated laboratory. For more detailed discussion, please consider the following literature and references therein [87–90].

3. Imaging of the Domain Structure and Beyond

The hysteresis properties, as defined in the previous paragraph, represent bulk characteristics or integrated amounts of the magnetic character of a sample as a whole. On the other hand, the magnetization is in principle a vector (and not a scalar quantity) and could be locally identified at the mesoscale. In addition, magnetic nanoparticles for both fundamental studies and technologies have been intensively explored for the past two decades. Different from the bulk counterparts, nanoparticles (NPs) with magnetic properties show unique magnetism characteristics, enabling the tuning of their magnetism by methodological nanoscale approaches. In a recent review, Wu et al. [91] have summarized the major synthetic procedures and the application fields of various magnetic NPs, including metals, metallic alloys, metal oxides, and multifunctional NPs.

Microscopies dedicated to magnetic investigation are perfectly suited for unravelling all these local magnetic properties, making available on the basis of the magnetism principles a set of information that can be adopted for improving and designating new materials or for the creation/modification of magnetic domains (spintronics) as well. For example, when dimensions of a system go down to the nanoscale the uniform magnetization hypothesis is unreliable and properties become exclusive. Local regions with uniform magnetization (i.e., magnetic domains), their boundary structure (i.e., domain walls), the directions of magnetization and the magnetization extent can be revealed under both static or dynamic conditions (i.e., magnetization reversal and time-reversal behaviors) from the cryogenic cooling to high temperature [92–95].

Electrons, photons, neutrons and X-rays are four kinds of modern probes offering a plethora of opportunities for the investigation of the magnetic properties (Figure 8). Besides other exciting properties, the aforementioned probes make it possible to visualize magnetic domains and structures, obtaining often quantitative results. The imaging of the magnetic domain structure is, for a fact, a way to reveal the local magnetic properties of materials from the macroscale down to the nanoscale. Since the publication of the comprehensive review on "Magnetic domains" by Craik and Tebble [31], dated 1961, new techniques and materials became available, while new insights were evidenced and perspectives appeared over the years. The method of observation can provide today useful insights also into the origin and reversal mechanisms of the magnetization, the sample magnetic vortex structures, that are all of a great importance from a fundamental point of view of the application, for example spintronics and other technologies. Shrewd interpretation of the variety of magnetic visualization techniques is used in this section both to begin the readers who are not familiar with these techniques and to help those more expert to figure out their own matter.

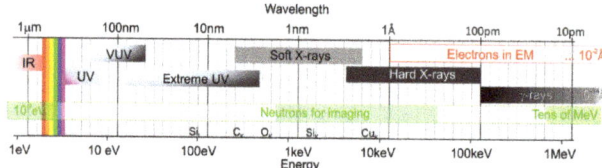

Figure 8. Electromagnetic spectrum from infrared (IR) to γ-rays and some probes for the magnetic imaging.

This review section is intended to be a survey around the topical methods for magnetic domain/domain structure observations and some of the recent findings. The appraisal of the today's significance in the corresponding field is discussed. The theories and principles of magnetism, including Weber's molecular and domain, and magnetic field theories are the background to this discussion.

Therefore, all aspects of magnetic structural characterization are not entirely covered, but a few techniques to be employed in modern investigations of magnetic systems in a wide dimensional range, from macro/microstructures (i.e., materials), to nanoparticles, molecular assemblies and single-molecule magnets are discussed. As each technique has its own advantages and disadvantages, the present review aims at helping researchers to select the most appropriate method for their purposes, but before discussing the techniques in more detail, a set of general comments for the imaging methods is needed. Firstly, almost all of the family of the visualization methods are firmly restricted to the topmost layer(s), a few micro- or nanometers deep in the matter, an area dominated by magnetic characteristics located at the surface. Nothing is associated with the properties of the inner portion that should be investigated by other characterization techniques more precisely dedicated to bulk magnetic properties (see the first section of this review). Secondly, some methods usually need a cleaned, smoothed and defect-free surface, and/or the operational conditions (i.e., vacuum) could be far from working conditions. Thirdly, the imaging of the domain walls in some methods could be affected by preparation (i.e., use of magnets to collect the particles that can affect their in-the-plane or transversal orientations). Fourthly, a compromise between sensitivity and resolution, must be taken under consideration and a multi-technique approach would be valuable.

As for characterization techniques, however, it is not surprising that in most papers a multi-technique approach is reported. The complementary investigation, typically adopted in materials science by means of microscopies or other visualization techniques with spectroscopies and textural analysis, overcomes the intrinsic limitation of individual methods and offers complementary details for assessing the correlation between morphology, structure and other properties of complex or nanostructured materials [11,21,44,96–108].

For the sake of comparison, a summary of the visualization techniques discussed in this review is detailed in Table 1. Such table should be intended as an overview of the possible investigation approaches and it offers a broad outline of the different characterization methods.

Table 1. Magnetic imaging methods and their characteristics.

Method(s)	Quantitative	Resolution (nm) Best	Resolution (nm) Typical	Complexity *	Acquisition Time	Vacuum Req.	Sample Thickness (nm)	Info on Depth (nm)	External Magn. Field	Selected Depth Info	High, Low Temperature ***	Types of Specimen ****	Ref.
Bitter	N	~100	500	L	<s	-	No limit	500	Yes	L	No	S, TF, NPs	[109]
Magneto-optic	Y	300	>500	L	<s	-	No limit	20	Yes	L	Cryo/VT	S, TE, NPs	[24,110,111]
L-TEM	Y	20	50	M	<s	HV	<100–150	ST**	Yes	No	VT	S, TE, NPs	[24,112,113]
DPC	Y	3	20	H	<s	HV					VT	S, TE, NPs	[24,114]
Electron holography	Y	5	20	H	1–10 s	HV	<150	ST**	Yes	No	VT	S, TE, NPs	[113,115]
SP-SEM	Y	10	100	H	1–100 min	UHV	No limit	1–2	Low	H	VT	S, TE, NPs	[24,116]
XMCD	Y	300	500	H	1–10 min	UHV	No limit	<5–10	-	H	VT	S, TE, NPs	[95,117,118]
TXMCD	Y	30	60	H	1–10 s	-	<150	ST**	Yes	H	VT	S, TE, NPs	[95]
SPLEEM	Y	3	40	H	1 s–10 min	UHV	No limit	1–2	-	VH	VT	S, TE, NPs	[95]
MFM	N	20–30	100	L/M	5–30 min	Air, HV	No limit	20–500	Yes	No	Cryo/VT.	S, TE, NPs	[24,119,120]
SP-STM	N/Y	≤1	2–3	M	10–100 min	(air), HV	No limit	<1–2	Yes	VH	Cryo/VT.	S, TF, NPs of metals/semic.	[24,121]
N-V Diamond	Y	400		L	<s	No		No	Yes	No	No	S, TF, LM	[122,123]
TXMs	Y	10–15	50	M/Y	<10 s	-	No limit	ST**	Yes	H	VT	S, TE, NPs, B	[117,121,124–127]
Neutron techniques		Tens of μm		VH	s	No	5–100	2–3	Yes	No	VT	LM; S, TF (pol. neutron reflectom.). NPs (SANS)	[117,128–130]

* L = low, M = medium, H = high, VH = very high; ** ST = sample thickness; *** Cryo = cryogenic temperature, VT = variable temperature; **** S = surfaces, TF = thin films, LM = living matter, NPs = nanoparticles, B = bulk.

3.1. Powder Pattern Imaging

It is easy to observe characteristic domain structures in materials by the Bitter powder imaging technique. Bitter patterns, firstly observed by Hamos and Thiessen [131], and Bitter [132] in 1931, involve a ferrofluid (or a colloidal suspension of small Fe/iron oxides particles), placed on the surface of a ferromagnetic material, which outlines the boundaries of the magnetic domains. The magnetic walls, having a higher magnetic flux than the regions located in the domains, can be visualized by the optical microscopy (Figure 9a) or better implemented into a device, where a coil generating a magnetic field helps for the observation (Figure 9b).

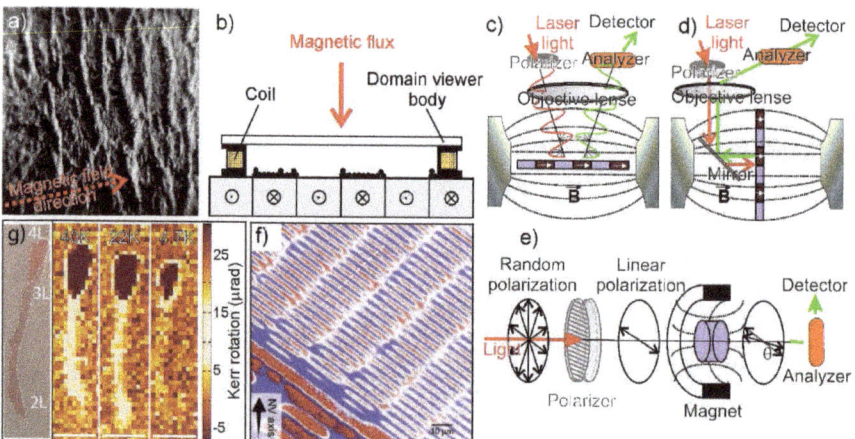

Figure 9. (a) Patterns obtained from Fe_2O_3 nanoparticles on an iron-silicon alloy (Bitter technique) in large fields (16× magnification). The observed striations were found perpendicular to the applied magnetic field without any relationship to the grain structure for large magnetizations. Reproduced with permission from [132], published by American Physical Society, 1931; (b) representation of the magnetic domain viewer (modified Bitter technique). The technique employs a coil to generate a magnetic field perpendicular to the surface of the sample. Magnetization directions are shown into (X) and out of (O) the plane, respectively; (c–e) representations of the longitudinal, polar magneto-optical Kerr and magneto-optical Faraday effect configurations, respectively; (f) magneto-optical imaging of a thin ferromagnetic film using a diamond-based sensor. Reprinted from [122] (published by Nature Publishing Group, 2016) licensed under CC BY 4.0; (g) optical (left panel) and Kerr images (right panels) of a few-layered $Cr_2Ge_2Te_6$ crystal exfoliated on SiO_2/Si obtained at different T. The layer number (2L, 3L, or 4L) is indicated by arrows in (g). Scale bars in (g) are 10 µm. Image adapted from [133], published by Springer Nature, 2017.

Nonetheless, no relationships to the grain structure can be obtained. Although the method is quick and simple, the investigation of the domain structure by the conventional Bitter pattern technique could be difficult for soft magnetic materials, complex domain structures, and even impossible for the smaller domain structures [134,135]. Furthermore, as the method is firmly a stray field decoration technique, patterns direction and magnitude of the magnetization are not available. Lastly, the Bitter powder pattern of Néel-type domain walls was found much more visible than that of a Bloch-walled structure [136]. Finally, as the method commonly gives more reliable results with high coercivity magnets or perpendicularly magnetized specimens, a small magnetic field applied perpendicularly to the sample surface is frequently used to increase the image contrast. Although old, Bitter technique has been recently improved. In a recent paper, Simpson et al. [122] have reported the magneto-optical response of negatively charged nitrogen-vacancy (NV) centers in diamond to quantitatively image the stray magnetic fields in thin ferromagnetic films with time and spatial resolutions of 20 ms and

440 nm. The technique, potentially working with any magnetic material with a stray magnetic field, can be operated under ambient conditions by placing the specimen in contact with the diamond imaging chip (see also in Section 3.10). In another paper, Sonntag et al. [137] investigated a deformed steel sample (S235JR) with composition: Fe-0.073C–0.47Mn–0.19Si–0.26Cu–0.12Cr–0.069Mo–0.18Ni to image macroscopic deformation gradients. The authors also concluded that geometrical features of the stress curves can be determined quantitatively by Bitter contrast images, including dimensions of different deformation regions and asymmetries.

3.2. Magneto-Optical Imaging

The interaction between polarized light and the magnetization of a material is generally a weak signal, but informative enough to attain the visualization of the domain. Though the principles of the magneto-optical interactions could be relatively difficult, the method is based on the change in polarization of incident light as it is reflected (or transmitted) by a magnetic material at the most elementary level. The magneto-optical effect in reflection mode is usually referred to as Kerr effect (MOKE), while in transmission mode as the Faraday effect. Precisely, the MOKE technique can be also divided into three different geometries: Longitudinal, polar, and transverse magneto-optical Kerr effect [138]. In the longitudinal MOKE geometry (Figure 9c), the magnetic field is applied in the plane of the sample and domains will tend to be oriented along the magnetic field direction. The light engendered by a laser passes firstly through a polarizer, then in an objective lens focusing the sample. On the other hand, the incident polarized light reflects off the sample surface, being rotated by the interaction between the polarized beam and the magnetic domain structure of the sample. The magnitude of the rotation is proportional to the local magnetization. The reflected signal passes through an analyzer so that the Kerr rotation signal can be measured. The degree of Kerr rotation can be used to determine the orientation and magnitude of the local magnetic domain. In the polar MOKE geometry, out-of-plane with respect to the sample plane the magnetic field is applied, and the resulting magnetization is transverse to the sample plane (Figure 9d). The incident laser excitation, perpendicularly aligned to the sample plane, is usually adopted for collecting the maximum signal. As in the longitudinal geometry, the polarization of incident laser light is rotated by a small degree when it reflects off from the surface of the magnetic sample. The magnitude of the Kerr rotation is related to the strength and orientation of the local magnetic domain. In fact, when the light beam is incident perpendicular to the surface in the Faraday or Kerr configurations, domains magnetized normal to the surface plane are imaged.

For the visualization, a light microscope with a working distance (WD) of several centimeters is required. The optical path allows to allocate a polarizing filter between the sample and objective lens, but owing to the relatively large WD, the resolution is slightly limited. To gain Kerr microscopy images closer to the limit resolution (300–500 nm), a relatively small distance between sample and objective lens, typically one millimeter or less (even possible if the sample is in the air), is required. Notwithstanding these techniques can be very sensitive, usually no quantitative determinations of the magnetization are provided. Nevertheless, the combination with a dedicated setup (e.g., the application of a bias potential) can provide quantitative results [122] (Figure 9f). Recent experimental studies on low-dimensional nanostructures (e.g., transition metal dichalcogenides, and graphene) show the potential of the magneto-optic effects to gain magnetic information on 2D systems. Gong and co-authors [133] reported the magnetic imaging of few-layered $Cr_2Ge_2Te_6$ crystals, as obtained at 40 K, 22 K, and 4.7 K (Figure 9g). Taking into consideration these images, the authors concluded that, although the magnetic properties are strongly suppressed by thermal fluctuations, due to the magnetic anisotropy a long-range ferromagnetic order can be observed in 2D crystals. Furthermore, Kerr rotation images were recently used by Lee et al. [139] to describe magneto-electric properties of 1-layer MoS_2-based devices to be employed as integrated photonic and spintronic devices. Together with the domains visualization under static or dynamic conditions, the motion of magnetic domain walls driven by magnetic field pulses or current pulses can be studied by time-resolved or transient

MOKE spectroscopy [140]. MO Faraday technique has been adopted by Kustov et al. [141] to obtain thermal imaging with high sensitivity, with micrometer as spatial and with millisecond as temporal resolutions, by using $BiLu_2Fe_4GaO_{12}$ ferrimagnetic film as a pyro-magneto-optical detection layer. Dikson et al. [111] adopted high-resolution optical techniques (confocal microscopy and scanning near-field optical microscopy: SNOM) for imaging the magnetic domains in ferromagnetic thin films based on gadolinium–iron–cobalt (GdFeCo) and terbium–iron–cobalt (TrFeCo). The authors showed that the magneto-optical resolution in the near-field measurements depends on the film thickness and it is affected by the diffraction on magnetic domains throughout the film.

3.3. Scanning Electron Microscopy (SEM): Type-I, Type-II, and Type-III Magnetic Contrast-Based Methods

Scanning electron microscopy (SEM) technique combined with energy-dispersive X-ray spectroscopy (EDX) is commonly adopted for the chemical and structural analysis. When the sample is probed by a focused beam of electrons some signals can be obtained. Among all, secondary and backscattered electrons (SEs and BSEs, respectively) coming from the sample are collected. BSEs have energies close to that of the primary beam and give rise to a chemical contrast that is proportional to the Z of elements, while SEs are more sensitive to the topological features. Furthermore, other kinds of imaging, such as conductive and magnetic mapping [101,142], are possible.

3.3.1. Secondary Electrons (SEs), Backscattered Electrons (BSEs) Imaging

The imaging of the magnetic domain structure of a material is possible by scanning electron microscope (SEM) techniques based on the deflection of the electron beams (i.e., Lorentz force), which are exposed to a magnetic field or from the electron polarization [143]. Three different kinds of contrast mechanisms are known: Secondary electrons (SEs), backscattered electrons (BSEs), and SE polarization, known as type-I, type-II, and type-III magnetic contrast [33], firstly observed by Banbury and Nixon [144], Philibert and Tixier [145], and Pierce and Celotta [146], respectively. Nonetheless, no full description of the experimental conditions required to obtain the magnetic contrast was given and no explanation of the observed effect was described. After the first observations, the principles of the involved processes were then provided. Now it is clear that the electron deflection can be caused by the magnetic field outside (SEs) or inside (BSEs) the sample (Figure 10a,b). In both cases, no special care of the sample is required, but the surface polishing is recommended to avoid unnecessary contributions by topographic features. The main difficulty of these two techniques is related to the separation between magnetic contrast and other kinds of signals and the spatial resolution limited to the intrinsic principles of these techniques. Furthermore, the simultaneous SE and BSE-based imaging is often possible, when the two detectors are simultaneously available. While the main modification of a conventional SEM has to be dedicated to the geometry (i.e., positions of detectors) for maximizing the signal collection, the acquisition potentials should be set as low as 10 keV (i.e., maximize the emission) for SEs and as high as possible (i.e., ca. 20–200 keV) with a tilted geometric surface of about 50–55° for BSEs to obtain enough image contrast. Several studies have shown that the tilting and rotation controls are usually required to maximize the contrast for both type-I and type-II magnetic contrast imaging.

3.3.2. Scanning Electron Microscopy with Polarization Analysis (SEMPA)

The SEM with polarization analysis (SEMPA) known also as spin-polarized SEM, was first developed by Koike et al. [147] in 1984. The technique is a more advanced method for the observation of magnetic domains and is based on the spin-polarization of SE that are emitted from the local region (ca. 50 nm) of a sample surface when electrons of medium energy (10–50 keV range) probe the sample. The main difference with traditional SEM lies in the fact that electrons arrive at the sample position with a nearly zero kinetic energy. The principles of the visualization technique are based on the measurement of the spin-polarization of SEs ejected from a ferromagnetic surface after being irradiated with a high-energy electron-beam (Figure 10c). The ejected polarized electrons are detected by spin detectors allowing all three vector components of magnetization (more commonly the electron polarization along the two projected orthogonal axes parallel to the sample surface, i.e., in-plane magnetization) to be attained with a very high spatial resolution (ca. 10 nm) suitable also for 3D surfaces [116]. Spin-polarized SEM, if combined with other imaging methods currently adopted in SEM (i.e., Auger, EDX, BSE diffraction microscopy), provides insightful information on chemical elements, crystal direction and topography with an increased spatial resolution down to about 3 nm. An example is reported in Figure 10d, where a NdFeB permanent magnet is imaged. Besides magnetic domain visualization, topography and elemental distributions of Nb and Fe for the same area were obtained by scanning Auger electron microscopy by Koike [116]. By comparing the different images, the author concluded that the Nd-rich areas were found to coincide with the grain boundary regions.

A set of spin-polarized SEM characteristics merits a particular attention compared to that of other magnetic imaging techniques: (i) spin-polarized SEM directly reveals the sample magnetization, whereas the majority of imaging techniques is sensitive to the magnetic fields out/inside of the sample; (ii) the quantitative magnetic interpretation from images is possible; (iii) despite the fact that SEs are generated in large quantity and the magnetization signal is pretty high, the low efficiency of the detectors makes the measurements more problematic; (iv) a very high spatial resolution can be achieved (below 10 nm); (v) both, magnetization and SE intensity properties, can be simultaneously monitored; (vi) both, the magnetic and topographic features, can be obtained and compared, providing information on the characteristics of the magnetic domain structure.

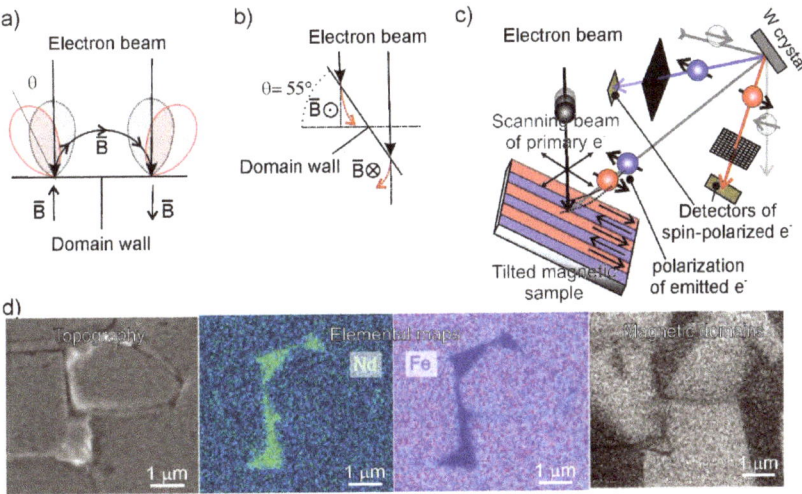

Figure 10. (**a**–**c**) representations of type-I, type-II, and type-III magnetism in SEM, corresponding to SE, BSE, and polarized electron detections, respectively; (**d**) Nd, Fe elemental distributions (left panels), topography and magnetic domain images (right panels) of a NdFeB permanent magnet. Element maps were acquired by Auger spectroscopy. Panels in (**d**) are adapted from [116], published by Oxford Academic, 2013.

Lastly, spin-polarization SEM signal, that is essentially sensitive to topmost few-layers, can be rapidly attenuated by the presence of non-magnetic coatings, overlays, molecularly adsorbed species, thus ion-beam together with ultra-high vacuum conditions, are gainfully required. Surface preparation under ultra-high vacuum conditions is necessary. Surfaces for spin-polarized SEM imaging can be prepared by depositing an ultra-thin overlayer of high spin polarized material (e.g., Fe). It is so assumed that, due to the very thin film thickness, the magnetic character of the underlaying sample is not significantly affected.

Depending on the emitted electrons, a lateral resolution as low as 10–50 nm can be obtained, without any contribution of the sample topography. However, the probing depth that corresponds to the quick escaping depth of SE, is of the order of a few nanometers. The typical image acquisition time depends on the image size and it is in the 1–100 min interval. SEMPA offers a series of advantages, including both the large field depth and the easily variable magnification, allowing to raster regions from a few millimeters to a few hundred nanometers. On the contrary, conductive samples are required to prevent charging effects and stray magnetic fields (above 10 Oe) must be avoided.

In recent experiments it was found that excited electrons in SEM could escape from the tip-target junction, helping in building new electronic systems based on polarized spin at nanoscale. Bertolini and co-authors [148] reported for the first time the scanning field emission microscopy with polarization analysis (SFEMPA). In SFEMPA a STM tip is held on one side and the spin polarization generated by electron scattering, thus providing the entire magnetic detail. However, due to the low efficiency of a spin polarimeter, spin-polarized images have been demonstrated to be limited to fewer pixels of the STM images during the imaging of Fe dots deposited on a W(110)-single crystal surface. The 1 nm spatial resolution, which primarily depends on the distance between the pixels, can be improved by increasing the efficiency of the spin polarimeter and by optimizing the electron optical parameters.

3.4. Transmission Electron Microscopy (TEM)

In the past, the spatial resolution of the TEM instruments has been primarily limited by the performances of the magnetic objective lens (and not by the incident electrons wavelength). Both, the primary image and diffraction pattern, which are responsible for the ultimate image quality, are by far affected by the lens aberration characteristics first, and are later magnified by all the other lenses. Magnetic objective lenses designed with small aberration coefficients (i.e., Cs: spherical and Cc: chromatic aberrations, respectively) for achieving atomic-scale spatial resolutions have been one primary achievement in the modern electron microscopy. Recent TEMs operate with high-power, high-quality magnetic lenses to produce magnified images. In order to obtain a short focal length, which is an essential condition to achieve atomic-resolution imaging, a relatively strong magnetic field as high as about 2–3 T is usually required inside the magnetic objective lens, where samples are placed. However, under these conditions, the magnetic and physical structures can be largely altered or even destroyed. The question can be unraveled by using newly developed magnetic objective lenses causing a magnetic field-free condition at the sample position [149]. This technical advancement, combined with the high-order aberration correction, permits to achieve a sub-angstrom spatial resolution (ca. 0.4 Å at 300 kV as an acceleration potential) [150] even under a very small residual magnetic field (below 0.2 mT) at the sample position. The direct atom-resolved imaging of magnetic materials can be attained enabling a new stage in atomic resolution electron microscopy with no alteration by high magnetic fields. In more detail, the standard double-tilt sample holders together with the interaction of the high energy beam with the sample result in characteristic spectroscopies, including electron energy loss spectroscopy (EELS) and energy dispersive X-ray (EDX) spectroscopy, which are both very informative about chemical bonds, structures, and compositions at the nanoscale. In the field of materials science, a multi-technique approach is often adopted by combining microscopies and spectroscopies together with other techniques. In this regard, the electron microscopies are more commonly used to determine the morphology and the structure, while the magnetic properties are revealed by complementary techniques [91,96,108,151].

3.4.1. Lorentz Transmission Electron Microscopies: Fresnel, Foucault and Differential Phase Contrast (DPC) Imaging

The Lorentz transmission electron microscopy (LTEM) is based on the Lorentz force causing the deflection of the electron paths of a beam perpendicularly oriented to the magnetic field. Lorentz microscopy operates in a conventional TEM or in a scanning TEM (STEM) instrument through a thin sample by means of three different techniques of imaging identified as Fresnel, Foucault (Figure 11) and differential phase contrast (DPC) microscopy (Figure 12). Lorentz microscopy in Fresnel mode (Figure 11a–d) is by far the most widely used also because the defocused electron beam transmitted through the sample is informative of the magnetic properties via the magnetic contrast located at the position of the domain walls (Figure 11d). Depending on the direction of magnetization of each domain, the beam is slightly deflected toward one or the other wall by the Lorentz force. Therefore, the domain, delimited between one bright and one adjacent dark region corresponding to the wall positions, can be visualized (Figure 11e,f). On the other hand, the domain structure in the Foucault mode is imaged in focus and the electrons, deflected by the magnetic field in one way, can be blocked by the aperture displacing (Figure 11c). The resulting contrast within a single domain thus depends on the direction of its magnetization. In both operational conditions, images directly display the domain wall structure and the dimensions can be determined quantitatively from these images.

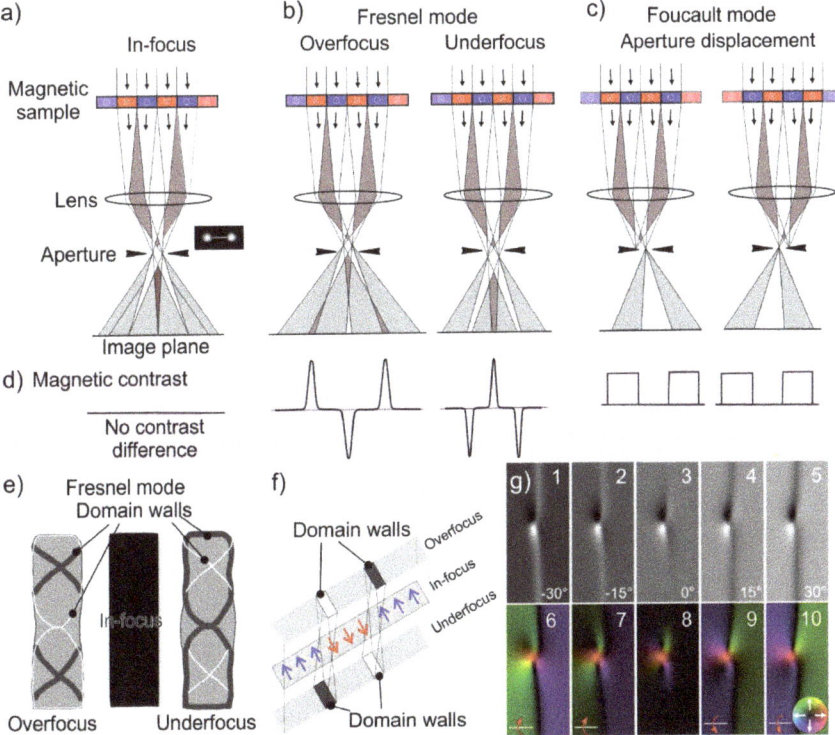

Figure 11. Lorentz microscope (**a**) in-focus; (**b**) overfocus and underfocus Fresnel mode TEM imaging; (**c**) Foucault mode; (**d**) the related magnetic contrast imaging; (**e**) domain wall contrast inversion in Fresnel mode; (**f**) scheme of the inverse bright/dark contrast corresponding at the positions of the magnetic domain walls in defocused Fresnel; and (**g**) bright-field Fresnel-mode L-TEM image sequence (1–5) of an isolated DW skyrmion collected with a different tilting angle and the corresponding in-plane magnetic induction maps (6–10) as obtained by TIE. Panel g reprinted with permission from [152], copyright by American Physical Society, 2019.

The maximum resolution attainable for imaging the magnetic structure largely depends on several factors, including the sample type, the delocalization value of the Lorentz lens, the spherical aberration of the instrument, and it is typically in the 2–20 nm range. As far as the sample is concerned, only thin specimens (less than about 150 nm) can be analyzed. In other words, the deflection is caused by the magnetic field within the sample and the information depth comes from the full thickness of the sample and with the same weighting for all depths. TEM instruments for this appropriate practice are already available, nevertheless, the moderately high complexity of methods, together with the preparation of thin samples that may affect the domain structure, have to be considered for this kind of imaging techniques. As experimental Lorentz TEM images in Fresnel-mode do not provide a direct information regarding the magnetic induction directions, the reconstruction of the phase is typically adopted using the transport of intensity equation (TIE). Such equation when applied to a focal series of the Fresnel images, may allow in reconstructing the magnetic phase shift of the electron wave (i.e., in-plane magnetic induction properties) and the quantitative mapping of the magnetic induction is also possible. In a recent paper by Cheng et al. [152], the Fresnel mode L-TEM was used to image a DW skyrmion, a kind of topological magnetic excitation that is characterized by vortex-like magnetic structure (a distinctive 360° transition) formed by the magnetic moments of electron spins. The sample tilting procedure was performed to allow magnetic contrast to be displayed in Fresnel mode (i.e., in the absence of the tilting stage, Néel walls do not appear) and the in-the-plane magnetic component emerges from the perpendicular induction of neighboring domains giving rise to contrast at the positions of Néel walls. TIE equation is also employed to calculate the integrated in-plane magnetic induction.

Differential phase contrast (DPC) microscopy is a scanning TEM-based technique rastering athwart the specimen (Figure 12a). The local Lorentz deflection is obtained at the electron probe position and observed by utilizing a quadrant detector able to record signals from detector segments located in opposite positions, thus providing the two components of the Lorentz deflection angle (β_L). Semi-quantitative maps of the magnetization directions and on the domain wall structures with a spatial resolution nearly equal to the size of the electron probe or 10–20 nm for more typical applications can be obtained. In this domain, with the aberration corrected TEM technique utilized in field-free Lorentz STEM imaging of magnetic specimens a spatial resolution lower than 1 nm for direct imaging of magnetic structure by EM is possible [153,154]. One example of the aberration corrected L-STEM is shown in Figure 12b–e. In these images a $Fe_{60}O_{30}C_{10}$ ferromagnetic nanostructure, 50 × 500 nm in size, is imaged. The in-plane magnetic characteristics of this nanostructure show a dual vortex structure (i.e., a close loop of the in-plane induction direction). From this image it is clear that the extent of the vortex core size can be easily measured across the vortex core direction and the associated deflection angle (±25 μrad) can be assumed proportional to the integrated magnetic induction (±40 T nm) [153].

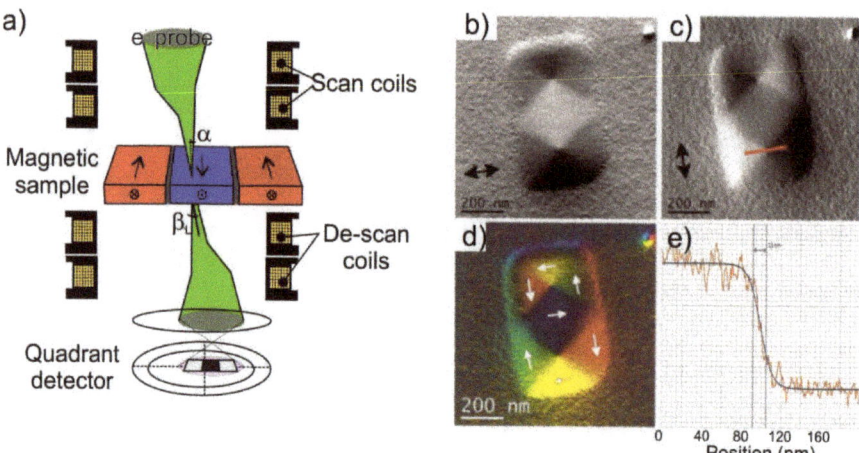

Figure 12. (**a**) scheme of the differential phase contrast (DPC) technique in L-STEM; (**b**,**c**) L-STEM images of a Fe nanostructure (from Fe exacarbonyl by EBID) with its in-plane components; (**d**) DPC color image (representation of the color wheel shown in the top-right inset) and (**e**) the profile of the line selected in (**c**) showing the induction profile, as measured from the deflection angle at the vortex core. β_L is Lorentz deflection angle. Panels b–e are reprinted with permission from [153], copyright by Elsevier, 2015.

3.4.2. Electron Holography

The principles of the phase and of the electron shifts are used in the electron holography (EH) (Figure 13).

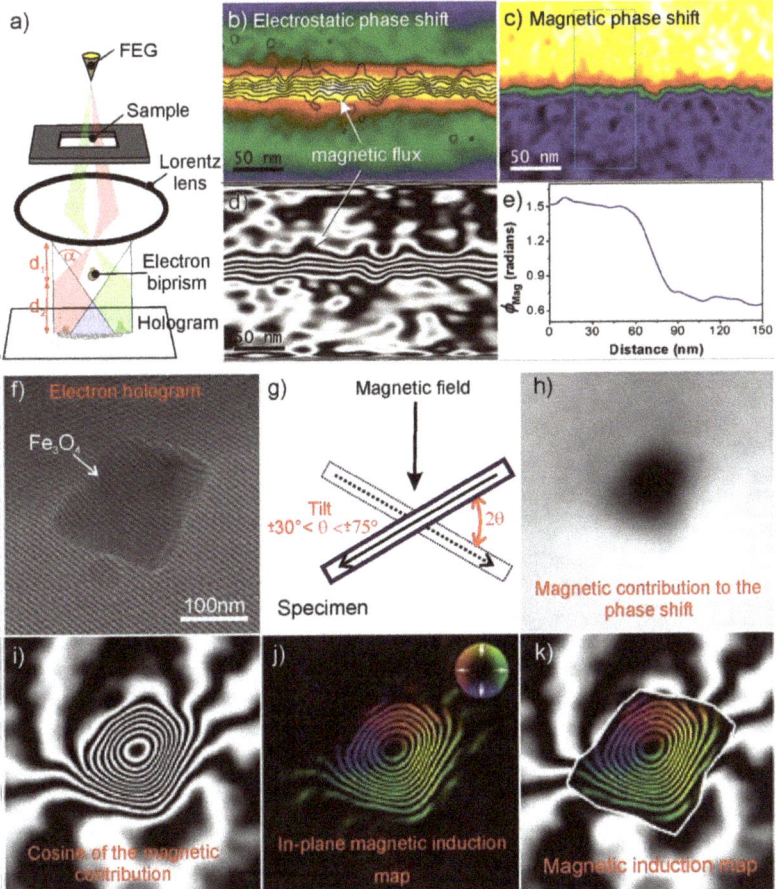

Figure 13. (a) Schematic diagram showing interference fringe formation from the overlap of the sample and reference waves by the electron biprism, adapted from [155], published by Wiley, 2010. Electron holography (EH) images of a Co nanowire 9 nm thick and 30 nm long: (b) electrostatic and (c) magnetic contributions to the electron beam phase shift; the magnetic flux deduced from the magnetic phase image is superimposed over the electrostatic phase shift image; (d) the cosine of 38 times the magnetic phase image that allows a better visualization of the magnetic flux within and out of the nanowire; (e) phase shift profile extracted from the right part of panel (d) where the magnetic induction is constant; (b–e) reproduced with permission [156], copyright (2011) by American Chemical Society. (f) Electron hologram of a Fe_3O_4 nanoparticle exposing well-resolved interference fringes; (g) representation of the tilting operation to obtain the in-plane magnetization component to the total phase shift; (h) magnetic contribution to the total phase shift; (i) Cosine of the magnetic contribution to the phase shown in (g) to produce contours with a spacing of 0.53 radians; (j) direction of the projected in-plane magnetic induction in the Fe_3O_4 nanoparticle as obtained with the addition of colors; and (k) the final magnetic induction map that shows the strength and direction of the magnetic signal in the Fe_3O_4 nanoparticle and the stray magnetic field. (f–k): panels are adapted from [157], published by Wiley, 2019.

Several types of EH are known, including TEM and scanning TEM. The most used form is the off-axis electron holography that involves an electron biprism inserted in the aperture plane of the column of the microscope perpendicularly to the electron beam direction. A field emission gun (FEG) is mandatory for delivering a coherent source of electrons. Moreover, by applying a voltage to the biprism, two trajectories of electrons are found in the back focal plane to constitute the electron hologram. By increasing the biprism voltage, the electron sources are pushed further apart, which will increase the width of the hologram and decrease the fringe spacing [155]. Several studies have shown that for medium-resolution EH, alternatively to the conventional objective lens, a Lorentz lens can be used to provide a useful field of view in the range 200 nm to 2 µm.

The off-axis electron holography approach makes possible both, the direct magnetic domain imaging and quantitative measurements of the magnetic flux (B).

For these reasons, together with the high energy required (i.e., 200 keV) an interference pattern is formed from electrons in TEMs allowing to play with electron holography. When electron holography is attained for the magnetic domain imaging, an interference pattern is generated by the splitting of the beam into two components through the negatively biased electron biprism: One electron trajectory passing through the thin specimen and another being unperturbed. When the two components are recombined on the detector plane, an interference figure is thus obtained (Figure 13a). Furthermore, the obtained additional phase shift is immediately related to the total magnetic flux enclosed by the two paths of electrons. Interestingly, interference pattern lines could be immediately interpreted as the magnetic flux lines. Additionally, a procedure can also be adopted as an alternative. In other words, as both electron trajectories pass through the magnetic sample with somewhat different paths, the observed difference in phase is proportional to the flux within the dual beam regions, but it is sensitive to the domain walls. Different approaches may be tackled for extracting the shift of the phase signal related to the magnetically induced electron wave, which is neighboring the sample. The magnetic figure can be fully quantitatively understood enabling the reconstruction of the electrostatic and magnetic phase shifts. In a paper by Serrano-Ramon et al. [156] an example of the off-axis electron holography is provided (Figure 13b–e). In the images, a 9 × 30 nm Co nanowire is illustrated together with electrostatic and magnetic phase shifts. In another paper, Almeida et al. [157] reported the magnetic structure of vortex-state Fe_3O_4 grains (Figure 13f–k). In these images, from the EH image (Figure 13f) by tilting the sample of equivalent and opposite angles in the ±30°–75° range (Figure 13g) the magnetic contribution was separated from the phase shift (Figure 13h). A cosine image (Figure 13i) of a chosen integer multiplied by the magnetic contribution to the phase shift shown in Figure 13j was used to generate the magnetic phase outlines (Figure 13k) within the Fe_3O_4 grain from the surrounding stray magnetic field. In fact, from electron holograms recorded with the particles magnetized in opposite directions, the mean inner potential (MIP) contribution to the phase could be subtracted from the total recorded phase to gain the magnetic contribution to the phase.

A spatial resolution of 10 nm is possible with electron holography with specimens about 50 nm in thickness, but thicker samples (below 150 nm) even conductive can be investigated. In any case, this technique should be intended as extremely powerful, but complex, due to the wide range of applications and difficulties in the preparation of specimens, image reconstruction and analysis time.

3.5. Spin-Polarized Low Energy Electron Microscopy (SPLEEM)

The SPLEEM has been used to address phenomena such as domain wall structures in thin magnetic films, micromagnetic configurations in surface supported nanostructures, spin reorientation transition, magnetic coupling in multilayers, phase transitions and finite-size effects. Its application has several advantages, such as real-time observation and the possibility to combine crystallography with magnetic information. The surface sensitivity limits its usefulness to samples prepared in situ (or grown elsewhere) and protected by a removable capping layer. The main disadvantage in the use of SPLEEM is its strong sensitivity to applied magnetic fields, which distress the trajectory of electrons and degrade the image quality. Modest fields of a few hundred gauss can be applied only in the surface normal direction, so that the Lorentz force is geometrically minimized. This limitation affects important fields of research, such as dynamics on domain walls and exotic magnetic states of matter. The usual image contrast is augmented by magnetic contrast generated by the exchange interaction between incident spin-polarized electrons and spin-polarized electrons in the magnetic material [95]. An important family of applications of SPLEEM employs the vector imaging capability to resolve domain wall (DW) spin textures. Zhou et al. [158] have recently reported a series of thin films, including Co/W(110), Co/Cu(001) and (Co/Ni)/W(110), obtained at zero contrast between magnetic domains. Under these conditions, the authors observed the appearance of magnetic contrast outlining the DWs.

3.6. Scanning Probe Techniques

In scanning probe methods, a sharp tip laterally scans over the sample surface. The interaction region and the resolution are primarily determined by the sizes and the properties of the tip. A feedback loop is commonly used to correct the height position of the tip and measures the tip-sample interaction. Depending on the type of the tip and sample characteristics, different scanning probe methods can be distinguished. Atomic force microscopy (AFM), takes advantage from either repulsive (contact-AFM mode) or attractive (non-contact AFM mode) interactions between the tip and the sample. The van der Waals forces are so used to regulate the tip-sample distance for the imaging from the microscale to the atomic resolution [159]. Scanning tunneling microscopy (STM) technique is based on the tunnel current between the tip and a conducting sample. The extent of the tunneling current represents the tip-sample interaction. Both AFM and STM are by far among the imaging techniques with the best lateral resolution able to detect the structure variations (i.e., conformation, adsorption geometry, bond-order relations) and visualize the charging state, when single electrons are added to or removed from small molecules (e.g., pentacene, porphine, azobenzene) or with ionic state variations from neutral to anionic and dianionic states of tetracyanoquinodimethane [160].

3.6.1. Magnetic Force Microscopy (MFM)

Magnetic force microscopy (MFM) is a relatively young technique, firstly demonstrated in 1987, but now is by far one of the most diffused investigation techniques for studying the magnetic structure at the mesoscale [161]. This technique is based on magnetic forces acting between a sharp ferromagnetic tip probe and the magnetic characteristics of a sample surface (Figure 14a,b). When the magnetic tip scans a magnetic surface, the small attraction (or repulsion) forces acting between the tip and surface are detected, thus making possible to image magnetic characteristics of the sample surface. The tip radius of 20–40 nm is used, which determines the limit of lateral resolution in the magnetic imaging.

In the typical MFM technique one (i.e., single-pass mode) or two consecutive scans (i.e., dual-pass mode) are involved. While in the single-pass mode both topography and magnetic properties are simultaneously acquired, in the dual-pass mode the topography of the surface is acquired first, by taking advantage of the van der Waals forces (i.e., short-range interactions) acting between the probe and surface, by using the conventional tapping-mode atomic force microscopy (AFM). Then, in a second scan, the probe preventively magnetized by a permanent magnet, is lifted above the surface and repeats the scanning of the surface at a constant lift height at a distance where van der Waals contributions are negligible and the magnetic tip probes are exposed only to the magnetic interactions (i.e., long-range forces). Magnetic force microscopy usually works in AC mode, in which the tip oscillates at a certain frequency near to its resonance frequency (ω_0). Importantly, ω_0 shifts when a force is applied on the cantilever, thus the resonance frequency shift ($\Delta\omega$) depends on the force gradient, causing a variation in the phase and amplitude (Figure 14c,d, respectively).

Two key advantages of MFM are the high spatial resolution that is of the order of 40–50 nm, and the flexibility to work in environmental conditions and with applied magnetic fields. Single magnetic nanoparticles acquired in the air are shown in Figure 14e–g [65]. The chitosan-pyrolized magnetic nanoparticles (Fe_3O_4, γ-Fe_2O_3) preventively separated by a permanent magnet and deposited on a freshly cleaved mica surface were AFM imaged in a first scan (Figure 14e) and their sizes determined in the 22–30 nm range from the related height profiles.

The phase shift signal, imaged in the second scan by using the same tip probe operating in lift mode at four selected heights (90, 100, 110, 120 nm), was found to be effective in minimizing the topographic features (i.e., short-range interactions) and in revealing the magnetic characteristics (i.e., long-range forces). It is worth mentioning that negative phase shifting (darker regions in Figure 14f), depends on the operating distance and it is associated with attractive interactions between the sample and the magnetic probe in combination with a positive shift of the amplitude signal in both forward and backward scans (Figure 14g). The method is relatively simple for bulk materials, films or for nanoparticles when they are placed on flat supports, like mica or highly oriented pyrolytic graphite (HOPG). The acquisition could be much more difficult for nanoparticles dispersed into a porous carbon texture [162] or in a polymer phase [163,164]. Other key advantages offered by the technique are the possibility to work under vacuum, with or without externally applied magnetical and electrical fields and variable temperature conditions. High-resolution MFM imaging can be obtained in vacuum conditions (10^{-4} Pa) to benefit from higher sensitivity (i.e., higher Q factor of the cantilever) and the (thermal) stability.

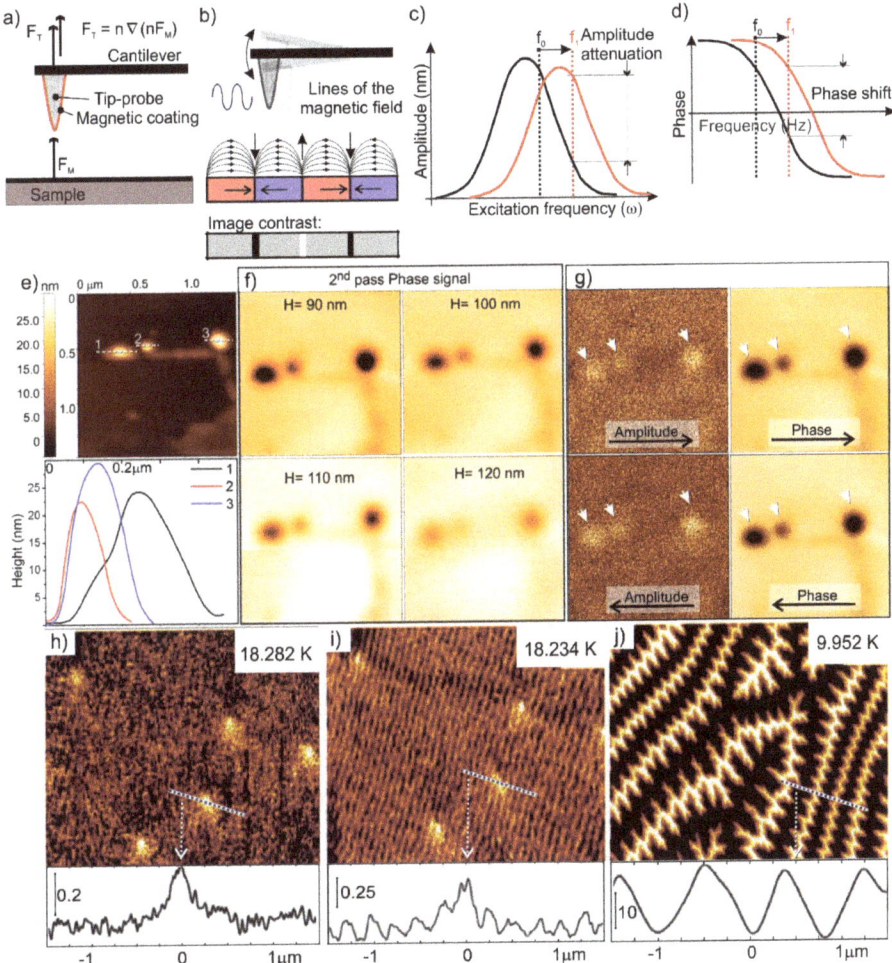

Figure 14. (**a**) Principles and (**b**) forces acting between the tip and the magnetic surface in MFM; (**c**) phase and (**d**) amplitude shifts associated with the long-range magnetic interactions; (**e**) topography signal (top panel) and height profiles (bottom panel) of pyrolyzed magnetic chitosan nanoparticles imaged on the freshly cleaved mica support; (**f**) magnetic force microscopy (MFM) phase shift images obtained at various lift heights (H = 90, 100, 110, and 120 nm); (**g**) forward and backward amplitude and phase maps obtained for H = 90 nm above the sample surface (obtained at constant height of 90 nm). Reproduced with permission from [65], copyright (2016) by American Chemical Society; (**h–j**) MFM maps acquired at T = 18.28 K, T = 18.23 K, and T = 9.95 K without a magnetic field. In the bottom insets, profiles of the phase signal along selected lines of the three structures are illustrated. Panels (**h–j**) are adapted from [165], published by American Association for the Advancement of Science, 2018.

In a recent paper, Geng et al. [166] showed that the magnetoelectric force microscopy (MeFM) that is a combination of magnetic force microscopy (MFM) with in situ modulated electric fields (**E**), can be employed to detect the **E**-induced magnetization (ME) and to show the multiferroic domain structure in hexagonal (h-)ErMnO$_3$. Along with multiferroic materials (e.g., Cu$_2$OSeO$_3$) [167], they exhibit the so-called magnetoelectric effect, in which an external magnetic field can cause electric polarization and an electric field can cause magnetic ordering. The field of the manipulation of magnetic structures by electrical fields is beyond the scope of this review, but it is highly desirable for technological applications.

The interplay between superconductivity and magnetism in single crystal EuFe$_2$(As$_{0.79}$P$_{0.21}$)$_2$ magnetic force microscopy (MFM) operated at low temperature and without applied magnetic field has been recently reported by Stolyarov and co-authors [165]. The MFM phase shift maps (Figure 14h,i,l) are representative of three temperature regions: TFM $\lesssim T < T_c$, $T \lesssim$ TFM $< T_c$, and $T <$ TFM $< T_c$ (in which TFM and T_c represent the ferromagnetic transition and the critical temperatures, respectively). Above TFM (ca. 18.28 K), as expected for a superconductor, the conventional Meissner state was found homogeneous (Figure 14h). Below TFM (ca. 18.23 K), the Meissner state first became striped with the domain width in the 100–200 nm range (Figure 14i), then a new phase exhibiting domain vortex-antivortex state (DVS), made of larger domains of ca. 350 nm, 40 times stronger than the magnetic contrast, was observed (Figure 14j).

The indirect magnetic force microscopy (ID-MFM) has been recently reported by Sifford et al. [108] as a novel method for the detection of magnetic domains. The newly developed method has been demonstrated for superparamagnetic iron oxide nanoparticles (SPIONs) immobilized on silicon nitride TEM windows and it has the potential to be applied in cells and tissue sections, without contaminating the magnetic probe.

Notably, due to the possible mutual interaction between tip and sample, the interpretation of MFM images is a debated topic and only qualitative data restricted to the more classical ferromagnetic samples (i.e., bulk materials, thin films, or nanostructures) are available.

3.6.2. Spin Polarized Scanning Tunneling Microscopy (SP-STM)

The spin-polarized scanning tunneling microscopy (SP-STM) is a special application of the STM technique that can provide an insightful detail of magnetic characteristics at the subatomic scale in addition to the topographic signal that is usually obtained by STM. Domain walls in antiferromagnetic/ferromagnetic systems obtained under static or dynamic magnetic and thermal conditions can be precisely determined by this technique. A magnetic-coated sharp tip scans over a sample surface under a potential applied between, thus promoting electrons to tunnel between the tip and sample and resulting in an electrical current. When the tip is magnetized the electrons having the same spin orientations (i.e., parallel orientation) at the sample surface provide a higher tunneling current than that given by spin with the antiparallel orientation (Figure 15a,b), while in the absence of magnetic interaction this current is indicative of the local electronic properties.

The magnetic probe is the most important component for the atomic spatial resolution: Both the geometry and the magnetic properties of the atomically shaped probe are crucial for the good interaction (i.e., high spin polarization, relatively small stray magnetic field and spin orientation properties at the tip apex). Three operational modes can be adopted: (i) constant current, (ii) spectroscopy, or modulated tip magnetization (exclusive for SP-STM). A particular sample (and probe) preparation is required as well as ultra-high vacuum (UHV) conditions to prevent oxidization and mislay of magnetization at the sample/tip surfaces. The separation of the magnetic features from topographic and electronic features is required also selecting the most proper operational mode.

The field of molecular magnets has become extremely active since 1990s, when it was shown that transition metal coordination molecules (i.e., [Mn$_{12}$O$_{12}$(OAc)$_{16}$(H$_2$O)$_4$] (Mn$_{12}$Ac) compounds) can retain magnetization for long time in the absence of external magnetic fields at liquid-helium temperatures. Over the past years, a large family of single-molecule magnets have been reported as well reviewed by Woodruff [51] and Guo [168]. In a recent paper [169], a single-molecule magnet, a dysprosium metallocene cation [(Cp^{iPr5})Dy(Cp*)]$^+$ (Cp^{iPr5}: penta-iso-propylcyclopentadienyl; Cp*: pentamethylcyclopentadienyl) that displays magnetic hysteresis above liquid-nitrogen temperatures, has been reported for the first time. An example of SP-STM imaging down to single molecule magnet is shown in Figure 15c–e. In a paper by Iacovita et al. [170] reported the spin polarization parallel and antiparallel of Co phtalocianine laying on magnetic nanolead support by adopting −0.32 V and

−0.16 V as working potentials, respectively. A spatially resolved molecule is illustrated together with the molecular structure model superimposed (Figure 15e). In a more recent paper, Schwobel and co-authors [171] reported an atomically resolved single molecule imaged by SP-STM technique (Figure 15f,g). In these images, topographies (Figure 15f,g top panels) and spin-resolved differential conductance maps (Figure 15f,g bottom panels) of bis(phthalo cyaninato)terbium(III) (TbPc$_2$) are shown. The topographic images showed only a slight difference, while the spin-resolved maps of differential conductance, taken with parallel or antiparallel alignments by applying a potential of −0.5 V, showed a remarkable magnetic contrast corresponding to an eight-lobe or a cross-shaped structure for parallel or antiparallel alignments, respectively. It is worth mentioning that either parallel (blue/dark colors) or antiparallel (red/grey colors) orientations can be obtained by applying an external magnetic field ($B = \pm 1$ T) and that Co islands did not rotate their orientations due to the hard magnet characteristics of Co.

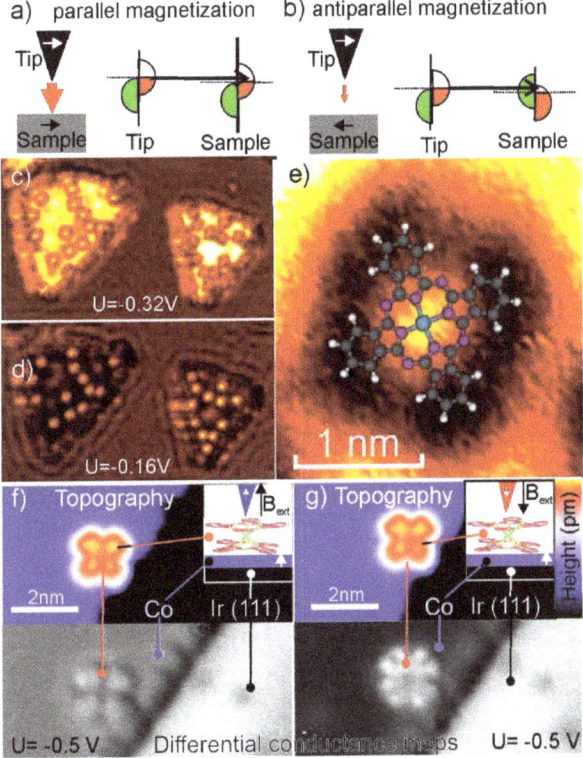

Figure 15. (**a**,**b**) representation of the spin polarized scanning tunneling microscopy (SP-STM) principle. In magnetic materials the density of states largely depends on the two different spin orientations: Parallel (higher tunneling current) or antiparallel (lower tunneling current) one; (**c**,**d**) spin-polarized tunneling conductance maps for Co phatolocyanine taken at −0.32 V and −0.16 V (image size: 40×20 nm^2); (**e**) map of a selected region in (**c**) (2.6×2.6 nm) with molecular structure model superimposed. Reprinted with permission from [170], copyright 2008 by American Physical Society; (**f**,**g**) topographies (top panels) and the associated spin-resolved differential conductance maps (bottom panels) of a single TbPc$_2$ molecule laying on a metal Co-layered Ir (111) surface for parallel (left panels) and antiparallel (right panels) magnetization alignments. In the top insets of (**f**) and (**g**) schemes of the operational conditions with the superimposed molecular structure model of TbPb$_2$, are illustrated. Adapted from [171], published by Nature Publishing Group, 2012.

3.6.3. Scanning-SQUID Microscopy

The scanning-superconducting quantum interference device (SQUID) microscopy is an ultra-sensitive technique for quantitative measurement of weak and local magnetic fields at the mesoscale. The probe, consisting of a superconducting quantum interference device (SQUID) (vide supra), allows the spatial scanning above the sample surface of hundreds of nanometers. The principal advantage and disadvantage of the scanning-SQUID microscopy with respect to the other magnetic scanning probe microscopies are the ultimate sensitivity of magnetic fields down to the nT scale and the limited spatial resolution as low as in the submicron scale only [172], although there are some promising new approaches with much higher spatial resolution. In any case, the last developed instruments allow cryogenic SQUID measurements [173]. By using microfabricated sensors, the scanning SQUID microscopy makes it possible various studies, including the ferromagnetism in novel materials and heterostructures [174] and fractional vortex formation in p-phase shift structures showing high-T_c superconducting state [175].

3.6.4. Scanning Hall Probe Microscopy (SHPM)

The scanning Hall probe microscopy (SHPM) is a variety of SPM techniques that incorporates the accurate sample approach and positioning of STM instruments equipped with a semiconductor Hall sensor. The combination of these two characteristics permits the magnetic induction mapping over a scanned surface. Although the technique has been demonstrated for domain imaging, it was originally designed for magnetic flux measurement in superconductor materials.

The SHPM technique can be considered as an improved magnetic imaging technique for a variety of reasons. Firstly, the technique can be combined with other scanning methods such as STM, but different from other methods, a small force, with negligible influence on the magnetic structure, is applied during the Hall probe analysis. Secondly, samples do not need to be electrically conductive (except for using a STM height control system). Thirdly, measurements can be performed from cryogenic to high temperatures under ultra-high vacuum (UHV) conditions, without special care of the surface or preparation. Lastly, the magnetic field sensitivity is very high (0.1 µT–10 T range). However, SHPM users have to take under consideration some limitations and difficulties, such as when they are acquiring high-resolution scans that become difficult due to the thermal noise of extremely small hall probes and that the obtained image is affected by the scanning height. Furthermore, a minimum height of the scanning distance is possible due to the hall probe assembling. Some other characteristics are in common with the probe scanning methods, for example the fact that a large scanning area is time-consuming and the importance of protecting from electromagnetic and acoustic noises (with a Faraday cage and an anti-vibrating platform, respectively) and static charge (ionizing units).

Ghirri end co-authors [176], reported the direct magnetic measurements for monolayers of molecular nanomagnets investigated by SHPM. In the paper, the magnetic response of $Cs_{0.7}Ni[Cr(CN)_6]_{0.9}$ (Prussian blue analogue) molecular structure is illustrated by studying the dependency of the magnetic images on the temperature and the applied fields. Dede et al. [177] reported three dimensional (3D) scanning Hall probe microscopy (3D-SHPM) of a magnetic patterned surface down to 700 nm as spatial resolution.

3.7. X-Ray Imaging Techniques

3.7.1. X-Ray Magnetic Circular Dichroism (XMCD) and Photoemission Electron Microscopy (X-PEEM) Techniques

The magnetic structures can be investigated at the nanoscale by means of X-ray methods. The spectroscopic and microscopic approaches can serve to identify fingerprints of the magnetic matter and quantitative information as well. Along this theme, the magnetic domain visualization with a spatial resolution of a few tens of nm, even in 3D, can be obtained with various X-ray microscopy techniques. In a recent paper [121], P. Fischer reviewed the X-ray imaging methods of the magnetic structures. The authors can largely refer to this review and references therein for the sake of completeness.

The X-ray magnetic circular dichroism (XMCD) and the X-ray photoemission electron microscopy (X-PEEM) techniques have been recently developed in synchrotron facilities offering tunable and intense radiations [118] (and references therein). In this contest, the dichroism definition is essentially associated with the polarization dependency of the absorption of X-ray photons with the magnetization orientation of the sample. The different X-rays absorption process, known as X-ray dichroism, is expected to be maximized when the material magnetization and the angular momentum of photons are antiparallelly or parallelly oriented. X-ray photons entirely transfer the angular momentum to photoelectrons (i.e., angular moment conservation during the absorption process) that are excited from a splitted spin–orbit core level. The magnetic properties are then given in a second step, when two X-ray absorption spectra (XAS) collected in a magnetic field, with the left- and right-handed circularly polarized lights, respectively, are compared. Such difference in the spectra can be informative of transitions that are too weak to be detected in conventional optical absorption spectra of the atomic site, including symmetry of the electronic levels, metal ion sites spin and orbit magnetic moments and paramagnetic properties.

In PEEM equipped with soft X-rays as excitation sources (X-PEEM), XAS and X-ray photoelectron spectroscopy (XPS) can be also implemented, thus offering a set of complementary information with a lateral resolution of the order of a few nanometers. On the other hand, an electron beam can be used as a probe as an alternative (PEEM), but samples with enough conductivity should be investigated for avoiding the surface charging and a lateral resolution of about 40 nm is expected, being the sensitivity confined to the topmost layer (i.e., 1–3 external atomic layers). The X-PEEM technique has been recently used by Kleibert et al. [178] for the direct observation of superparamagnetic and unconventional magnetization state in single 3d transition metal (i.e., Fe, Co, and Ni) nanoparticles (Figure 16a–f).

Nanoparticles were imaged by means of elemental maps obtained by collecting two images for the sample site: Firstly, an edge image was recorded with the photon energy tuned at the L_3 X-ray absorption edge of the element. Secondly, a pre-edge image having the photon energy tuned a few eV below the L_3 X-ray absorption edge energy was recorded. Ruiz-Gomex et al. [179] investigated single-crystal magnetite islands exhibiting spin structures in ultrathin, magnetically soft magnetite by X-ray spectromicroscopy (Figure 16g–k). In these images, the triangularly-shaped island with edges aligned along the magnetite [110] directions is shown (Figure 16g). XAS image illustrated in Figure 16h exhibits a contrast with a brighter region located at the magnetite position, as obtained by collecting spatially resolved low-energy (ca. 2 eV) SEs emitted from the sample upon X-rays irradiation with a photon energy corresponding to the L_3 absorption edge of Fe. X-PEEM image shown in Figure 16i illustrates an unambiguous magnetic contrast in the magnetite island, whose hexagonal pattern in the diffraction pattern, shown in Figure 16j, is indicative of an iron oxide with a spinel structure. The difference between the XAS spectra acquired with opposite light helicities from a single domain gives the corresponding XMCD spectrum shown in Figure 16k [179].

The versatility of XMCD in providing detailed information on the electronic and magnetic structure of a wide variety of systems, including nanoparticles, molecular magnets, organometallic complexes, and crystals has been recently demonstrated [118,180].

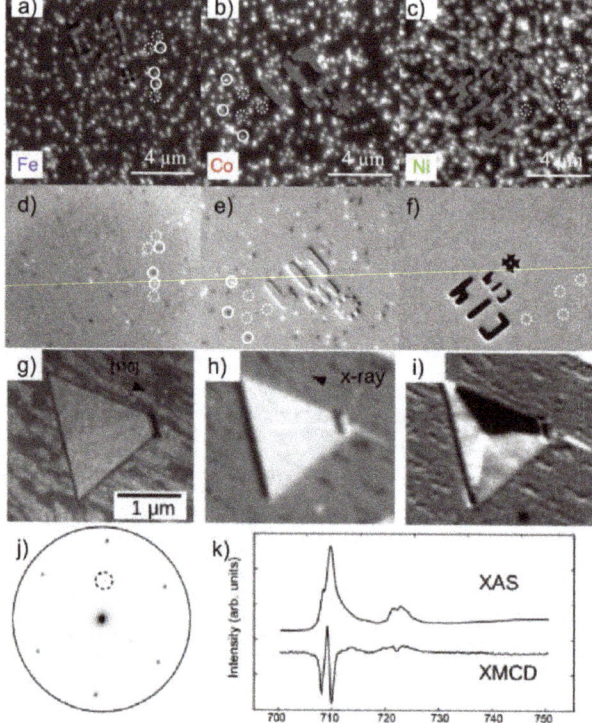

Figure 16. (**a**–**f**) elemental maps of the NPs: (**a**) Fe, (**b**) Co, and (**c**) Ni nanoparticles; (**d**–**f**) the corresponding magnetic contrast images. Dashed and solid circles represent NPs in the superparamagnetic and magnetically blocked states, respectively. Au markers shown in the images are used for the particle identification. Reproduced with permission from [178], copyright (2017) by American Physical Society. (**g**–**k**) multitechnique imaging of a Fe_3O_4 (magnetite) island on an FeO wetted layer on the Ru(0001) surface: (**g**) LEEM image (electron energy = 8 eV), (**h**) XAS image of the same area acquired at a photon energy close to the maximum of the L_3 Fe absorption edge, (**i**) X-ray magnetic circular dichroism image of the same area, (**j**) low-energy ED pattern acquired from the island at 30 eV. One of the ED spot, characteristic of the spinel phase is shown by a circle. (**k**) XAS spectrum of the island showing the L_3 and the L_2 Fe edges and XMCD spectrum taken in the black domain in (**i**). Reproduced with permission from [179], published by The Royal Society of Chemistry, 2018.

3.7.2. Scanning and Transmission X-Ray Microscopies (TXMs)

Full-field and scanning transmission X-ray microscopy (STXM) are two techniques to obtain a microscopy image by monitoring the intensity of focused X-rays transmitted through a relatively thin specimen. Since these techniques require collimated and high-intensity X-rays, the use of X-ray microscopies has been typically limited to synchrotrons. In the scanning operational mode, the X-ray beam is reduced to a small spot size at the specimen position by a focusing lens used to focus, and the sample is raster-scanned by measuring the transmitted intensity at each point. Together with transmitted X-rays, other signals (i.e., diffracted X-rays, photoelectrons and fluorescent photons) are usually simultaneously detected to map chemical and structural properties of the specimen [121,124].

On the other hand, full-field X-ray microscopy is analogous to that of conventional microscopies and the real-space image of the illuminated specimen within the field of view is directly captured. A tiny and lithographically fabricated concentric ring structure (known as zone plate) is used to focus X-rays from the synchrotron beam, working as an objective lens for X-rays. The concentric rings of the zone plate lead to the X-ray diffraction, and the observed diffraction angles depend on the wavelength, while the focal length is proportional to the energy of X-ray photons. Furthermore, the final spatial resolution is determined by the zone plate and by the width of its outmost concentric ring. With the actual lithographic feature, a spatial resolution of 10–20 nm can be attained. The technique is fast and allows the three-dimensional imaging and the study of biological specimens, when operated at near pressure conditions [125]. The energy-resolved TXM mapping can be performed with the complement provided by the X-ray absorption near-edge structure (XANES).

Comparing to PEEM, the XMCD signal monitored by TXM differs by the facts that the transmission configuration is not sensitive to external magnetic fields and it makes possible the imaging of reversal processes of magnetization, the resolution is typically higher, but the field of view cannot be easily changed. The main difference between X-PEEM and XTXM, is that X-PEEM is a more surface dependent (penetration depth is of about 2 nm), while XTXM is a transmission method through a thicker interaction region (below 50–100 nm) [124].

The photon energy can be tuned around the absorption edge of a specific element. The spatial resolution, i.e., the focusing size of the X-rays in the soft X-ray STXM is typically 20–100 nm. It is in principle determined by the diffraction limit of the lithographically fabricated FZPs.

A newly developed scanning X-ray microscope equipped with an 8T superconducting magnet as a tool for investigating the magnetic domain structures in inhomogeneous magnetic materials with a spatial resolution of 90 nm under high magnetic fields have been recently illustrated by Kotani and co-authors [181]. The performance and features were demonstrated by magnetic domain observations of Nd–Fe–B sintered magnet.

In order to shed light on both quantitative composition and 2D/3D chemical and magnetic mapping, recent improvements in soft X-ray synchrotron-based scanning transmission microscopy have been carried out (see also Section 3.9). Images with a spatial resolution better than 5 nm, obtained by a special setup including ptychographic coherent diffraction imaging in STXM platforms [182] have been obtained, as demonstrated in proton conductive ionomer in fuel cells and magnetotactic bacteria [183]. Using hard X-ray magnetic nanotomography, Donnelly and coauthors [126] have determined the 3D magnetization configuration in soft magnetic $GdCo_2$ microcylinders. The method, based on circular left polarized X-rays with hard-X-ray dichroic ptychography, showed a spatial resolution of about 100–130 nm. Interestingly, it was shown that the two ferrimagnet sublattices are clearly coupled antiparallel to each other.

In a recent paper, Blanco-Roldan et al. [184] reported an imaging method based on the angular dependence of the magnetic contrast in a series of high resolution transmission X-ray microscopy images. For ferromagnetic $NdCo_5$ layers 55–120 nm in thickness and $NdCo_5$ film covered with permalloy, the authors observed a quantitative character of the magnetization (i.e., canting angles relative to the surface plane). Furthermore, the proposed method allows identifying complex topological defects (e.g., merons or $\frac{1}{2}$ skyrmions).

3.8. Neutron Magnetic Imaging Techniques

The imaging techniques driven by neutrons have been documented to be very informative for the study of the inward properties of the matter, including those that cannot be investigated by X-ray sources, such as hydrogen-rich water and organic-based systems. Thermal neutrons, often with low energy (even 10^{-2} eV), or fast neutrons with high energies (e.g., few MeV or above 10 MeV) can be used. The energy profile covers a wide interval depending on the dedicated facilities. Fast neutrons possess a low attenuation in the most matter, thus allowing them to be transmitted through the matter. Furthermore, some new methods of imaging make use of pulsed neutron beams based on accelerator facilities as developed also at compact accelerator-driven neutron sources, which open new applications in the field of the neutron imaging. The first instrument in the world fed by pulsed neutron sources and dedicated to the imaging was realized in Japan [128].

The most relevant elements of an NTT instrument are the neutron sources, as follows: (i) radionuclide-based spontaneous fission, (ii) low Z matrix, such as Be with α emitters (i.e., Am or Po); (iii) photoneutrons, such as D or Be with a high γ-emitter, and (iv) neutron generators from deuterium–deuterium or deuterium–tritium reactions. The other principal elements are collimators and detectors. Detectors are the most relevant constituent for the high spatial resolution: Cold neutron radiography and tomography (3D) have the higher spatial resolution: about 100 μm (1–25 s as exposure time) and about 300–500 μm (10–500 ms).

As far as the magnetic imaging is concerned, neutrons have no electric charge, allowing them to be penetrated deeply into the matter. Furthermore, neutrons have a magnetic moment, making them sensitive to magnetic fields via the Zeeman interaction. Some neutron scattering techniques based on thermal neutrons have been developed to investigate the magnetic structure, as follows: (i) Polarized small-angle neutron scattering (SANS); (ii) polarized neutron reflectometry, and polarized neutron radiography and tomography. SANS is a technique perfectly appropriated for the investigation of nanoparticles, including their internal structure via the measurement of the magnetic form factor or the magnetic interactions between magnetic nanoparticles (polarized SANS technique). Polarized neutron reflectometry is more suitable for magnetic thin films (5–100 nm in thickness) by measuring directions and the magnitude of magnetic induction in heterostructures with a depth resolution of about 2–3 nm. Polarized neutron radiography and tomography technique can be adopted for films or bulk materials for mapping the distribution of induction fields by a 2D detector. The spatial distribution of magnetic fields of bulk samples have been recently developed utilizing polarized neutrons and the technique may allow to reveal the 3D magnetic field distribution in the solid matter [129].

3.9. 3D Imaging of Magnetic Domains

Three-dimensional analysis of the domain structure in bulk materials is an important issue for understanding ultimate magnetic properties and for developing materials and devices.

While bidimensional imaging of magnetic domains on surfaces can be obtained using the aforementioned techniques, 3D visualization methods have eluded scientists for many years, due to the intrinsic problems of the more conventional imaging techniques. In these years, physicists were able to study the effect of domains on the magnetic properties of materials, but they were not able to make 3D images of domains deeply inside the matter.

In the past, the magnetism was conventionally associated with the imaging and analysis of magnetic domains in ferromagnetic systems. More recently, it has been demonstrated that the magnetic structures at the nanoscale can have a more complex magnetic texture and a deeper investigation is required. A new age in the magnetism has begun thanks to the magnetic imaging techniques addressing the challenge towards the more complex 3D magnetic textures. In this context, the magnetic X-ray transmission tomography (i.e., tomographic reconstruction method) is an excellent method for rising the challenge of this new field, capable of both, probing and 3D reconstructing the magnetic textures [185]. In fact, in recent years, some approaches for the 3D visualization the magnetic domains become available. Among these, Streubel et al. [127] have shown the 3D visualization and reconstruction of magnetic domain structures in curved magnetic thin films with tubular shape by means of full-field soft X-ray microscopy with a lateral resolution of a few tens of nm. It was shown that the 3D magnetization is obtained from a 2D projection sequence by investigating the magnetic contrast that is varying with the projection angle.

In another paper, Manke, Kardjilov and colleagues' work [186], have 3D imaged the magnetic domains by using a new technique called Talbot-Lau neutron tomography. Due to the fact that the sample slicing to investigate the interior domains could be useless, because a new domain pattern structure related to the newly developed surface can be rearranged, the tomographic approach has been developed to identify domain walls in wedge-shaped crystals basing on the refraction of neutrons (Figure 17a).

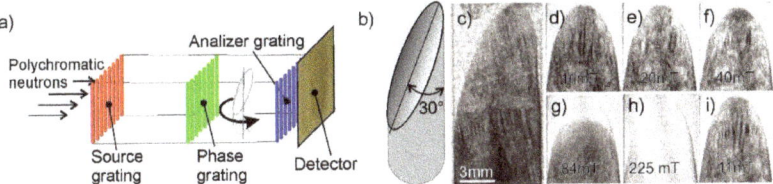

Figure 17. (a) representative scheme of the Talbot Lau neutron tomography setup; (b) representation of the investigated wedge-shaped FeSi single crystal, radiographic projection images as obtained by Talbot-Lau neutron dark-field radiography; (c–i) the same radiographic projection images obtained under an external magnetic field (parallel to the cylinder axis) of 11 mT, 20 mT, 40 mT, 84 mT, 225 mT, and 11 mT. The dark stripes are caused by domain walls that are nearly parallel to the incident beam. Adapted from [186], published by Nature Publishing Group, 2010.

For the image contrast a double crystal diffractometer can demonstrate that domain walls oriented almost parallel to the neutron beam direction can be visualized. The setup is based on grating-based shearing interferometer consisting of three gratings: (i) source grating (G_S), is an array of slits within an absorbing Gd mask with a periodicity of 790 μm; (ii) The phase grating (G_{Ph}) with a much smaller period (7.96 μm), causes a phase shift of $\lambda/2$ between incoming and produces; (iii) an analyzer grating (G_A), is an absorption grating with a period of 4 μm (~$P_{Ph}/2$). G_A is used to detect the interference pattern. The tomographic approach has been adopted for FeSi (Fe 12.8 at % Si single crystals) magnetic sample with a cylindrical shape (Figure 18b–i) [186]. An X-ray tomographic technique, more recently developed by Suzuki et al. [187], was employed to investigate the internal magnetic domain structure in ferromagnetic samples at the microscale. The technique, based on a scanning hard X-ray nanoprobe using XMCD, allows 3D reconstruction of the magnetic vector components with a spatial resolution of 360 nm. The authors argued that the method is applicable to practical magnetic materials and can be extended to 3D visualization of the magnetic domain formation process under external magnetic fields. Very recently, Wolf et al. [188] have shown the 3D magnetic induction mapping of a layered Cu/Co nanowire after reconstruction of both the magnetic induction vector field and 3D chemical composition of the material with sub-10 nm spatial resolution using the holographic vector field electron tomography approach.

3.10. Towards Imaging in Living Matter

The development of non-invasive measurements probing magnetic fields in organisms, living matter, even down to smaller scales, possibly to single cell sizes (i.e., microns), or better, inside the cell, is a matter of great interest for the scientific community [189]. It is known that magnetic imaging methods have either low spatial resolution or they are not applicable for imaging the cell structures and living biological samples. Grinolds et al. [190] demonstrated that microscopy based on NV-center (or nitrogen–vacancy, N–V) may allow to quantitatively measure the stray field of a sample with a sensitivity down to a single electron spin. From this, the magnetic imaging of magnetotactic bacteria under ambient conditions with a nm scale resolution it was reported, using an optically detected magnetic field imaging array consisting of a N–V color centre implanted at the surface of a diamond chip (Figure 18a). Such N–V color center, absorbs the light from green and emits to the red light, thus making local magnetic field investigation possible [123]. Then, magnetotactic bacteria are placed on the diamond, and the N–V quantum spin states are optically probed, thus making possible the reconstruction of magnetic-field vector components (Figure 18b–e).

In a recent paper by Thiel et al. [191], the probing of magnetism also in 2D materials at the nanoscale with single-spin microscopy was reported. In the paper, scanning single-spin magnetometry based on diamond NV centers was used to image the magnetization, localized defects and magnetic domains of atomically thin crystals of the van der Waals magnet CrI_3.

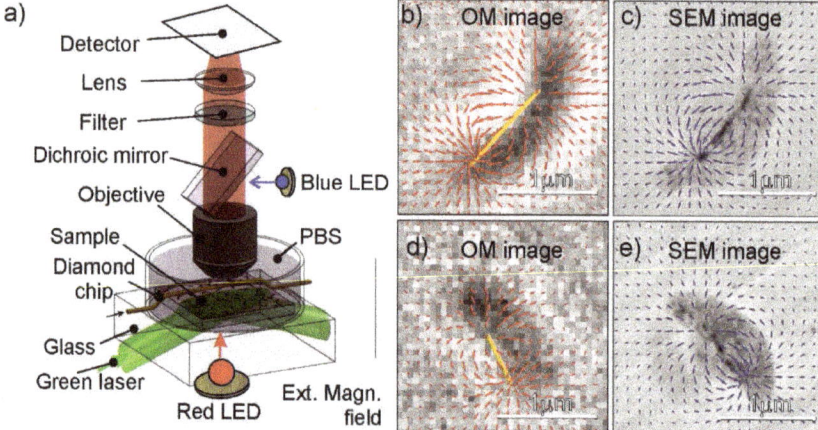

Figure 18. (a) Representation of the wide-field magnetic imaging microscope field fluorescence microscope setup vector plots of the measured (red arrows, left panel) and simulated (blue arrows, right panel) magnetic field projections in the xy plane are reported superimposed; (**b,d**) optical and (**c,e**) backscattered electron (BSE) images, respectively. Adapted from: [123], published by Nature Publishing Group, 2013.

4. Characterization of Magnetic Systems for Biomedical Applications

In the previous two sections of this review, a shrewd description of a variety of representative methods currently adopted for the investigation of magnetic materials, nanoparticles and molecular structures (i.e., single-molecule magnets) is provided. In the following the authors aim at illustrating how the (inorganic) magnetic systems for biomedical uses could be designed and characterized. The field is very broad, but very relevant due to its many implications in medicine; it is difficult to summarize all methods and adopted strategies, which can be very different. Some of the most relevant techniques are discussed for the sake of brevity in the following paragraphs. It is not surprising that due to the multidisciplinarity of the fields a multi-technique approach is required to address suitable strategies to detect and monitor specific diseases in human's body [192–196].

In order to evaluate potential applications of magnetic nanosystems in the biomedical field, some specific characteristics have to be carefully determined. First of all, morphology, size distribution as well as surface zeta potential must be investigated (Figure 19). Then solubility and stability in aqueous solutions have to be pursued. Possibly, relaxivity measurements could be performed to evaluate the possible application as magnetic resonance imaging (MRI) contrast agents. Finally, eventual toxicity effects exerted towards cell cultures must be assessed. In the following paragraphs some of the common techniques nowadays available to obtain precise information about magnetic systems, in view of possible biomedical applications, are enumerated and briefly discussed. For more details, the readers can refer to the more dedicated literature.

4.1. Microscopies, X-Ray Diffraction: Morphology, Composition, and Shape

Morphology and homogeneity of the sample are two fundamental aspects that should be deeply investigated in the light of in vivo administration of magnetic systems [197]. Both the morphology and the size of the system, in fact, influence its distribution, blood circulation time and possible toxicity [198–200]. Iron oxide nanoparticles, for instance, can be classified according to their size into USPIOs (ultrasmall superparamagnetic iron oxides), SPIOs (superparamagnetic iron oxides) or MPIOs (micrometer-sized iron oxide particles), displaying a hydrodynamic size smaller than 50 nm, about 50–150 nm or around 1 µm, respectively (coating included) [201,202]. While USPIOs are generally able to avoid the early and massive uptake by the macrophages from the mononuclear phagocyte system (especially spleen and liver macrophages), resulting in long blood circulation time and possible targeting of macrophages in deep compartments, SPIOs are generally taken up readily by macrophages in liver/spleen and provide passive targeting for reticuloendothelial system, whereas MPIOs are mostly intended to label and efficiently track macrophages [203]. Moreover, morphology and shape have been reported to affect toxicity [204]. Lee et al. [205] reported higher toxicity for rod-shaped particles in comparison to spherical ones, probably due to different wrapping mode on cell surface. Internalization through nonspecific cellular uptake was observed for rod-shaped nanoparticles more than sphere-shaped ones, resulting in faster cell penetration, dispersion throughout the cells and consequent higher pro-inflammatory cytokines production.

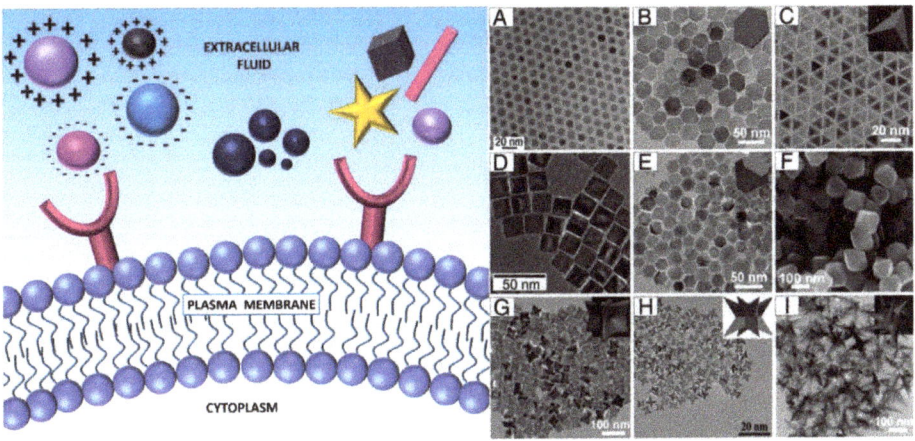

Figure 19. (**Left**) Physicochemical factors that affect cellular uptake of nanoparticles: Surface charge, size and shape. (**Right**) Different morphologies of iron oxide nanoparticles. Reproduced with permission from: [206], published by Ivyspring, 2018.

In order to study and report on shape and size of magnetic systems various microscopy cutting-edge techniques are employed, as reviewed in the previous paragraph. However, valid alternative and complementary methods are nowadays available [104]. Information about system structure, for instance, can be acquired by X-ray diffraction, using both conventional and synchrotron radiation sources, thermal analysis, Mössbauer [207] and infrared (IR) spectroscopy [208,209]. Conventional X-ray diffraction (XRD) is carried out to obtain information about the crystalline structure of the particles, especially about the proportion of iron oxide formed in a mixture, by comparing experimental peak and reference peak intensities [36,209]. Moreover, the crystal and particle size can also be extrapolated from the peak broadening in the XRD pattern using the Scherrer equation [104,210–212]. The advantage of using synchrotron radiation sources, based on a collimated light source with high intensity, instead of conventional ones, lies in the possibility of obtaining more accurate information about phase composition, crystallite size, strain and defects and of performing time-resolved studies [213]. Both types of XRD are performed on solid samples, but with energy dispersive X-ray diffraction is possible to obtain information also on samples in suspension, thus allowing knowledge of fine structural details [104,201].

4.2. Thermogravimetry and Differential Scanning Calorimetry: Thermal Properties and Composition

Another approach widely employed to the complementary characterization of magnetic systems is the thermal analysis, term that includes a group of techniques in which a physical property of a substance is measured as a function of temperature or time while the substance is subjected to a controlled temperature program. Thermal analysis provides information about sample thermal stability, adsorption of water molecules or specific coating molecules onto the magnetic core, phase transition and crystallinity of the system [214–217]. For instance, Catalano et al. [218] used thermal analysis to characterize novel hydrogel chitosan-coated iron oxide nanoparticles for cancer therapy: Through this analysis it was possible to accurately determine the chitosan content in the particle and gain information about temperature degradation of this innovative system. The most widely used thermal analysis is differential thermal analysis, DTA, in which the difference in temperature, ΔT, between the sample and a reference material is recorded while both are subjected to the same heating program. It could provide information about transformations that have occurred, such as phase transitions, crystallization, melting and sublimation. Conversely, in power-compensated differential scanning calorimetry (pc-DSC), the sample and a reference material are maintained at the same temperature; the differences in independent power supplies needed by the sample and the reference to keep this temperature constant are recorded and plotted against the programmed temperature or time [219]. Instead, thermo-gravimetric analysis (TGA), mainly exploited to determine the composition of materials and to predict their thermal stability, consists in measurements of weight/mass change (loss or gain) and the rate of weight change as a function of temperature, time and atmosphere [64,220]. An interesting example of TGA measurement applied to magnetic systems was reported by Xu and co-workers: they estimated the amount of Fe_3O_4 in polyacrylamide-coated magnetic particles [221]. In addition, they could correlate, with reasonable precision, the amount of magnetite with saturation magnetization, applying a specific equation. In 2014 Mansfield et al. [222] compared classical TGA with microscale-TGA (μ-TGA) in the analysis of surface coating in gold nanoparticles. They demonstrated that μ-TGA is a valid method for quantitative determination of the coatings on nanoparticles, such as surface-bond ligand coverage, and in some cases, can provide purity and compositional data of the nanoparticles themselves.

4.3. Mössbauer (or Gamma-Resonance) and Infrared Spectroscopies

Another appealing technique, probably the most adequate method to describe magnetic nanoparticle dynamics is Mössbauer (or gamma-resonance) spectroscopy [223]. Due to its relevance, Mössbauer spectroscopy should be placed among the characterization techniques dedicated to magnetic materials as well, but for the sake of brevity the technique is here briefly discussed.

Magnetic nanoparticle dynamic is related to the relaxation rates of nanoparticle magnetization vectors as determined by the size of nanoparticles. This spectroscopy method considers a group of nanoparticles as a cluster of interacting single domain magnetic particles and gives information about magnetic dipole interactions that is essential for applications involving a magnetic field such as targeted drug delivery. With Mössbauer spectroscopy three types of nuclear interactions can be typically observed: Isomer shift, quadrupole splitting and magnetic hyperfine splitting [224]. The technique is well-known due to its very high sensitivity in terms of frequency resolution, thanks to the high energy and narrow line widths of employed gamma rays. Gabbasov et al. [225] deeply characterized iron oxide nanoparticles coated with an oleic polymer using XRD and Mössbauer spectroscopy. They claimed that in some cases, X-ray diffraction measurements were unable to estimate the size of the magnetic core and Mössbauer data were necessary for the correct interpretation of the experimental results. A further example of the practical importance of Mössbauer spectroscopy in materials science is the thorough discrimination among systems having almost same structures, e.g., magnetite (Fe_3O_4) from maghemite (γ-Fe_2O_3) phase by ^{57}Fe Mössbauer spectroscopy [96,226].

Finally, infrared spectroscopy (IR) is widely used for the complementary investigation in magnetic system characterization due to its simplicity and availability. This technique is based on the interaction between infrared radiation and matter and is extremely useful to study chemical bonds of the magnetic systems, possible surface modifications and the presence of coating molecules [227]. It can be carried out on solid samples or liquids.

However, most of the methods reported above can be employed only on dry samples, while magnetic systems for biomedical applications are usually suspended in aqueous media.

To measure size and homogeneity of hydrated samples dynamic light scattering (DLS), also known as photon correlation spectroscopy (PCS), can be envisaged.

4.4. Dynamic Light Scattering and Zeta Potential: Particle Size Distribution and Particle–Liquid Interface

The employment of DLS to measure the size of magnetic nanoparticles has been already extensively reviewed by Lim et al. [228]. Briefly, during DLS measurement a light beam (electromagnetic wave) passes through a particle suspension. As the incident light impinges on the particles, a process known as scattering takes place. This process consists in alteration of the direction and intensity of the light beam. The variation of the intensity with time can then be used to measure the diffusion coefficient of the particles, that are in constant random motion (Brownian motion) due to their kinetic energy (Figure 20) [229]. Depending on the shape of the magnetic nanoparticles, for spherical particles, the hydrodynamic radius of the particle can be calculated from its diffusion coefficient by the Stokes–Einstein equation [230]. However, magnetic particle radii measured by microscopy techniques and DLS do not corroborate well. As a rule, TEM gives the "true radius" of the particle, while DLS provides the hydrodynamic radius, defined as the radius of a sphere that has the same diffusion coefficient within the same viscous environment of the particles being measured (Figure 20).

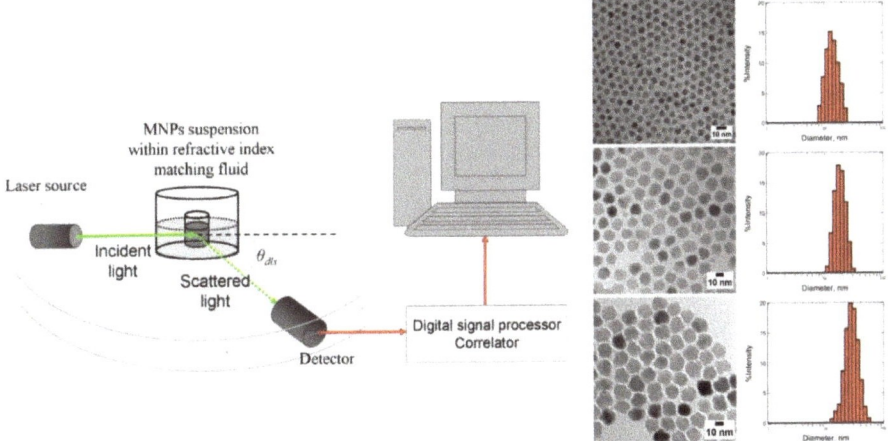

Figure 20. **Left**: schematic representation of the experimental setup for dynamic light scattering measurements. **Right**: TEM micrographs of Fe_3O_4 magnetic nanoparticles with the corresponding size distribution determined by DLS. The Z-average of magnetic nanoparticles measured by DLS is 16.9 ± 5.2 nm (top), 21.1 ± 5.5 nm (middle), and 43.1 ± 14.9 nm (bottom), respectively. Reproduced with permission from [228] (published by Springer Open, 2013) licensed under CC BY 2.0.

The advantage of using DLS in addition to other microscopy techniques are: (i) Wide sampling of the specimen (millions of particles), (ii) speed and ease of measurement, (iii) estimation of the radius of solvated particles (not dried and under vacuum like TEM), mimicking possible behavior in biological fluids, (iv) accessible and automated process, not requiring an extensive experience, (v) high sensitivity to the presence of small aggregates, useful to monitor sample colloidal stability [227,229,231]. The size range detectable by DLS is approximately included between 1 nm and 5 µm. The technique could also be employed to measure dimensions of non-spherical systems, such as rod-like particles, nanostars, nanotubes [232–234]. For example, Fang et al. [235] employed DLS to monitor the behavior of β-Ferric oxyhydroxide (β-FeOOH) nanorods depending on pH variations. They demonstrated that by DLS it was possible to monitor in situ self-assembling of this nanorods in a side-by-side fashion to form highly oriented 2D nanorod arrays which are further stacked in a face-to-face fashion to form the final 3D layered architectures. Results obtained well corroborate with TEM and SEM results. Moreover, through DLS is possible to obtain information about sample polydispersity. The polydispersity index (PDI), with $0 \leq PDI \leq 1$, is an indicator of sample homogeneity: The lower is PDI, the higher is particle homogeneity in terms of size distribution. PDI of 0.2 and below are most commonly deemed acceptable for polymer-based nanoparticle materials [236].

Most of the instruments nowadays available to measure DLS offer the possibility to further investigate Zeta potential. Zeta potential, also known as electrokinetic potential, is the potential at the slipping/shear plane of a colloid particle moving under an electric field [229]. More in details, when a charged particle is exposed to a fluid, an electric double layer (EDL) develops on its surface. The two layers that compose the EDL are the Stern layer, which is closer to the surface, and the diffuse layer [237]. The Stern layer is primarily formed by ions/molecules with opposite charge to that of the particle, while the diffuse layer is made up of both same and opposite charged ions/molecules (Figure 21) [238]. While the charge of the Stern layer is stable as it is due to direct chemical interactions with the particle, the peculiarity of the diffuse layer is that it is dynamic as its composition is influenced by various factors (e.g., concentration, ionic strength, and pH). Zeta potential is measured at the interface between the particles, moving under the electric field, and the layer of dispersant around it.

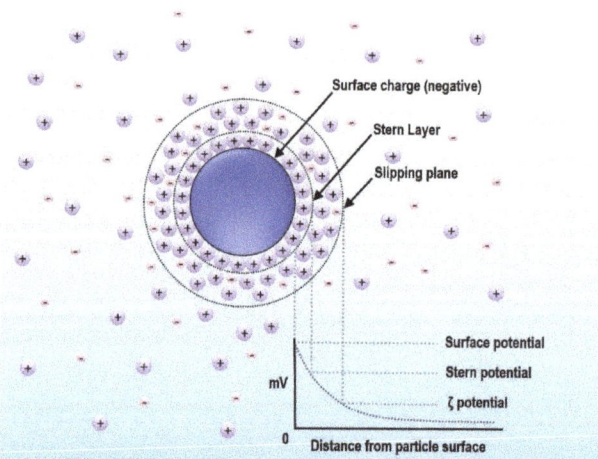

Figure 21. Schematic representation of the two layers forming the electric double layer (EDL) developed on a negatively charged particle. The inner layer, called the Stern layer, is composed by ions of charge opposed to the particle surface, i.e., positive ions in this case. The outer layer, also known as diffuse layer, is made up of both negative and positive charges. The Zeta Potential is the electrokinetic potential at the slipping plane.

The potential on the particle surface itself is known as the *Nernst potential* (ψ_0) and cannot be measured [239]. Investigation of the zeta potential of magnetic nanoparticles is of utmost importance to predict their protein absorption and subsequent biological behavior, toxicity effects, possible interactions with cells and other microorganisms, as well as particle stability. Schwegmann and co-workers studied the influence of the zeta potential on the sorption between microorganisms (*Saccharomyces cerevisiae* and *Escherichia coli*) and iron oxide nanoparticles, at two different pH values, in order to evaluate potential uptake and connected toxicity effects [240]. Sharma et al. deeply investigated and compared the influence of size and charge of various magnetic particles on systemic distribution. They noticed that the highest accumulation of iron occurred in the lungs, when positively charged particles were administered, while it occurred in the liver and spleen when nanoparticles possessing a negative surface potential were injected [241]. These results suggest that nanoparticle distribution in vivo is dependent not only on the size but most likely on a complex interplay between size and charge, thus highlighting the importance of evaluating these two parameters if looking for perspective biomedical applications.

4.5. Methods to Investigate Stability and Protein Corona Adsorption

Stability of magnetic systems is of utmost importance in view of possible in vivo applications. As the colloidal stability of bare iron oxide nanoparticles, for example, is poor, different types of natural and synthetic coating materials are generally added to the preparations. Coatings can be classified into three main groups: (i) monomeric, (ii) inorganics, (iii) polymer stabilizers. Among the monomeric coating, carboxylates and phosphates are the most diffuse. Citric acid stabilized magnetic systems have been widely reported and have been also under clinical investigation [242]. Inorganic materials employed as stabilizers include silica, gold or gadolinium, leading to nanoparticles endowed with an inner iron oxide core and an outer metallic shell. As far as polymer coatings are concerned, a wide range of polymer stabilizers is nowadays available on the market, offering the possibility of tailoring each magnetic system according to its specific application. Poly(ethylenimine) (PEI), for example, is a cationic polymer frequently used to design gene-transfer vehicles for both in vitro and in vivo applications, achieving high transfection efficiency [243]. Iron oxide nanoparticles coated with the cationic polymer PEI, in fact, can easily capture negatively charged molecules, such as DNA and RNA, giving rise to cell tracking probes and MRI detectable gene/drug delivery carriers. Polyethylene glycol (PEG) is definitely the most popular coating polymer, due to its excellent anti-fouling property (preventing opsonization) and high steric hindrance, ideal to stabilize various magnetic nanoparticles. Dextran is another (bio)polymer widely used as a stabilizer, mainly because of its high biocompatibility.

In order to be administered in vivo, however, the stability of magnetic systems should be favorable not only at 4 °C in aqueous buffers, but also at 37 °C in serum or other simulated biological fluids. In the presence of serum proteins, in fact, particles can rapidly aggregate if not appropriately formulated and stabilized [244]. Moreover, the growth of protein corona onto the surface of administered magnetic systems should be investigated in order to predict their in vivo behavior [245]. Among the methods available to study particle stability there are DLS, described in the previous paragraph, which gives an idea about size variations and therefore particle aggregation and/or interaction with proteins, nuclear magnetic resonance (NMR) relaxivity studies [246], that will be described in the next paragraph and can account for particle aggregation, instability of the formulation, release of metal ions and interaction with serum proteins, such as serum albumin. TEM, SEM, AFM and fluorescence microscopy can also be used to monitor magnetic system stability [247]. Lazzari et al. [248] employed DLS to investigate the stability of various nanoparticles in buffers, simulated biological fluids (e.g., saliva, gastric juice, intestinal fluid), serum and tissue homogenates (mice brain, liver and spleen). They found out that while poly-lactic acid (PLA) particles were stable in such biologically relevant conditions, poly-methyl methacrylate (PMMA) based systems were unstable and tended to aggregate over time. Such systematic DLS studies provided an in vitro tool to investigate the stability of various systems before in vivo studies. Khan et al. [249] instead, investigated by both DLS and Zeta potential hard and soft corona formations on gold nanospheres of different sizes (2–40 nm). Understanding the dynamics of the growth of protein corona on the nanoparticle surface, indeed, is important to predict how the NPs behave in vivo. Experiments were carried out in the presence of three different proteins: Human serum albumin, bovine serum albumin, and hemoglobin. Through the correlation of obtained DLS data with mathematical modelling, they extrapolated the adsorption kinetics, number of adsorbed proteins, and binding orientation. They demonstrated that the growth kinetics of a protein corona is exclusively dependent on both protein structure and surface chemistry of the nanoparticles. In 2013, Salvati et al. [250] studied the in vivo targeting ability of transferrin-functionalized silica nanoparticles: They pointed out that due to the adsorption of protein corona onto nanoparticle surface the receptor-targeting capability was lost (Figure 22). Hence, detailed investigation to understand the effect of protein corona on different magnetic systems will enhance their translational potential and DLS can be regarded as an effective tool in such studies, along with other techniques.

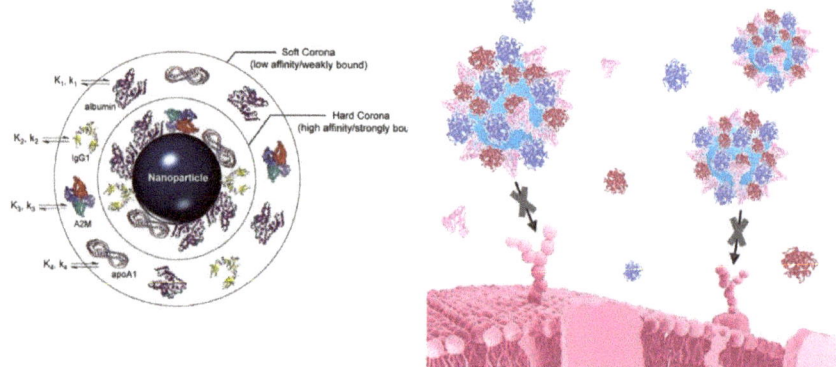

Figure 22. Left: Schematic of soft and hard corona formed on the surface of a NP. Reproduced with permission from Bhattacharjee et al. [229], copyright by Elsevier 2016. **Right**: schematic representation of loss of transferrin receptor targeting for transferrin-conjugated nanoparticles in the presence of corona proteins. Adapted from Salvati et al. [250], published by Springer Nature, 2013.

4.6. Relaxometric Properties and Magnetic Fluid Hyperthermia

Amongst the possible biomedical applications of magnetic systems there are (i) separation techniques, like cell isolation and cellular proteomics [8], (ii) diagnosis, such as magnetic resonance imaging (MRI), cell tracking, biosensing, and (iii) therapy, basically including hyperthermia and drug delivery (Figure 23) [251].

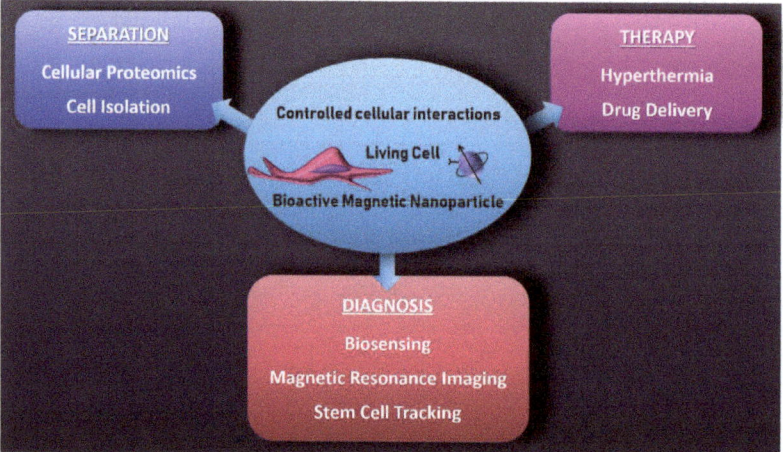

Figure 23. Biomedical applications of magnetic nanoparticles.

In this paragraph the use of magnetic systems as MRI contrast agents or for hyperthermia mediated treatment will be discussed. To be eligible as a MRI contrast agent, the magnetic system must be superparamagnetic. Superparamagnetism is a magnetic behavior occurring in small ferrimagnetic or ferromagnetic nanoparticles. If the core diameter of these NPs is lower than 3–50 nm, depending on the material, magnetization can randomly flip direction under the influence of temperature. The mean time between two flips is called the Néel relaxation time [252]. In the absence of an external magnetic field, when the NPs magnetization measurement time is greater than the Néel relaxation time, the average value of NPs magnetization appears to be zero and these nanoparticles are said to be in the superparamagnetic state. In this state it is possible to magnetize the nanoparticles with an external magnetic field, similar to a paramagnet. However, the magnetic susceptibility of these NPs is much greater than that of paramagnets [253]. To investigate the possibility of using magnetic suspensions as MRI contrast agents longitudinal (R_1) and transverse (R_2) relaxivities should be measured. Relaxivity is defined as the increase of the relaxation rate of the solvent (water) induced by 1 mmol/L of the active ion. In the case of magnetite, the relaxivity is the relaxation rate enhancement observed for an aqueous solution containing 1 mmol of iron/L (Equation (2)):

$$R_{i\,(obs)} = \frac{1}{T_{i\,(obs)}} = \frac{1}{T_{i\,(diam)}} + r_i C \; i = 1 \text{ or } 2 \qquad (2)$$

where $R_{i(obs)}$ is the global relaxation rate of the aqueous system (s^{-1}), $T_{i(obs)}$ is the global relaxation time of the aqueous system (s), $T_{i(diam)}$ is the relaxation time of the system before the addition of the contrast agent (s), C is the concentration of the superparamagnetic center (mmol·L^{-1}), and r_i is the relaxivity (s^{-1}·$mmol^{-1}$ L) [104]. Since relaxivity is temperature-dependent, it is usually measured at 298 K or 310 K. Superparamagnetic systems generally behave like small movable magnets, creating a strong magnetic inhomogeneity in their vicinity and considerably reducing the T_2 relaxation time of nearby water protons. They are also known as signal "killers" as a negative enhancement (corresponding to image darkening) can be seen in regions in which they accumulate [254]. Recently, the use of iron oxide also as positive contrast agents, resulting in signal enhancement in regions of accumulation, has been reported by many groups [255,256]. Positive contrast agents, based on Gadolinium, are currently the mainstream clinical MRI contrast agents. However, some gadolinium-based contrast agents have shown a long-term toxicity effect (nephrogenic systemic fibrosis, NSF) and Gd depositions in the brain. On this basis, The NSF has triggered a Food and Drug Administration (FDA) black-box warning and a contraindication, as well as long term adverse effect monitoring, of some Gd-based contrast [257,258]. Newly developed ultrasmall iron oxide nanoparticles have then recently attracted much attention to serve as safer alternatives to gadolinium-based T_1 contrast agents.

As a rule, to evaluate if a MRI contrast agent has higher T_2 versus T_1 shortening effects the ratio between the relaxivity R_2 and R_1 (R_2/R_1) must be calculated. The higher is this ratio, the higher is the predicted T_2 effect. If this ratio is lower than 1, the contrast agent is predicted to be more suitable for T_1 shortening purposes. Magnetic nanoparticles used in vivo are mainly composed of magnetite (Fe_3O_4) and maghemite (γ-Fe_2O_3) coated with protective shells (e.g., dextran, polyethylene glycol, starch) to prevent agglomeration. It has been reported that the coating thickness of superparamagnetic systems can significantly affect the R_2, and the R_2/R_1 ratio. Moreover, both R_1 and R_2 can vary in case of system instability (metal ions release, aggregation, coating degradation) and following interactions with proteins (e.g., protein corona or specific target proteins) [259,260].

To individuate the best magnetic field at which the system can be used, the acquisition of a nuclear magnetic relaxation dispersion (NMRD) profile, which gives the evolution of the longitudinal relaxivity R_1 with respect to the external magnetic field, is envisaged (Figure 24).

The model to fit NMRD experimental data, proposed by Roch et al. [261] and further developed by Gossuin et al. [262], can provide information about the nanomagnet crystals: Their average radius, their Néel relaxation time, their anisotropy energy and their specific magnetization. Combining these data with the results obtained by magnetometry is possible to obtain a very complete description of the morphology and physical properties of a magnet [261,263]. Ruggiero et al. [264] compared NMRD profiles of newly-synthesized iron oxide nanoparticles coated with Poly(lactic-co-glycolic acid) (PLGA) to the NMRD profile of Endorem™ (commercially available SPIO particles used as MRI contrast agent). From these profiles, displaying a shape of the curve typical of the relaxation induced by superparamagnetic particles, they deduced that loading iron oxide nanoparticles into the PLGA matrix seems not to affect their overall magnetic properties. Moreover, they pointed out that the different profile shapes observed for the various preparations depend on the particle properties (such as: size, clustering, Néel relaxation time and saturation magnetization) [264].

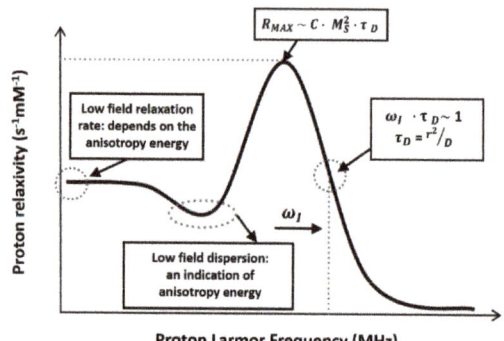

Figure 24. Nuclear magnetic relaxation dispersion (NMRD) profile of magnetite particles in colloidal solution.

PLGA particles reported by Ruggiero et al. [264] were designed for magnetic fluid hyperthermia (MFH) applications. The use of nanomaterials for MFH was first investigated in 1957 by Gilchrist et al. [265]. The authors reported the possibility of heating various tissue samples with particles of γ-Fe_2O_3 (20–100 nm in size) exposed to a 1.2 MHz magnetic field. Since then, the use of different magnetic materials, particle design and delivery methods, as well as various field strengths and frequencies have been investigated for MFH purposes [266]. Briefly, when exposed to a fast switching magnetic field, the magnetization of superparamagnetic nanoparticles can quickly flip direction. The friction caused by the rotation of the particle itself, known as Brownian relaxation, and the spin fluctuations within the crystal, known as Néel relaxation, lead to the loss of magnetic energy and the generation of thermal energy [252]. The heating power of a certain magnetic system, expressed as the specific absorption rate (SAR, W/g), mainly depends on the mean size of the system, on its saturation magnetization (M_s) and magnetic anisotropy (K) as well as on the alternating magnetic field amplitude (H_{max}) and frequency (f). MFH effect can be exploited in cancer treatment, to raise the temperature at the tumor site above the systemic one, or to induce a controlled drug release through the thermal stimulus [267]. This field is also known as theranostics: A discipline that combines diagnostic, as superparamagnetic nanoparticles can be used as MRI contrast agents, and localized therapy, through controlled temperature increase or drug release. For practical therapeutic applications, a temperature of 41–46 °C must be reached with hyperthermia. Temperatures greater than 50 °C cause thermoablation and can induce several side effects. Therefore, it is of crucial importance for the development of MFH dedicated systems to rely on analytical methods able to characterize new magnetic nanoparticles and to predict their heating capacity in physiological conditions. Besides NMRD profiles, that, as previously described, allow to predict magnetic systems performance in both MRI and MFH applications, other particle characteristics can account for MFH efficiency. Size, volume, and morphology of the NPs can be determined by DLS, SEM, and TEM; stability of the preparation and zeta potential should be investigated. Thermogravimetric analysis, X-ray diffraction and magnetometry can be exploited to obtain valuable additional information. Finally, heating efficiency, reported in terms of SAR, can be measured by exposing the magnetic fluid to a time varying magnetic field [268]. Therefore, all the techniques described in this last section can be exploited to fully characterize magnetic systems for MFH purpose.

Given the variety of information that can be gained with the here reported techniques, for the sake of clearness, in Table 2 a recapitulatory comparison between all the reported techniques (even if we know that the list is not exhaustive), useful to characterize magnetic materials for biomedical applications, is provided. More in details, advantages and limitations of each technique are reported, as well as the specific information that can be obtained with the different instrumentations.

Table 2. Main properties of the techniques reported in the previous paragraph, useful to characterize magnetic systems designed for biomedical applications.

Method(s)	Information Provided	Acq. Time	Complexity	Sample Form	Advantages	Drawbacks	Ref.
X-ray Diffraction	Crystalline structure; crystal size, strain, defects; particle size; sample purity; iron oxide proportion	<20 min	L	Solid, liquid	Qualitative and quantitative analysis; powerful and rapid; minimal sample preparation required	Size limit (better results are obtained with large crystals); high amount of material needed (g); possible misleading results interpretation (peaks overlay)	[269]
Thermal Analysis (DTA, pc-DSC, TGA, μ-TGA)	Thermal stability; water adsorption; molecule adsorption; phase transition temperature; crystallinity and purity	min–hours	M-H	Solid, liquid	Qualitative and quantitative analysis; low amount of sample needed (mg–g)	Challenging results interpretation; challenging results comparison between laboratories; strong dependence of DTA signals from experimental conditions	[270,271]
Mössbauer Spectroscopy	Size of magnetic core; magnetic interactions; precise identification of iron oxides; magnetite and maghemite discrimination; blocking temperature	min–hours	H	Solid	Low amount of sample required (1–5 mg); valence state of iron in minerals detectable; best technique to identify different iron oxides	Low spatial resolution; thinly spread powders needed; the optimal amount of sample should be selected each time; data analysis techniques complex and variable; low number of elements suitable to be investigated; very low temperatures required	[104,223]
IR Spectroscopy; FTIR spectroscopy	Material structure; chemical Bonds; surface coating/functionalization of magnetic NPs	min	L	Solid, liquid	Simplicity and availability; non expensive technique; qualitative and quantitative analysis; fast mean of identification in case of magnetite; high sensitivity (μg)	Complex mixtures are hardly analyzed; functional groups cannot be exactly located	[272]
Dynamic light scattering (DLS)	Hydrodynamic radius of NPs; polydispersity of the sample; aggregation of NPs; adsorption of protein corona	<5 min	L	Liquid	Wide sampling of the specimen; speed and ease of measurement; estimation of the radius of solvated particles; accessible and automated process; high sensitivity to small aggregates; small amount of sample required (μL)	Highly polydispersed, diluted or fluorescent samples cannot be measured accurately	[227–229,231]
Zeta Potential	Apparent surface potential of NPs; adsorption of protein corona	<5 min	L	Liquid	Wide sampling of the specimen; speed and ease of measurement; accessible and automated process; small amount of sample required (μL)	Highly polydispersed samples cannot be measured accurately; measurements influenced by pH, ionic strength, dispersion media; measurements affected by sample sonication; measurements affected by fast metal/ion release	[273]

Table 2. *Cont.*

Method(s)	Information Provided	Acq. Time	Complexity	Sample Form	Advantages	Drawbacks	Ref.
SEM TEM	Size of the magnetic core; morphology; size distribution; homogeneity; surface structure	min	L/M	Solid	Accurate information provided on size of magnetic core; images of NPs are displayed; very high magnification possible	Long and complex sample preparation; expensive technique; cryo-EM is needed for samples sensitive to temperature; low sampling of the specimen; absence of information about sample aggregation in liquid suspensions; highly qualified personnel needed; possible artifacts formation; vacuum needed	[274]
NMRD Profile Fitting	Nanomagnet crystals (N.C.) average radius; N.C. Néel relaxation time; N.C. anisotropy energy; N.C. specific magnetization; possible NPs clustering; longitudinal Relaxivity R_1 at different magnetic fields; MRI efficiency prediction	hours	H	Liquid	Multiplicity of information provided in one fitting; prediction of MRI efficiency	Highly specific instrumentation required; samples must be in liquid form and stable during the acquisition time; staff must be highly trained	[261,262]

Note: L = low, M = moderate, H = high.

4.7. Toxicity Assays

Toxicity effects exerted towards cell cultures and blood components must be investigated before testing the magnetic systems directly in vivo. Obviously, it must be taken into account that these assays can be highly operator-dependent and that the lack of standard protocols make more difficult the comparison between different systems. However, the higher is the number of tests performed, the safer can be considered the preparation. Classical and less expensive cytotoxicity tests include MTT assay, Alamar Blue and Trypan blue exclusion assay. The MTT (3-[4,5-dimethylthiazol-2-yl]-2,5 diphenyl tetrazolium bromide) colorimetric assay is based on the cleavage by mitochondrial activity of the yellow soluble tetrazolium salt, MTT, to form an insoluble dark blue formazan product. As mitochondrial activity is generally proportional to the number of viable cells, this test can be used to estimate the cytotoxic effect of various compounds in vitro. Cells are first incubated with different concentrations of the system to be tested and for different time ranges, then the system is washed away, cells are then incubated with MTT for a few hours and the presence of violet formazan can be analyzed using a plate reader. The samples must be always compared to controls to determine the percentage of viable cells. In order to obtain accurate and reliable results particular attention should be devoted to cell density, culture medium and MTT exposure time [275]. Similar to MTT, the Alamar Blue test is based on the reduction of the blue non-fluorescent dye resazurin, upon entering living cells, to resofurin, a pink-colored highly fluorescent compound. The amount of fluorescent compound produced can be quantified and correlated to the number of viable cells. Resazurin is then considered a reliable indicator of the oxidation–reduction (REDOX) processes that occur both during aerobic and anaerobic respirations in cells incubated with various molecules, systems and/or drugs [276]. The trypan blue dye exclusion test, instead, is based on the different permeability of cell membranes towards certain molecules (e.g., trypan blue, eosin or propidium): While these substances can easily penetrate in dead cells, they cannot cross the membrane of living cells. To perform this test the cells are mixed with a solution of trypan blue and then the number of blue colored cells (died cells) versus the number of cells having a clear cytoplasm (living cells) are immediately quantified by visual inspection. The percentage viable cells in incubated samples will be then compared to the percentage of viable cells in control samples. All these assays can be used not only to assess cell viability, but also cell proliferation. To this aim, after incubation with variable amounts of the magnetic system and for variable time ranges, cells are washed and incubated in culture medium for different days. The number of viable cells measured at 24, 48, and 72 h post incubation allows estimating the cell proliferation time, which is then compared to that of control cells. A biocompatible system should not affect this parameter. Other convenient assays are the metabolic NAD/NADH fluorimetric assay, the protease viability marker assay, the Comet assay, useful to quantify DNA damages induced by the preparation, and the ATP (Adenosine 5′-triphosphate) quantification test, using firefly luciferase. As each cell viability assay displays its own advantages and drawbacks the most appropriate to each situation should be selected. MTT and Alamar Blue tests for example are less expensive than the ATP detection assay. However, the ATP detection assay is characterized by fewer steps, a minimum amount of interferences and is by far the most sensitive [277].

Another assay, widely used to evaluate potential cytotoxicity of magnetic particles, is the reactive oxygen species (ROS) production test. This test is based on the use of specific kits able to detect and quantify ROS, as an indicator of cell oxidative stress. ROS include hydrogen peroxide, anions and hydroxyl radicals. The production of free radicals, challenged in cells by a "detoxification" mechanism, which involves enzymes such as superoxide dismutase and glutathione peroxidase, is considered one of the primary mechanism of nanoparticle induced toxicity. The overproduction of ROS has been associated to DNA strand breaks, alteration in gene transcription, lipid peroxidation and generation of protein radicals [278]. Incubation with various magnetic systems could also induce apoptosis, a well-controlled, tightly regulated physiological process, in which the cells participate in self-destruction. Numerous assays are nowadays available to measure the number of apoptotic cells: light microscopy (Trypan blue staining), fluorescence microscopy (acridin orange/ethidium bromide and annexin V/propidium iodide staining) and agarose gel electrophoresis of fragmented genomic DNA. Trypan blue and Hoechst staining detect apoptotic cells on the basis of their reduced DNA content and morphological changes that include nuclear condensation; annexin-V and propidium iodide can distinguish between early and late apoptotic events based on plasma membrane composition and function. The cleavage of many cellular proteins, carried out by apoptosis-activated caspases, results in fragments that can serve as apoptosis markers [279].

Finally, the hemocompatibility properties of magnetic systems can be investigated by means of a standard hemolysis test [280]. Following the incubation of animal (e.g., rabbit, mouse, rat) or human donor erythrocytes with the system, the amount of released hemoglobin is measured by means of UV–vis spectroscopy (λ = 415 nm). Released hemoglobin is directly related to the number of destroyed erythrocytes that could then be easily estimated. Normal saline and deionized water are generally used as negative (0% hemolysis) and positive (100% hemolysis) controls, respectively. The percentage hemolytic rate % HR can be calculated as (Equation (3)):

$$\% \ HR = \frac{\text{mean } OD \text{ of sample to be tested } - \text{ mean } OD \text{ of negative control}}{\text{mean } OD \text{ of positive control } - \text{ mean } OD \text{ of negative control}} \times 100 \qquad (3)$$

where OD stands for optical density. If the hemolytic rate is less than 5%, the material will have no hemolytic effect and conform to the requirements of the hemolysis test for medical materials [281].

In Table 3 a comparison between the most diffuse toxicity assays, useful to evaluate potential toxic effects exerted by magnetic NPS, is reported.

Table 3. Comparison between some of the most diffuse toxicity assays, useful to evaluate potential toxic effects exerted by magnetic NPs.

Assay	Parameter Measured	Information Provided	Measurem. Method	Advantages	Drawbacks	Ref.
MTT	Mitochondrial activity	Possible cytotoxic effects of compounds/drugs/NPs	Fluorescence Absorbance	Inexpensive; high throughput screening; low amount of sample needed	Operator and procedure dependent (e.g., cell density, culture medium and MTT exposure time); Substrate conversion needed	[277]
Alamar Blue (Resazurin)	Cellular metabolic activity	Possible cytotoxic effects of compounds/drugs/NPs	Fluorescence Absorbance	Inexpensive, High throughput screening; low amount of sample needed; more accurate than MTT	Operator and procedure dependent (e.g., cell density, culture medium and MTT exposure time); substrate conversion needed	[277]
Trypan Blue Exclusion Test	Membrane integrity	Possible cytotoxic effects of compounds/drugs/NPs Number of dead/viable cells	Optical microscopy	Inexpensive; extremely rapid assay; no substrate conversion needed	Operator dependent; possible interference in case of incubation with colored/fluorescent samples	[277]
ATP detection	ATP synthesis	Possible cytotoxic effects of compounds/drugs/NPs; number of viable cells	Luminescence	Fast assay; high sensitivity; less artifacts occurrence; no substrate conversion needed	Expensive	[277]
ROS production	Generation of ROS *	Possible cytotoxic effects of compounds/drugs/NPs; cell oxidative stress	Fluorescence	Inexpensive	Substrate conversion needed	[278]
Hemolysis Assay	Red blood cell lysis	Hemolytic power of compounds/drugs/NPs	UV-vis spectroscopy	Extremely useful in view of (pre)clinical applications; possibility to run the test in plasma	Red blood cells of the chosen specie needed	[280]

* ROS: Reactive Oxygen Species.

5. Conclusions

Magnetic systems and particles are extremely appealing materials that found applications in a large variety of advanced technological fields. In the last decades, the scientific literature reported a plethora of research studies and documents characterized by different physicochemical, morphological, and biological methods (depending on the use), fundamental for pointing out their magnetic response/properties. However, due to the multidisciplinary approach proper of these studies, in most cases such very specific characterization techniques are little known (or fully unknown) to most of the users.

Therefore, in order to overcome this lack, the main goal of this review is to summaries (in a detailed, but hopefully concise manner) a large number of techniques that are currently available to characterize magnetic systems, highlighting the ordinary uses together with the main relevant advantages and disadvantages.

More in details, characterization techniques were classified into three sections, and properly discussed with examples from the literature. Part I is dedicated to the definitions of magnetism and magnetization (hysteresis) techniques. Part II is dedicated to the morphological aspects, thus illustrating all the different visualization methods of magnetic domains. Finally, Part III is dedicated to the principal physicochemical characterizations, with a last section particularly devoted to the biomedical applications.

With this review, authors hope of having provided a real toolbox that acts as guidelines for helping in the choice of the most suitable method(s) as appropriate, as well as in the comprehension of the magnetic properties/responses of these promising materials.

Author Contributions: The review was written through contributions of all authors. All authors have read and agreed to the published version of the manuscript.

Funding: This work was supported by MIUR (Ministero dell'Istruzione, dell'Università e della Ricerca), INSTM Consorzio and NIS (Nanostructured Interfaces and Surfaces) Inter-Departmental Centre of University of Torino.

Conflicts of Interest: The authors declare no conflict of interest.

Abbreviations

AFM: atomic force microscopy; ATP: Adenosine 5′-triphosphate B: magnetic flux; BSE: backscattered electrons; Cc: chromatic aberration; Cs: spherical aberration; DLS: dynamic light scattering; DPC: differential phase contrast; DTA: differential thermal analysis; DW: domain walls; ED pattern: electron diffraction pattern; EDL: electric double layer; EDX: energy dispersive X-ray; EELS: electron energy loss spectroscopy; EH: electron holography; FDA: Food and Drug Administration; FEG: field emission gun; HOPG: highly ordered pyrolytic graphite; HR: Hemolytic rate; HV: high vacuum; IR spectroscopy: Infrared spectroscopy; L-MOKE: longitudinal MOKE; LTEM: Lorentz transmission electron microscopy; MFH: magnetic fluid hyperthermia; MFM: magnetic force microscopy; MIP: mean inner potential; MOKE: magneto-optical Kerr effect; MPIOs: micrometer-sized iron oxide particles; MRI: magnetic resonance imaging; MTT: (3-[4,5-dimethylthiazol-2-yl]-2,5 diphenyl tetrazolium bromide; NMR: Nuclear magnetic resonance; NMRD: Nuclear magnetic relaxation dispersion; NPs: nanoparticles; NSF: nephrogenic systemic fibrosis; OD: optical density; pc-DSC: power-compensated differential scanning calorimetry; PCS: photon correlation spectroscopy; PDI: polydispersity index; PEEM: photoemission electron microscopy, PEG: polyethylene glycol; PEI: poly(ethylenimine) PLGA: poly(lactic-co-glycolic acid); P-MOKE: polar MOKE; R_1: longitudinal relaxivity; R_2: transversal relaxivity; ROS: reactive oxygen species; SAM: Auger electron microscopy; SANS: small-angle neutron scattering; SAR: specific absorption rate; SE: secondary electrons; SEM: scanning electron microscopy; SEMPA: spin-polarized analysis SEM; SHPM: scanning Hall probe microscopy; SPIOs: superparamagnetic iron oxides; SPLEEM: spin-polarized low energy electron microscopy; SPM: scanning probe microscopy; SP-STM: spin polarized scanning tunneling microscopy; SQUID: superconducting quantum interference device; STEM: scanning transmission electron microscopy; STM: scanning tunneling microscopy; T_1: longitudinal relaxation time; T_2: transversal relaxation time; TEM: transmission electron microscopy; TGA: thermogravimetric analysis; TIE: transport-of-intensity equation; T-MOKE: transversal MOKE; TXMCD: transversal X-ray magnetic circular dichroism; UHV: ultra-high vacuum; USPIOs: Ultrasmall superparamagnetic iron oxides; XANES: X-ray absorption near-edge structure XANES; XAS: X-ray absorption spectroscopy; XMCD: X-ray magnetic circular dichroism; XRD: X-ray diffraction.

References

1. Nisticò, R. Magnetic materials and water treatments for a sustainable future. *Res. Chem. Intermed.* **2017**, *43*, 6911–6949. [CrossRef]
2. Bozorth, R.M. Magnetism. *Rev. Modern Phys.* **1947**, *19*, 29–86. [CrossRef]
3. Singh, R. Unexpected magnetism in nanomaterials. *J. Magn. Magn. Mater.* **2013**, *346*, 58–73. [CrossRef]
4. Mehta, R.V. Synthesis of magnetic nanoparticles and their dispersions with special reference to applications in biomedicine and biotechnology. *Mat. Sci. Eng. C* **2017**, *79*, 901–916. [CrossRef]
5. Gupta, A.K.; Gupta, M. Synthesis and surface engineering of iron oxide nanoparticles for biomedical applications. *Biomaterials* **2005**, *26*, 3995–4021. [CrossRef]
6. Tietze, R.; Zaloga, J.; Unterweger, H.; Lyer, S.; Friedrich, R.P.; Janko, C.; Pottler, M.; Durr, S.; Alexiou, C. Magnetic nanoparticle-based drug delivery for cancer therapy. *Biochem. Biophys. Res. Commun.* **2015**, *468*, 463–470. [CrossRef]
7. Serrà, A.; Gimeno, N.E.; Gómez, E.; Mora, M.; Sagristá, M.L.; Vallés, E. Magnetic Mesoporous nanocarriers for drug delivery with improved therapeutic efficacy. *Adv. Funct. Mater.* **2016**, *26*, 6601–6611.
8. Kurgan, E.; Gas, P. Magnetophoretic Placement of Ferromagnetic Nanoparticles in RF Hyperthermia. In Proceedings of the 2017 Progress in Applied Electrical Engineering (PAEE), Koscielisko, Poland, 25–30 June 2017; pp. 1–4. [CrossRef]
9. Sheng, Y.; Li, S.; Duan, Z.; Zhang, R.; Xue, J. Fluorescent magnetic nanoparticles as minimally-invasive multi-functional theranostic platform for fluorescence imaging, MRI and magnetic hyperthermia. *Mater. Chem. Phys.* **2018**, *204*, 388–396. [CrossRef]
10. Pankhurst, Q.A.; Connolly, J.; Jones, S.K.; Dobson, J. Applications of magnetic nanoparticles in biomedicine. *J. Phys. D Appl. Phys.* **2003**, *36*, R167–R181. [CrossRef]
11. Garello, F.; Vibhute, S.; Gunduz, S.; Logothetis, N.K.; Terreno, E.; Angelovski, G. Innovative design of Ca-sensitive paramagnetic liposomes results in an unprecedented increase in longitudinal relaxivity. *Biomacromolecules* **2016**, *17*, 1303–1311. [CrossRef]
12. Liu, W.; Wong, P.K.J.; Xu, Y. Hybrid spintronic materials: Growth, structure and properties. *Progr. Mater. Sci.* **2019**, *99*, 27–105. [CrossRef]
13. Tudu, B.; Tiwari, A. Recent developments in perpendicular magnetic anisotropy thin films for data storage applications. *Vacuum* **2017**, *146*, 329–341. [CrossRef]
14. Peyer, K.E.; Zhang, L.; Nelson, B.J. Bio-inspired magnetic swimming microrobots for biomedical applications. *Nanoscale* **2013**, *5*, 1259–1272. [CrossRef]
15. Li, J.; Barjuei, E.S.; Ciuti, G.; Hao, Y.; Zhang, P.; Menciassi, A.; Huang, Q.; Dario, P. Magnetically-driven medical robots: An analytical magnetic model for endoscopic capsules design. *J. Magn. Magn. Mater.* **2018**, *452*, 278–287. [CrossRef]
16. Fu, F.; Dionysiou, D.D.; Liu, H. The use of zero-valent iron for groundwater remediation and wastewater treatment: A review. *J. Hazard. Mater.* **2014**, *267*, 194–205. [CrossRef] [PubMed]
17. Ivanets, A.I.; Srivastava, V.; Roshchina, M.Y.; Sillanpää, M.; Prozorovich, V.G.; Pankov, V.V. Magnesium ferrite nanoparticles as a magnetic sorbent for the removal of Mn^{2+}, Co^{2+}, Ni^{2+} and Cu^{2+} from aqueous solution. *Ceram. Int.* **2018**, *44*, 9097–9104. [CrossRef]
18. Nistico, R.; Cesano, F.; Franzoso, F.; Magnacca, G.; Scarano, D.; Funes, I.G.; Carlos, L.; Parolo, M.E. From biowaste to magnet-responsive materials for water remediation from polycyclic aromatic hydrocarbons. *Chemosphere* **2018**, *202*, 686–693. [CrossRef]
19. Palma, D.; Bianco Prevot, A.; Celi, L.; Martin, M.; Fabbri, D.; Magnacca, G.; Chierotti, M.; Nisticò, R. Isolation, characterization, and environmental application of bio-based materials as auxiliaries in photocatalytic processes. *Catalysts* **2018**, *8*, 197. [CrossRef]
20. Neamtu, M.; Nadejde, C.; Hodoroaba, V.D.; Schneider, R.J.; Verestiuc, L.; Panne, U. Functionalized magnetic nanoparticles: Synthesis, characterization, catalytic application and assessment of toxicity. *Sci. Rep.* **2018**, *8*, 6278. [CrossRef]
21. Sun, Z.; Zhou, X.; Luo, W.; Yue, Q.; Zhang, Y.; Cheng, X.; Li, W.; Kong, B.; Deng, Y.; Zhao, D. Interfacial engineering of magnetic particles with porous shells: Towards magnetic core—Porous shell microparticles. *Nano Today* **2016**, *11*, 464–482. [CrossRef]

22. Yermolenko, I.Y.; Ved, M.V.; Sakhnenko, N.D.; Shipkova, I.G.; Zyubanova, S.I. Nanostructured magnetic films based on iron with refractory metals. *J. Magn. Magn. Mater.* **2019**, *475*, 115–120. [CrossRef]
23. Chen, X.-Z.; Hoop, M.; Mushtaq, F.; Siringil, E.; Hu, C.; Nelson, B.J.; Pané, S. Recent developments in magnetically driven micro- and nanorobots. *Appl. Mater. Today* **2017**, *9*, 37–48. [CrossRef]
24. Bedanta, S.; Petracic, O.; Kleemann, W. Supermagnetism. In *Handbook of Magnetic Materials*; Bruck, E., Ed.; Elsevier: North Holland, The Netherlands, 2015; Volume 23, pp. 1–83.
25. Karimi, Z.; Karimi, L.; Shokrollahi, H. Nano-magnetic particles used in biomedicine: Core and coating materials. *Mat. Sci. Eng. C* **2013**, *33*, 2465–2475. [CrossRef] [PubMed]
26. Li, D.; Li, Y.; Pan, D.; Zhang, Z.; Choi, C.-J. Prospect and status of iron-based rare-earth-free permanent magnetic materials. *J. Magn. Magn. Mater.* **2019**, *469*, 535–544. [CrossRef]
27. Rossi, L.M.; Costa, N.J.S.; Silva, F.P.; Wojcieszak, R. Magnetic nanomaterials in catalysis: Advanced catalysts for magnetic separation and beyond. *Green Chem.* **2014**, *16*, 2906–2933. [CrossRef]
28. Sundaresan, A.; Rao, C.N.R. Ferromagnetism as a universal feature of inorganic nanoparticles. *Nano Today* **2009**, *4*, 96–106. [CrossRef]
29. Hubert, A.; Schäfer, R. *Magnetic Domains, The Analysis of Magnetic Microstructures*; Springer-Verlag: Berlin/Heidelberg, Germany, 1998. [CrossRef]
30. Tumanski, S. *Handbook of Magnetic Measurements*; CRC Press, Taylor & Francis Group: Boca Raton, FL, USA, 2011.
31. Craik, D.J.; Tebble, R.S. Magnetic domains. *Rep. Prog. Phys.* **1961**, *24*, 116–166. [CrossRef]
32. Jiles, D.C.; Atherton, D.L. Theory of ferromagnetic hysteresis. *J. Magn. Magn. Mater.* **1986**, *61*, 48–60. [CrossRef]
33. Newbury, D.E.; Joy, D.C.; Echlin, P.; Fiori, C.E.; Goldstein, J.I. *Advanced Scanning Electron Microscopy and X-Ray Microanalysis*; Springer Science: New York, NY, USA, 1986. [CrossRef]
34. Jiles, D.C. Theory of the magnetomechanical effect. *J. Phys. D Appl. Phys.* **1995**, *28*, 1537–1546. [CrossRef]
35. Chien, C.L. Magnetism and Magnetic Measurement, Introduction. In *Characterization of Materials*, 2nd ed.; Kaufmann, E.N., Ed.; John Wiley & Sons: Hoboken, NY, USA, 2012. [CrossRef]
36. Thanh, N.T.K. *Magnetic Nanoparticles: From Fabrication to Clinical Applications*; CRC Press, Taylor & Francis Group: Boca Raton, FL, USA, 2012.
37. Quarterman, P.; Sun, C.; Garcia-Barriocanal, J.; Dc, M.; Lv, Y.; Manipatruni, S.; Nikonov, D.E.; Young, I.A.; Voyles, P.M.; Wang, J.P. Demonstration of Ru as the 4th ferromagnetic element at room temperature. *Nat. Commun.* **2018**, *9*, 2058. [CrossRef]
38. Jordán, D.; González-Chávez, D.; Laura, D.; León Hilario, L.M.; Monteblanco, E.; Gutarra, A.; Avilés-Félix, L. Detection of magnetic moment in thin films with a home-made vibrating sample magnetometer. *J. Magn. Magn. Mater.* **2018**, *456*, 56–61. [CrossRef]
39. IEEE Magnetics Society. Magnetic Units. Available online: http://www.ieeemagnetics.org/index.php?option=com_content&view=article&id=118&Itemid=107 (accessed on 25 September 2019).
40. Sung, H.W.F.; Rudowicz, C. Physics behind the magnetic hysteresis loop—A survey of misconceptions in magnetism literature. *J. Magn. Magn. Mater.* **2003**, *260*, 250–260. [CrossRef]
41. Kurgan, E.; Gas, P. Methods of Calculation the Magnetic Forces Acting on Particles in Magnetic Fluids. In Proceedings of the 2018 Progress in Applied Electrical Engineering (PAEE), Koscielisko, Poland, 18–22 June 2018; pp. 1–5. [CrossRef]
42. Gerstein, G.; L'Vov, V.A.; Żak, A.; Dudziński, W.; Maier, H.J. Direct observation of nano-dimensional internal structure of ferromagnetic domains in the ferromagnetic shape memory alloy Co–Ni–Ga. *J. Magn. Magn. Mater.* **2018**, *466*, 125–129. [CrossRef]
43. James, R.D. Magnetic alloys break the rules. *Nature* **2015**, *521*, 298–299. [CrossRef]
44. Lu, A.H.; Salabas, E.L.; Schuth, F. Magnetic nanoparticles: Synthesis, protection, functionalization, and application. *Angew. Chem. Int. Ed. Engl.* **2007**, *46*, 1222–1244. [CrossRef]
45. Ando, K. Seeking room-temperature ferromagnetic semiconductors. *Science* **2006**, *312*, 1883–1885. [CrossRef]
46. Fitta, M.; Czaja, P.; Krupiński, M.; Lewińska, G.; Szuwarzyński, M.; Bałanda, M. Magnetic properties of bilayer thin film composed of hard and soft ferromagnetic Prussian Blue analogues. *Chem. Select* **2017**, *2*, 7930–7934. [CrossRef]
47. Devi, E.C.; Soibam, I. Tuning the magnetic properties of a ferrimagnet. *J. Magn. Magn. Mater.* **2019**, *469*, 587–592. [CrossRef]

48. Matsuura, K.; Sagayama, H.; Uehara, A.; Nii, Y.; Kajimoto, R.; Kamazawa, K.; Ikeuchi, K.; Ji, S.; Abe, N.; Arima, T.-H. Magnetic excitations in the orbital disordered phase of MnV$_2$O$_4$. *Phys. B Cond. Matter* **2018**, *536*, 372–376. [CrossRef]
49. Gignoux, D.; Schmitt, D. Chapter 2 Magnetism of compounds of rare earths with non-magnetic metals. In *Handbook of Magnetic Materials*; Buschow, K.H.J., Ed.; Elsevier: North Holland, The Netherlands, 1997; Volume 10, pp. 239–413.
50. Rhyne, J.J.; Erwin, R.W. Magnetism in artificial metallic superlattices of rare earth metals. In *Handbook of Magnetic Materials*; Buschow, K.H.J., Ed.; Elsevier: North Holland, The Netherlands, 1995; Volume 8, pp. 1–57.
51. Woodruff, D.N.; Winpenny, R.E.; Layfield, R.A. Lanthanide single-molecule magnets. *Chem. Rev.* **2013**, *113*, 5110–5148. [CrossRef]
52. München, D.D.; Veit, H.M. Neodymium as the main feature of permanent magnets from hard disk drives (HDDs). *Waste Manag.* **2017**, *61*, 372–376. [CrossRef] [PubMed]
53. Ucar, H.; Choudhary, R.; Paudyal, D. An overview of the first principles studies of doped RE-TM5 systems for the development of hard magnetic properties. *J. Magn. Magn. Mater.* **2020**, *496*, 165902. [CrossRef]
54. Chikazumi, S.S.; Graham, C.D. *Physics of Ferromagnetism*; University Press: Oxford, UK, 2009; Volume 94.
55. Preller, T.; Menzel, D.; Knickmeier, S.; Porsiel, J.C.; Temel, B.; Garnweitner, G. Non-aqueous sol–gel synthesis of FePt nanoparticles in the absence of in situ stabilizers. *Nanomaterials* **2018**, *8*, 297. [CrossRef] [PubMed]
56. Simon, M.D.; Geim, A.K. Diamagnetic levitation: Flying frogs and floating magnets. *J. Appl. Phys.* **2000**, *87*, 6200–6204. [CrossRef]
57. Palagummi, S.; Yuan, F.G. Magnetic levitation and its application for low frequency vibration energy harvesting. In *Structural Health Monitoring (SHM) in Aerospace Structures*; Yuan, F.G., Ed.; Woodhead Publishing and Elsevier: North Holland, The Netherlands, 2016; pp. 213–251. [CrossRef]
58. Paulo, V.I.M.; Neves-Araujo, J.; Revoredo, F.A.; Padrón-Hernández, E. Magnetization curves of electrodeposited Ni, Fe and Co nanotubes. *Mater. Lett.* **2018**, *223*, 78–81. [CrossRef]
59. Venkata Ramana, E.; Figueiras, F.; Mahajan, A.; Tobaldi, D.M.; Costa, B.F.O.; Graça, M.P.F.; Valente, M.A. Effect of Fe-doping on the structure and magnetoelectric properties of (Ba$_{0.85}$Ca$_{0.15}$)(Ti$_{0.9}$Zr$_{0.1}$)O$_3$ synthesized by a chemical route. *J. Mater. Chem. C* **2016**, *4*, 1066–1079. [CrossRef]
60. Peixoto, E.B.; Carvalho, M.H.; Meneses, C.T.; Sarmento, V.H.V.; Coelho, A.A.; Zucolotto, B.; Duque, J.G.S. Analysis of zero field and field cooled magnetization curves of CoFe$_2$O$_4$ nanoparticles with a T-dependence on the saturation magnetization. *J. Alloys Comp.* **2017**, *721*, 525–530. [CrossRef]
61. Nisticò, R.; Magnacca, G.; Antonietti, M.; Fechler, N. "Salted silica": Sol–gel chemistry of silica under hypersaline conditions. *Z. Anorg. Allg. Chem.* **2014**, *640*, 582–587. [CrossRef]
62. Nisticò, R.; Scalarone, D.; Magnacca, G. Sol-gel chemistry, templating and spin-coating deposition: A combined approach to control in a simple way the porosity of inorganic thin films/coatings. *Microp. Mesop. Mater.* **2017**, *248*, 18–29. [CrossRef]
63. Wang, S.; Huang, K.; Hou, C.; Yuan, L.; Wu, X.; Lu, D. Low temperature hydrothermal synthesis, structure and magnetic properties of RECrO$_3$ (RE = La, Pr, Nd, Sm). *Dalton Trans.* **2015**, *44*, 17201–17208. [CrossRef]
64. Cesano, F.; Fenoglio, G.; Carlos, L.; Nisticò, R. One-step synthesis of magnetic chitosan polymer composite films. *Appl. Surf. Sci.* **2015**, *345*, 175–181. [CrossRef]
65. Nisticò, R.; Franzoso, F.; Cesano, F.; Scarano, D.; Magnacca, G.; Parolo, M.E.; Carlos, L. Chitosan-derived iron oxide systems for magnetically guided and efficient water purification processes from polycyclic aromatic hydrocarbons. *ACS Sustain. Chem. Eng.* **2016**, *5*, 793–801. [CrossRef]
66. Bianco Prevot, A.; Baino, F.; Fabbri, D.; Franzoso, F.; Magnacca, G.; Nistico, R.; Arques, A. Urban biowaste-derived sensitizing materials for caffeine photodegradation. *Environ. Sci. Pollut. Res. Int.* **2017**, *24*, 12599–12607. [CrossRef] [PubMed]
67. Palma, D.; Bianco Prevot, A.; Brigante, M.; Fabbri, D.; Magnacca, G.; Richard, C.; Mailhot, G.; Nistico, R. New insights on the photodegradation of caffeine in the presence of bio-based substances-magnetic iron oxide hybrid nanomaterials. *Materials* **2018**, *11*, 1084. [CrossRef]
68. Miller, J.S. Organic- and molecule-based magnets. *Mater. Today* **2014**, *17*, 224–235. [CrossRef]
69. Shirakawa, N.; Tamura, M. Low temperature static magnetization of an organic ferromagnet, β-*p*-NPNN. *Polyhedron* **2005**, *24*, 2405–2408. [CrossRef]

70. Shum, W.W.; Her, J.H.; Stephens, P.W.; Lee, Y.; Miller, J.S. Observation of the pressure dependent reversible enhancement of Tc and loss of the anomalous constricted hysteresis for [Ru$_2$(O$_2$CMe)$_4$]$_3$[Cr(CN)$_6$]. *Adv. Mater.* **2007**, *19*, 2910–2913. [CrossRef]
71. Fu, L.; Zhang, K.; Zhang, W.; Chen, J.; Deng, Y.; Du, Y.; Tang, N. Synthesis and intrinsic magnetism of bilayer graphene nanoribbons. *Carbon* **2019**, *143*, 1–7. [CrossRef]
72. Ominato, Y.; Koshino, M. Orbital magnetism of graphene nanostructures. *Sol. State Commun.* **2013**, *175–176*, 51–61. [CrossRef]
73. Tajima, K.; Isaka, T.; Yamashina, T.; Ohta, Y.; Matsuo, Y.; Takai, K. Functional group dependence of spin magnetism in graphene oxide. *Polyhedron* **2017**, *136*, 155–158. [CrossRef]
74. Calle, D.; Negri, V.; Munuera, C.; Mateos, L.; Touriño, I.L.; Viñegla, P.R.; Ramírez, M.O.; García-Hernández, M.; Cerdán, S.; Ballesteros, P. Magnetic anisotropy of functionalized multi-walled carbon nanotube suspensions. *Carbon* **2018**, *131*, 229–237. [CrossRef]
75. Kim, D.W.; Lee, K.W.; Lee, C.E. Defect-induced room-temperature ferromagnetism in single-walled carbon nanotubes. *J. Magn. Magn. Mater.* **2018**, *460*, 397–400. [CrossRef]
76. Tucek, J.; Hola, K.; Bourlinos, A.B.; Blonski, P.; Bakandritsos, A.; Ugolotti, J.; Dubecky, M.; Karlicky, F.; Ranc, V.; Cepe, K.; et al. Room temperature organic magnets derived from sp^3 functionalized graphene. *Nature Commun.* **2017**, *8*, 14525. [CrossRef] [PubMed]
77. Georgakilas, V.; Perman, J.A.; Tucek, J.; Zboril, R. Broad family of carbon nanoallotropes: Classification, chemistry, and applications of fullerenes, carbon dots, nanotubes, graphene, nanodiamonds, and combined superstructures. *Chem. Rev.* **2015**, *115*, 4744–4822. [CrossRef] [PubMed]
78. Wlodarski, Z. Analytical description of magnetization curves. *Phys. B Cond. Matter* **2006**, *373*, 323–327. [CrossRef]
79. Stiles, M.D.; McMichael, R.D. Coercivity in exchange-bias bilayers. *PRB* **1995**, *63*, 064405. [CrossRef]
80. Fiorillo, F. DC and AC magnetization processes in soft magnetic materials. *J. Magn. Magn. Mater.* **2002**, *242–245*, 77–83. [CrossRef]
81. Yoshida, T.; Nakamura, T.; Higashi, O.; Enpuku, K. Effect of viscosity on the AC magnetization of magnetic nanoparticles under different AC excitation fields. *J. Magn. Magn. Mater.* **2019**, *471*, 334–339. [CrossRef]
82. Panina, L.V.; Dzhumazoda, A.; Evstigneeva, S.A.; Adama, A.M.; Morchenko, A.T.; Yudanova, N.A.; Kostishyna, V.G. Temperature effects on magnetization processes and magnetoimpedance in low magnetostrictive amorphous microwires. *J. Magn. Magn. Mater.* **2018**, *459*, 147–153. [CrossRef]
83. Franse, J.J.M.; Hién, T.D.; Ngân, N.H.K.; Dúc, N.H. Magnetization and AC susceptibility of Tb$_x$Y$_{1-x}$Co$_2$ compounds. *J. Magn. Magn. Mater.* **1983**, *39*, 275–278. [CrossRef]
84. Boutaba, A.; Lahoubi, M.; Varazashvili, V.; Puc, S. Magnetic, magneto-optical and specific heat studies of the low temperature anomalies in the magnetodielectric DyIG ferrite garnet. *J. Magn. Magn. Mater.* **2018**, *476*, 551–558. [CrossRef]
85. Kodama, K. Measurement of dynamic magnetization induced by a pulsed field: Proposal for a new rock magnetism method. *Front. Earth Sci.* **2015**, *3*, 5. [CrossRef]
86. Jackson, M. Magnetization, Isothermal Remanent. In *Encyclopedia of Geomagnetism and Paleomagnetism*; Gubbins, D., Herrero-Bervera, E., Eds.; Springer: Dordrecht, The Netherlands, 2007. [CrossRef]
87. Clarke, J. SQUID Concepts and Systems. In *Superconducting Electronics. NATO ASI Series (Series F: Computer and Systems Sciences)*; Weinstock, H., Nisenoff, M., Eds.; Springer: Berlin/Heidelberg, Germany, 1989; Volume 59.
88. Conta, G.; Amato, G.; Coisson, M.; Tiberto, P. Experimental insight into the magnetic and electrical properties of amorphous Ge$_{1-x}$Mn$_x$. *Sci. Technol. Adv. Mater.* **2017**, *18*, 34–42. [CrossRef]
89. Drung, D.; Cantor, R.; Peters, M.; Ryhanen, T.; Koch, H. Integrated DC SQUID magnetometer with high dV/dB. *IEEE Trans. Magnet.* **1991**, *27*, 3001–3004. [CrossRef]
90. Meredith, D.J.; Pickett, G.R.; Symko, O.G. Application of a SQUID magnetometer to NMR at low temperatures. *J. Low Temp. Phys.* **1973**, *13*, 607–615. [CrossRef]
91. Wu, L.; Mendoza-Garcia, A.; Li, Q.; Sun, S. Organic Phase Syntheses of Magnetic Nanoparticles and Their Applications. *Chem. Rev.* **2016**, *116*, 10473–10512. [CrossRef]
92. Gaul, A.; Emmrich, D.; Ueltzhoffer, T.; Huckfeldt, H.; Doganay, H.; Hackl, J.; Khan, M.I.; Gottlob, D.M.; Hartmann, G.; Beyer, A.; et al. Size limits of magnetic-domain engineering in continuous in-plane exchange-bias prototype films. *Beilstein J. Nanotechnol.* **2018**, *9*, 2968–2979. [CrossRef]

93. Coïsson, M.; Barrera, G.; Celegato, F.; Tiberto, P. Rotatable magnetic anisotropy in Fe$_{78}$Si$_9$B$_{13}$ thin films displaying stripe domains. *Appl. Surf. Sci.* **2019**, *476*, 402–411. [CrossRef]
94. Staňo, M.; Fruchart, O. Magnetic Nanowires and Nanotubes. In *Handbook of Magnetic Materials 2018*; Elsevier: North Holland, The Netherlands, 2018; pp. 155–267. [CrossRef]
95. Sala, A. Sala Imaging at the Mesoscale. 2018. Available online: https://arxiv.org/abs/1812.01610 (accessed on 28 December 2019).
96. Dehsari, H.S.; Ksenofontov, V.; Möller, A.; Jakob, G.; Asadi, K. Determining Magnetite/Maghemite Composition and Core–Shell Nanostructure from Magnetization Curve for Iron Oxide Nanoparticles. *J. Phys. Chem. C* **2018**, *122*, 28292–28301. [CrossRef]
97. Pagoto, A.; Stefania, R.; Garello, F.; Arena, F.; Digilio, G.; Aime, S.; Terreno, E. Paramagnetic phospholipid-based micelles targeting VCAM-1 receptors for MRI visualization of inflammation. *Bioconj. Chem.* **2016**, *27*, 1921–1930. [CrossRef]
98. Garello, F.; Arena, F.; Cutrin, J.C.; Esposito, G.; D'Angeli, L.; Cesano, F.; Filippi, M.; Figueiredo, S.; Terreno, E. Glucan particles loaded with a NIRF agent for imaging monocytes/macrophages recruitment in a mouse model of rheumatoid arthritis. *RSC Adv.* **2015**, *5*, 34078–34087. [CrossRef]
99. Garello, F.; Pagoto, A.; Arena, F.; Buffo, A.; Blasi, F.; Alberti, D.; Terreno, E. MRI visualization of neuroinflammation using VCAM-1 targeted paramagnetic micelles. *Nanomedicine* **2018**, *14*, 2341–2350. [CrossRef] [PubMed]
100. Garello, F.; Stefania, R.; Aime, S.; Terreno, E.; Delli Castelli, D. Successful entrapping of liposomes in glucan particles: An innovative micron-sized carrier to deliver water-soluble molecules. *Mol. Pharm.* **2014**, *11*, 3760–3765. [CrossRef]
101. Cesano, F.; Rattalino, I.; Bardelli, F.; Sanginario, A.; Gianturco, A.; Veca, A.; Viazzi, C.; Castelli, P.; Scarano, D.; Zecchina, A. Structure and properties of metal-free conductive tracks on polyethylene/multiwalled carbon nanotube composites as obtained by laser stimulated percolation. *Carbon* **2013**, *61*, 63–71. [CrossRef]
102. Cravanzola, S.; Cesano, F.; Magnacca, G.; Zecchina, A.; Scarano, D. Designing rGO/MoS$_2$ hybrid nanostructures for photocatalytic applications. *RSC Adv.* **2016**, *6*, 59001–59008. [CrossRef]
103. Groppo, E.; Lamberti, C.; Cesano, F.; Zecchina, A. On the fraction of Cr-II sites involved in the C$_2$H$_4$ polymerization on the Cr/SiO$_2$ Phillips catalyst: A quantification by FTIR spectroscopy. *PCCP* **2006**, *8*, 2453–2456. [CrossRef]
104. Laurent, S.; Forge, D.; Port, M.; Roch, A.; Robic, C.; Elst, L.V.; Muller, R.N. Magnetic Iron Oxide Nanoparticles: Synthesis, Stabilization, Vectorization, Physicochemical Characterizations, and Biological Applications. *Chem. Rev.* **2008**, *108*, 2064–2110. [CrossRef]
105. Bianco Prevot, A.; Arques, A.; Carlos, L.; Laurenti, E.; Magnacca, G.; Nisticò, R. Innovative sustainable materials for the photoinduced remediation of polluted water. In *Sustainable Water and Wastewater Processes*; Galanakis, C.M., Agrafioti, E., Eds.; Elsevier Inc.: Amsterdam, The Netherlands, 2019; pp. 203–238.
106. Nisticò, R.; Bianco Prevot, A.; Magnacca, G.; Canone, L.; Garcia-Ballesteros, S.; Arques, A. Sustainable magnetic materials (from chitosan and municipal biowaste) for the removal of Diclofenac from water. *Nanomaterials* **2019**, *9*, 1091. [CrossRef]
107. Nisticò, R.; Celi, L.R.; Bianco Prevot, A.; Carlos, L.; Magnacca, G.; Zanzo, E.; Martin, M. Sustainable magnet-responsive nanomaterials for the removal of arsenic from contaminated water. *J. Hazard. Mater.* **2018**, *342*, 260–269. [CrossRef]
108. Sifford, J.; Walsh, K.J.; Tong, S.; Bao, G.; Agarwal, G. Indirect magnetic force microscopy. *Nanoscale Adv.* **2019**, *1*, 2348–2355. [CrossRef] [PubMed]
109. Celotta, R.J.; Unguris, J.; Kelley, M.H.; Pierce, D.T. Techniques to Measure Magnetic Domain Structures. In *Characterization of Materials*, 2nd ed.; Kaufmann, E., Ed.; John Wiley & Sons: Hoboken, NJ, USA, 2012. [CrossRef]
110. Celotta, R.J.; Unguris, J.; Pierce, D.T. Magnetic Domain Imaging of Spintronic Devices. In *Magnetic Interactions and Spin Transport*; Chtchelkanova, A.; Wolf, S.Y.I., Eds.; Springer: Boston, MA, USA, 2003. [CrossRef]
111. Dickson, W.; Takahashi, S.; Pollard, R.; Atkinson, R.; Zayats, A.V. High-Resolution Optical Imaging of Magnetic-Domain Structures. *IEEE Trans. Nanotechnol.* **2005**, *4*, 229–237. [CrossRef]
112. Petford-Long, A.K.; Chapman, J.N. Lorentz Microscopy. In *Magnetic Microscopy of Nanostructures*; Hopster, H., Oepen, H.P., Eds.; Springer: Berlin/Heidelberg, Germany, 2005. [CrossRef]

113. Tanase, M.; Petford-Long, A.K. In situ TEM observation of magnetic materials. *Microsc. Res. Tech.* **2009**, *72*, 187–196. [CrossRef] [PubMed]
114. Shibata, N.; Findlay, S.D.; Matsumoto, T.; Kohno, Y.; Seki, T.; Sánchez-Santolino, G.; Ikuhara, Y. Direct Visualization of Local Electromagnetic Field Structures by Scanning Transmission Electron Microscopy. *Acc. Chem. Res.* **2018**, *50*, 1502–1512. [CrossRef] [PubMed]
115. Kovács, A.; Dunin-Borkowski, R.E. Magnetic Imaging of Nanostructures Using Off-Axis Electron Holography. In *Handbook of Magnetic Materials, Handbook of Magnetic Materials*; Brück, E., Ed.; Elsevier: North Holland, The Netherlands, 2018; Volume 27, pp. 59–153. [CrossRef]
116. Koike, K. Spin-polarized scanning electron microscopy. *Microscopy* **2013**, *62*, 177–191. [CrossRef]
117. Takeichi, Y. Scanning Transmission X-Ray Microscopy. In *Compendium of Surface and Interface Analysis*; Springer: Singapore, 2018; pp. 593–597. [CrossRef]
118. Van der Laan, G.; Figueroa, A.I. X-ray magnetic circular dichroism—A versatile tool to study magnetism. *Coord. Chem. Rev.* **2014**, *277–278*, 95–129. [CrossRef]
119. De Groot, L.V.; Fabian, K.; Bakelaar, I.A.; Dekkers, M.J. Magnetic force microscopy reveals meta-stable magnetic domain states that prevent reliable absolute palaeointensity experiments. *Nat. Commun.* **2014**, *5*, 4548. [CrossRef]
120. Ferri, F.A.; Pereira-da-Silva, M.A.; Marega, E. Magnetic Force Microscopy: Basic Principles and Applications. In *Imaging, Measuring and Manipulating Surfaces at the Atomic Scale*; Bellitto, V., Ed.; IntechOpen: London, UK, 2012. [CrossRef]
121. Fischer, P. X-Ray Imaging of Magnetic Structures. *IEEE Trans. Magn.* **2015**, *51*, 0800131. [CrossRef]
122. Simpson, D.A.; Tetienne, J.P.; McCoey, J.M.; Ganesan, K.; Hall, L.T.; Petrou, S.; Scholten, R.E.; Hollenberg, L.C. Magneto-optical imaging of thin magnetic films using spins in diamond. *Sci. Rep.* **2016**, *6*, 22797. [CrossRef]
123. Le Sage, D.; Arai, K.; Glenn, D.R.; DeVience, S.J.; Pham, L.M.; Rahn-Lee, L.; Lukin, M.D.; Yacoby, A.; Komeili, A.; Walsworth, R.L. Optical magnetic imaging of living cells. *Nature* **2013**, *496*, 486–489. [CrossRef]
124. Fischer, P. Viewing spin structures with soft X-ray microscopy. *Mater. Today* **2010**, *13*, 14–22. [CrossRef]
125. Ge, M.; Coburn, D.S.; Nazaretski, E.; Xu, W.; Gofron, K.; Xu, H.; Yin, Z.; Lee, W.-K. One-minute nano-tomography using hard X-ray full-field transmission microscope. *Appl. Phys. Lett.* **2018**, *113*, 083109. [CrossRef]
126. Donnelly, C.; Guizar-Sicairos, M.; Scagnoli, V.; Gliga, S.; Holler, M.; Raabe, J.; Heyderman, L.J. Three-dimensional magnetization structures revealed with X-ray vector nanotomography. *Nature* **2017**, *547*, 328–331. [CrossRef] [PubMed]
127. Streubel, R.; Kronast, F.; Fischer, P.; Parkinson, D.; Schmidt, O.G.; Makarov, D. Retrieving spin textures on curved magnetic thin films with full-field soft X-ray microscopies. *Nat. Commun.* **2015**, *6*, 7612. [CrossRef] [PubMed]
128. Kiyanagi, Y. Neutron Imaging at Compact Accelerator-Driven Neutron Sources in Japan. *J. Imaging* **2018**, *4*, 55. [CrossRef]
129. Kardjilov, N.; Hilger, A.; Manke, I.; Strobl, M.; Banhart, J. Imaging with Polarized Neutrons. *J. Imaging* **2018**, *4*, 23. [CrossRef]
130. Manke, I.; Kardjilov, N.; Schäfer, R.; Hilger, A.; Grothausmann, R.; Strobl, M.; Dawson, M.; Grünzweig, C.; Tötzke, C.; David, C.; et al. Three-Dimensional Imaging of Magnetic Domains with Neutron Grating Interferometry. *Phys. Procedia* **2015**, *69*, 404–412. [CrossRef]
131. Hámos, L.; Thiessen, P.A. Über die Sichtbarmachung von Bezirken verschiedenen ferromagnetischen Zustandes fester Körper. *Z. Phys.* **1931**, *71*, 442–444. [CrossRef]
132. Bitter, F. On Inhomogeneities in the Magnetization of Ferromagnetic Materials. *Phys. Rev.* **1931**, *38*, 1903–1905. [CrossRef]
133. Gong, C.; Li, L.; Li, Z.; Ji, H.; Stern, A.; Xia, Y.; Cao, T.; Bao, W.; Wang, C.; Wang, Y.; et al. Discovery of intrinsic ferromagnetism in two-dimensional van der Waals crystals. *Nature* **2017**, *546*, 265–269. [CrossRef]
134. Bozorth, R.M. Magnetic domain patterns. *J. Phys. Radium* **1951**, *12*, 308–321. [CrossRef]
135. Szmaja, W.; Balcerski, J. Domain investigation by the conventional Bitter pattern technique with digital image processing. *Czech. J. Phys.* **2002**, *52*, 223–226. [CrossRef]
136. Olson, A.L. On the Visibility of Bitter Powder Patterns on Ferromagnetic Films with Bloch and Néel-Type Domain Walls. *J. Appl. Phys.* **1967**, *38*, 1869–1871. [CrossRef]

137. Sonntag, N.; Cabeza, S.; Kuntner, M.; Mishurova, T.; Klaus, M.; Kling e Silva, L.; Skrotzki, B.; Genzel, C.; Bruno, G. Visualisation of deformation gradients in structural steel by macroscopic magnetic domain distribution imaging (Bitter technique). *Strain* **2018**, *54*, e12296. [CrossRef]
138. Arregi, J.A.; Riego, P.; Berger, A. What is the longitudinal magneto-optical Kerr effect? *J. Phys. D Appl. Phys.* **2017**, *50*, 03LT01. [CrossRef]
139. Lee, J.; Wang, Z.; Xie, H.; Mak, K.F.; Shan, J. Valley magnetoelectricity in single-layer MoS_2. *Nat. Mater.* **2017**, *16*, 887–891. [CrossRef] [PubMed]
140. Chen, J.Y.; Zhu, J.; Zhang, D.; Lattery, D.M.; Li, M.; Wang, J.P.; Wang, X. Time-Resolved Magneto-Optical Kerr Effect of Magnetic Thin Films for Ultrafast Thermal Characterization. *J. Phys. Chem. Lett.* **2016**, *7*, 2328–2332. [CrossRef] [PubMed]
141. Kustov, M.; Grechishkin, R.; Gusev, M.; Gasanov, O.; McCord, J. A Novel Scheme of Thermographic Microimaging Using Pyro-Magneto-Optical Indicator Films. *Adv. Mater.* **2015**, *27*, 5017–5022. [CrossRef]
142. Oatley, C.W. The early history of the scanning electron microscope. *J. Appl. Phys.* **1982**, *53*, R1. [CrossRef]
143. Robinson, V.N.E. Imaging with Backscattered Electrons in a Scanning Electron Microscope. *Scanning* **1980**, *3*, 15–26. [CrossRef]
144. Banbury, J.R.; Nixon, W.C. The direct observation of domain structure and magnetic fields in the scanning electron microscope. *J. Sci. Instr.* **1967**, *44*, 889. [CrossRef]
145. Philibert, J.; Tixier, R. Effets de contraste cristallin en microscopie électronique à balayage. *Micron* **1969**, *1*, 174–186. [CrossRef]
146. Pierce, D.T.; Celotta, R.J. Spin Polarization in Electron Scattering from Surfaces. *Adv. Electron. Electr. Phys.* **1981**, *56*, 219–289.
147. Koike, K.; Hayakawa, K. Spin Polarization of Electron-Excited Secondary Electrons from a Permalloy Polycrystal. *Jpn. J. Appl. Phys.* **1984**, *23*, L85–L87. [CrossRef]
148. Bertolini, G.; De Pietro, T.; Bähler, T.; Cabrera, H.; Gürlüab, O.; Pescia, D.; Ramsperger, U. Scanning Field Emission Microscopy with Polarization Analysis (SFEMPA). *J. Electr. Spectr. Rel. Phen.* **2019**, *2019*. in press. [CrossRef]
149. Shibata, N.; Kohno, Y.; Nakamura, A.; Morishita, S.; Seki, T.; Kumamoto, A.; Sawada, H.; Matsumoto, T.; Findlay, S.D.; Ikuhara, Y. Atomic resolution electron microscopy in a magnetic field free environment. *Nat. Commun.* **2019**, *10*, 2308. [CrossRef] [PubMed]
150. Morishita, S.; Ishikawa, R.; Kohno, Y.; Sawada, H.; Shibata, N.; Ikuhara, Y. Attainment of 40.5 pm spatial resolution using 300 kV scanning transmission electron microscope equipped with fifth-order aberration corrector. *Microscopy* **2018**, *67*, 46–50.
151. Cesano, F.; Cravanzola, S.; Rahman, M.; Scarano, D. Interplay between Fe-Titanate Nanotube Fragmentation and Catalytic Decomposition of C_2H_4: Formation of C/TiO_2 Hybrid Interfaces. *Inorganics* **2018**, *6*, 55. [CrossRef]
152. Cheng, R.; Li, M.; Sapkota, A.; Rai, A.; Pokhrel, A.; Mewes, T.; Mewes, C.; Xiao, D.; De Graef, M.; Sokalski, V. Magnetic domain wall skyrmions. *PRB* **2019**, *99*, 184412. [CrossRef]
153. McVitie, S.; McGrouther, D.; McFadzean, S.; MacLaren, D.A.; O'Shea, K.J.; Benitez, M.J. Aberration corrected Lorentz scanning transmission electron microscopy. *Ultramicroscopy* **2015**, *152*, 57–62. [CrossRef]
154. Nago, Y.; Ishiguro, R.; Sakurai, T.; Yakabe, M.; Nakamura, S.; Yonezawa, S.; Kashiwaya, S.; Takayanagi, H.; Maeno, Y. Evolution of supercurrent path in $Nb/Ru/Sr_2RuO_4$ dc-SQUIDs. *PRB* **2016**, *94*, 054501. [CrossRef]
155. Cooper, D.; Béché, A.; Hertog, M.D.; Masseboeuf, A.; Rouvière, J.-L.; Guillem, P.B.; Gambacorti, N. Off-Axis Electron Holography for Field Mapping in the Semiconductor Industry. *Microsc. Anal.* **2010**, *5*, 7.
156. Serrano-Ramon, L.; Cordoba, R.; Rodriguez, L.A.; Magen, C.; Snoeck, E.; Gatel, C.; Serrano, I.; Ibarra, M.R.; De Teresa, J.M. Ultrasmall functional ferromagnetic nanostructures grown by focused electron-beam-induced deposition. *ACS Nano* **2011**, *5*, 7781–7787. [CrossRef] [PubMed]
157. Almeida, T.P.; Muxworthy, A.R.; Williams, W.; Takeshi, K.; Kovács, A.; Dunin-Borkowski, R.E. Off-axis electron holography for imaging the magnetic behavior of vortex-state minerals. *Microsc. Anal.* **2019**, *8*, 5–8.
158. Zhou, C.; Chen, G.; Xu, J.; Liang, J.; Liu, K.; Schmid, A.K.; Wu, Y. Magnetic domain wall contrast under zero domain contrast conditions in spin polarized low energy electron microscopy. *Ultramicroscopy* **2019**, *200*, 132–138. [CrossRef] [PubMed]

159. Scarano, D.; Bertarione, S.; Cesano, F.; Spoto, G.; Zecchina, A. Imaging polycrystalline and smoke MgO surfaces with atomic force microscopy: A case study of high resolution image on a polycrystalline oxide. *Surf. Sci.* **2004**, *570*, 155–166. [CrossRef]
160. Fatayer, S.; Albrecht, F.; Zhang, Y.; Urbonas, D.; Peña, D.; Moll, N.; Gross, L. Molecular structure elucidation with charge-state control. *Science* **2019**, *365*, 142–145. [PubMed]
161. Kazakova, O.; Puttock, R.; Barton, C.; Corte-León, H.; Jaafar, M.; Neu, V.; Asenjo, A. Frontiers of magnetic force microscopy. *J. Appl. Phys.* **2019**, *125*, 060901. [CrossRef]
162. Cesano, F.; Cravanzola, S.; Brunella, V.; Damin, A.; Scarano, D. From Polymer to Magnetic Porous Carbon Spheres: Combined Microscopy, Spectroscopy, and Porosity Studies. *Front. Chem.* **2019**, *6*, 84. [CrossRef]
163. Krivcov, A.; Schneider, J.; Junkers, T.; Möbius, H. Magnetic Force Microscopy of in a Polymer Matrix Embedded Single Magnetic Nanoparticles. *Phys. Stat. Sol. A* **2018**, *216*, 1800753. [CrossRef]
164. Coïsson, M.; Celegato, F.; Barrera, G.; Conta, G.; Magni, A.; Tiberto, P. Bi-Component Nanostructured Arrays of Co Dots Embedded in $Ni_{80}Fe_{20}$ Antidot Matrix: Synthesis by Self-Assembling of Polystyrene Nanospheres and Magnetic Properties. *Nanomaterials* **2017**, *7*, 232. [CrossRef]
165. Stolyarov, V.S.; Veshchunov, I.S.; Grebenchuk, S.Y.; Baranov, D.S.; Golovchanskiy, I.A.; Shishkin, A.G.; Zhou, N.; Shi, Z.; Xu, X.; Pyon, S.; et al. Domain Meissner state and spontaneous vortex-antivortex generation in the ferromagnetic superconductor $EuFe_2(As_{0.79}P_{0.21})_2$. *Sci. Adv.* **2018**, *4*, eaat1061. [CrossRef] [PubMed]
166. Geng, Y.; Das, H.; Wysocki, A.L.; Wang, X.; Cheong, S.W.; Mostovoy, M.; Fennie, C.J.; Wu, W. Direct visualization of magnetoelectric domains. *Nat. Mater.* **2014**, *13*, 163–167. [CrossRef] [PubMed]
167. Seki, S.; Yu, X.Z.; Ishiwata, S.; Tokura, Y. Observation of Skyrmions in a Multiferroic Materials. *Science* **2012**, *336*, 198–201. [CrossRef] [PubMed]
168. Guo, F.S.; Bar, A.K.; Layfield, R.A. Main Group Chemistry at the Interface with Molecular Magnetism. *Chem. Rev.* **2019**, *119*, 8479–8505. [CrossRef] [PubMed]
169. Guo, F.S.; Day, B.M.; Chen, Y.C.; Tong, M.L.; Mansikkamäki, A.; Layfield, R.A. Magnetic hysteresis up to 80 kelvin in a dysprosium metallocene single-molecule magnet. *Science* **2018**, *362*, 1400–1403. [CrossRef] [PubMed]
170. Iacovita, C.; Rastei, M.V.; Heinrich, B.W.; Brumme, T.; Kortus, J.; Limot, L.; Bucher, J.P. Visualizing the spin of individual cobalt-phthalocyanine molecules. *PRL* **2008**, *101*, 116602. [CrossRef] [PubMed]
171. Schwobel, J.; Fu, Y.; Brede, J.; Dilullo, A.; Hoffmann, G.; Klyatskaya, S.; Ruben, M.; Wiesendanger, R. Real-space observation of spin-split molecular orbitals of adsorbed single-molecule magnets. *Nat. Commun.* **2012**, *3*, 953. [CrossRef]
172. Reith, P.; Renshaw Wang, X.; Hilgenkamp, H. Analysing magnetism using scanning SQUID microscopy. *Rev. Sci. Instrum.* **2017**, *88*, 123706. [CrossRef]
173. Walbrecker, J.O.; Kalisky, B.; Grombacher, D.; Kirtley, J.; Moler, K.A.; Knight, R. Direct measurement of internal magnetic fields in natural sands using scanning SQUID microscopy. *J. Magn. Reson.* **2014**, *242*, 10–17. [CrossRef]
174. Boschker, H.; Harada, T.; Asaba, T.; Ashoori, R.; Boris, A.V.; Hilgenkamp, H.; Hughes, C.R.; Holtz, M.E.; Li, L.; Muller, D.A.; et al. Ferromagnetism and Conductivity in Atomically Thin $SrRuO_3$. *Phys. Rev. X* **2019**, *9*, 011027. [CrossRef]
175. Kirtley, J.R. Probing the order parameter symmetry in the cuprate high temperature superconductors by SQUID microscopy. *Comptes Rendus Phys.* **2011**, *12*, 436–445. [CrossRef]
176. Ghirri, A.; Candini, A.; Evangelisti, M.; Gazzadi, G.C.; Volatron, F.; Fleury, B.; Catala, L.; David, C.; Mallah, T.; Affronte, M. Magnetic Imaging of Cyanide-Bridged Co-ordination Nanoparticles Grafted on FIB-Patterned Si Substrates. *Small* **2008**, *4*, 2240–2246. [CrossRef] [PubMed]
177. Dede, M.; Akram, R.; Oral, A. 3D scanning Hall probe microscopy with 700 nm resolution. *Appl. Phys. Lett.* **2017**, *109*, 182407. [CrossRef]
178. Kleibert, A.; Balan, A.; Yanes, R.; Derlet, P.M.; Vaz, C.A.F.; Timm, M.; Fraile Rodríguez, A.; Béché, A.; Verbeeck, J.; Dhaka, R.S.; et al. Direct observation of enhanced magnetism in individual size- and shape-selected 3d transition metal nanoparticles. *Phys. Rev. B* **2017**, *95*, 195404. [CrossRef]
179. Ruiz-Gomez, S.; Perez, L.; Mascaraque, A.; Quesada, A.; Prieto, P.; Palacio, I.; Martin-Garcia, L.; Foerster, M.; Aballe, L.; de la Figuera, J. Geometrically defined spin structures in ultrathin Fe_3O_4 with bulk like magnetic properties. *Nanoscale* **2018**, *10*, 5566–5573. [CrossRef]

180. Guo, F.; Li, Q.; Zhang, H.; Yang, X.; Tao, Z.; Chen, X.; Chen, J. Czochralski Growth, Magnetic Properties and Faraday Characteristics of CeAlO₃ Crystals. *Crystals* **2019**, *9*, 245. [CrossRef]
181. Kotani, Y.; Senba, Y.; Toyoki, K.; Billington, D.; Okazaki, H.; Yasui, A.; Ueno, W.; Ohashi, H.; Hirosawa, S.; Shiratsuchi, Y.; et al. Realization of a scanning soft X-ray microscope for magnetic imaging under high magnetic fields. *J. Synchrotron Radiat.* **2018**, *25*, 1444–1449. [CrossRef]
182. Sala, A. Multiscale X-ray imaging using ptychography. *J. Synchrotron Radiat.* **2019**, *25*, 1214–1221. [CrossRef]
183. Hitchcock, A. Advances in Soft X-Ray Spectromicroscopy. Imaging and Microscopy 2019. Available online: https://www.imaging-git.com (accessed on 29 December 2019).
184. Blanco-Roldán, C.; Quirós, C.; Sorrentino, A.; Hierro-Rodríguez, A.; Álvarez-Prado, L.M.; Valcárcel, R.; Duch, M.; Torras, N.; Esteve, J.; Martín, J.I.; et al. Nanoscale imaging of buried topological defects with quantitative X-ray magnetic microscopy. *Nat. Commun.* **2015**, *6*, 8196. [CrossRef]
185. Hierro-Rodriguez, A.; Gürsoy, D.; Phatak, C.; Quirós, C.; Sorrentino, A.; Álvarez-Prado, L.M.; Vélez, M.; Martín, J.I.; Alameda, J.M.; Pereiro, E.; et al. 3D reconstruction of magnetization from dichroic soft X-ray transmission tomography. *J. Synchr. Rad.* **2018**, *25*, 1144–1152. [CrossRef]
186. Manke, I.; Kardjilov, N.; Schafer, R.; Hilger, A.; Strobl, M.; Dawson, M.; Grunzweig, C.; Behr, G.; Hentschel, M.; David, C.; et al. Three-dimensional imaging of magnetic domains. *Nat. Commun.* **2010**, *1*, 125. [CrossRef]
187. Suzuki, M.; Kim, K.-J.; Kim, S.; Yoshikawa, H.; Tono, T.; Yamada, K.T.; Taniguchi, T.; Mizuno, H.; Oda, K.; Ishibashi, M.; et al. Three-dimensional visualization of magnetic domain structure with strong uniaxial anisotropy via scanning hard X-ray microtomography. *Appl. Phys. Express* **2018**, *11*, 036601. [CrossRef]
188. Wolf, D.; Biziere, N.; Sturm, S.; Reyes, D.; Wade, T.; Niermann, T.; Krehl, J.; Warot-Fonrose, B.; Büchner, B.; Snoeck, E.; et al. Holographic vector field electron tomography of three-dimensional nanomagnets. *Commun. Phys.* **2019**, *2*, 87. [CrossRef]
189. Hopper, D.A.; Shulevitz, H.J.; Bassett, L.C. Spin Readout Techniques of the Nitrogen-Vacancy Center in Diamond. *Micromachines* **2018**, *9*, 437. [CrossRef] [PubMed]
190. Grinolds, M.S.; Hong, S.; Maletinsky, P.; Luan, L.; Lukin, M.D.; Walsworth, R.L.; Yacoby, A. Nanoscale magnetic imaging of a single electron spin under ambient conditions. *Nat. Phys.* **2013**, *9*, 215. [CrossRef]
191. Thiel, L.; Wang, Z.; Tschudin, M.A.; Rohner, D.; Gutiérrez-Lezama, I.; Ubrig, N.; Gibertini, M.; Giannini, E.; Morpurgo, A.F.; Maletinsky, P. Probing magnetism in 2D materials at the nanoscale with single-spin microscopy. *Science* **2019**, *364*, 973–976. [CrossRef]
192. Turino, L.N.; Ruggiero, M.R.; Stefania, R.; Cutrin, J.C.; Aime, S.; Geninatti Crich, S. Ferritin Decorated PLGA/Paclitaxel Loaded Nanoparticles Endowed with an Enhanced Toxicity Toward MCF-7 Breast Tumor Cells. *Bioconjugate Chem.* **2017**, *28*, 1283–1290. [CrossRef]
193. Kudr, J.; Haddad, Y.; Richtera, L.; Heger, Z.; Cernak, M.; Adam, V.; Zitka, O. Magnetic Nanoparticles: From Design and Synthesis to Real World Applications. *Nanomaterials* **2017**, *7*, 243. [CrossRef]
194. Ito, A.; Shinkai, M.; Honda, H.; Kobayashi, T. Medical application of functionalized magnetic nanoparticles. *J. Biosci. Bioeng.* **2005**, *100*, 1–11. [CrossRef]
195. Garello, F.; Terreno, E. Sonosensitive MRI nanosystems as cancer theranostics: A recent update. *Front. Chem.* **2018**, *6*, 157. [CrossRef]
196. Anderson, S.D.; Gwenin, V.V.; Gwenin, C.D. Magnetic Functionalized Nanoparticles for Biomedical, Drug Delivery and Imaging Applications. *Nanoscale Res. Lett.* **2019**, *14*, 188. [CrossRef] [PubMed]
197. Hirt, A.M.; Sotiriou, G.A.; Kidambi, P.R.; Teleki, A. Effect of size, composition, and morphology on magnetic performance: First-order reversal curves evaluation of iron oxide nanoparticles. *J. Appl. Phys.* **2014**, *115*, 044314. [CrossRef]
198. Hoshyar, N.; Gray, S.; Han, H.; Bao, G. The effect of nanoparticle size on in vivo pharmacokinetics and cellular interaction. *Nanomedicine* **2016**, *11*, 673–692. [CrossRef] [PubMed]
199. Albanese, A.; Tang, P.S.; Chan, W.C. The effect of nanoparticle size, shape, and surface chemistry on biological systems. *Ann. Rev. Biomed. Eng.* **2012**, *14*, 1–16. [CrossRef] [PubMed]
200. Toy, R.; Peiris, P.M.; Ghaghada, K.B.; Karathanasis, E. Shaping cancer nanomedicine: The effect of particle shape on the in vivo journey of nanoparticles. *Nanomedicine* **2014**, *9*, 121–134. [CrossRef] [PubMed]
201. Marco, M.D.; Sadun, C.; Port, M.; Guilbert, I.; Couvreur, P.; Dubernet, C. Physicochemical characterization of ultrasmall superparamagnetic iron oxide particles (USPIO) for biomedical application as MRI contrast agents. *Int. J. Nanomed.* **2007**, *2*, 609–622.

202. Wu, Y.L.; Ye, Q.; Foley, L.M.; Hitchens, T.K.; Sato, K.; Williams, J.B.; Ho, C. In situ labeling of immune cells with iron oxide particles: An approach to detect organ rejection by cellular MRI. *Proc. Natl. Acad. Sci. USA* **2006**, *103*, 1852–1857. [CrossRef]
203. Williams, J.B.; Ye, Q.; Hitchens, T.K.; Kaufman, C.L.; Ho, C. MRI detection of macrophages labeled using micrometer-sized iron oxide particles. *J. Magn. Reson. Imaging* **2007**, *25*, 1210–1218. [CrossRef]
204. Feng, Q.; Liu, Y.; Huang, J.; Chen, K.; Huang, J.; Xiao, K. Uptake, distribution, clearance, and toxicity of iron oxide nanoparticles with different sizes and coatings. *Sci. Rep.* **2018**, *8*, 2082. [CrossRef]
205. Lee, J.H.; Ju, J.E.; Kim, B.I.; Pak, P.J.; Choi, E.K.; Lee, H.S.; Chung, N. Rod-shaped iron oxide nanoparticles are more toxic than sphere-shaped nanoparticles to murine macrophage cells. *Environ. Toxicol. Chem.* **2014**, *33*, 2759–2766. [CrossRef]
206. Xie, W.; Guo, Z.; Gao, F.; Gao, Q.; Wang, D.; Liaw, B.-S.; Cai, Q.; Sun, X.; Wang, X.; Zhao, L. Shape-, size- and structure-controlled synthesis and biocompatibility of iron oxide nanoparticles for magnetic theranostics. *Theranostics* **2018**, *8*, 3284–3307. [CrossRef]
207. Serna, C.J.; Bødker, F.; Mørup, S.; Morales, M.P.; Sandiumenge, F.; Veintemillas-Verdaguer, S. Spin frustration in maghemite nanoparticles. *Solid State Commun.* **2001**, *118*, 437–440. [CrossRef]
208. Hou, C.; Wang, Y.; Ding, Q.; Jiang, L.; Li, M.; Zhu, W.; Pan, D.; Zhu, H.; Liu, M. Facile synthesis of enzyme-embedded magnetic metal-organic frameworks as a reusable mimic multi-enzyme system: Mimetic peroxidase properties and colorimetric sensor. *Nanoscale* **2015**, *7*, 18770–18779. [CrossRef] [PubMed]
209. Franzoso, F.; Nisticò, R.; Cesano, F.; Corazzari, I.; Turci, F.; Scarano, D.; Bianco Prevot, A.; Magnacca, G.; Carlos, L.; Mártire, D.O. Biowaste-derived substances as a tool for obtaining magnet-sensitive materials for environmental applications in wastewater treatments. *Chem. Eng. J.* **2017**, *310*, 307–316. [CrossRef]
210. Cravanzola, S.; Jain, S.M.; Cesano, F.; Damin, A.; Scarano, D. Development of a multifunctional TiO$_2$/MWCNT hybrid composite grafted on a stainless steel grating. *RSC Adv.* **2015**, *5*, 103255–103264. [CrossRef]
211. Langford, J.I.; Wilson, J.C. Seherrer after Sixty Years: A Survey and Some New Results in the Determination of Crystallite Size. *J. Appl. Cryst.* **1978**, *11*, 102–113. [CrossRef]
212. Muhammed Shafi, P.; Chandra Bose, A. Impact of crystalline defects and size on X-ray line broadening: A phenomenological approach for tetragonal SnO$_2$ nanocrystals. *AIP Adv.* **2015**, *5*, 057137. [CrossRef]
213. Cheng, H.; Lu, C.; Liu, J.; Yan, Y.; Han, X.; Jin, H.; Wang, Y.; Liu, Y.; Wu, C. Synchrotron radiation X-ray powder diffraction techniques applied in hydrogen storage materials—A review. *Progr. Nat. Sci. Mater. Int.* **2017**, *27*, 66–73. [CrossRef]
214. Bautista, M.C.; Bomati-Miguel, O.; Morales, M.d.P.; Serna, C.J.; Veintemillas-Verdaguer, S. Surface characterisation of dextran-coated iron oxide nanoparticles prepared by laser pyrolysis and coprecipitation. *J. Magn. Magn. Mater.* **2005**, *293*, 20–27. [CrossRef]
215. Cesano, F.; Rahman, M.M.; Bardelli, F.; Damin, A.; Scarano, D. Magnetic Hybrid Carbon via Graphitization of Polystyrene-*co*-Divinylbenzene: Morphology, Structure and Adsorption Properties. *Chem. Sel.* **2016**, *1*, 2536–2541. [CrossRef]
216. Ichiyanagi, Y.; Kimishima, Y. Structural, magnetic and thermal characterizations of Fe$_2$O$_3$ nanoparticle Systems. *J. Therm. Anal. Calorim.* **2002**, *69*, 919–923. [CrossRef]
217. Von White, G., 2nd; Chen, Y.; Roder-Hanna, J.; Bothun, G.D.; Kitchens, C.L. Structural and thermal analysis of lipid vesicles encapsulating hydrophobic gold nanoparticles. *ACS Nano* **2012**, *6*, 4678–4685. [CrossRef] [PubMed]
218. Catalano, E.; Di Benedetto, A. Characterization of physicochemical and colloidal properties of hydrogel chitosan-coated iron-oxide nanoparticles for cancer therapy. *J. Phys. Conf. Ser.* **2017**, *841*, 012010. [CrossRef]
219. Klančnik, G.; Medved, J.; Mrvar, P. Differential thermal analysis (DTA) and differential scanning calorimetry (DSC) as a method of material investigation. *RMZ* **2010**, *57*, 127–142.
220. Tenório-Neto, E.T.; Jamshaid, T.; Eissa, M.; Kunita, M.H.; Zine, N.; Agusti, G.; Fessi, H.; El-Salhi, A.E.; Elaissari, A. TGA and magnetization measurements for determination of composition and polymer conversion of magnetic hybrid particles. *Polym. Adv. Technol.* **2015**, *26*, 1199–1208. [CrossRef]
221. Xu, Z.Z.; Wang, C.C.; Yang, W.L.; Deng, Y.H.; Fu, S.K. Encapsulation of nanosized magnetic iron oxide by polyacrylamide via inverse miniemulsion polymerization. *J. Magn. Magn. Mater.* **2004**, *277*, 136–143. [CrossRef]

222. Mansfield, E.; Tyner, K.M.; Poling, C.M.; Blacklock, J.L. Determination of nanoparticle surface coatings and nanoparticle purity using microscale thermogravimetric analysis. *Anal. Chem.* **2014**, *86*, 1478–1484. [CrossRef] [PubMed]
223. Cranshaw, T.E.; Longworth, G. Mössbauer Spectroscopy of Magnetic Systems. In *Mössbauer Spectroscopy Applied to Inorganic Chemistry*; Long, G.J., Ed.; Springer: Boston, MA, USA, 1984; Volume 1, pp. 171–194.
224. Fultz, B. Mössbauer Spectrometry. In *Characterization of Materials*, 2nd ed.; Kaufmann, E., Ed.; John Wiley: New York, NY, USA, 2012; pp. 1–21. [CrossRef]
225. Gabbasov, R.; Polikarpov, M.; Cherepanov, V.; Chuev, M.; Mischenko, I.; Lomov, A.; Wang, A.; Panchenko, V. Mössbauer, magnetization and X-ray diffraction characterization methods for iron oxide nanoparticles. *J. Magn. Magn. Mater.* **2015**, *380*, 111–116. [CrossRef]
226. Nisticò, R.; Carlos, L. High yield of nano zero-valent iron (nZVI) from carbothermal synthesis using lignin-derived substances from municipal biowaste. *J. Anal. Appl. Pyrol.* **2019**, *140*, 239–244. [CrossRef]
227. Lin, P.C.; Lin, S.; Wang, P.C.; Sridhar, R. Techniques for physicochemical characterization of nanomaterials. *Biotechnol. Adv.* **2014**, *32*, 711–726. [CrossRef]
228. Lim, J.; Yeap, S.P.; Che, H.X.; Low, S.C. Characterization of magnetic nanoparticle by dynamic light scattering. *Nanosc. Res. Lett.* **2013**, *8*, 381. [CrossRef] [PubMed]
229. Bhattacharjee, S. DLS and zeta potential—What they are and what they are not? *J. Control. Release* **2016**, *235*, 337–351. [CrossRef] [PubMed]
230. Miller, C.C. The Stokes-Einstein Law for diffusion in solution. *Proc. R. Soc. Lond. A* **1924**, *106*, 740. [CrossRef]
231. Fissan, H.; Ristig, S.; Kaminski, H.; Asbach, C.; Epple, M. Comparison of different characterization methods for nanoparticle dispersions before and after aerosolization. *Anal. Met.* **2014**, *6*, 7324–7334. [CrossRef]
232. Nair, N.; Kim, W.J.; Braatz, R.D.; Strano, M.S. Dynamics of surfactant-suspended single-walled carbon nanotubes in a centrifugal field. *Langmuir* **2008**, *24*, 1790–1795. [CrossRef] [PubMed]
233. Niu, W.; Chua, Y.A.; Zhang, W.; Huang, H.; Lu, X. Highly Symmetric Gold Nanostars: Crystallographic Control and Surface-Enhanced Raman Scattering Property. *JACS* **2015**, *137*, 10460–10463. [CrossRef]
234. Zhang, Q.; Huang, J.Q.; Qian, W.Z.; Zhang, Y.Y.; Wei, F. The road for nanomaterials industry: A review of carbon nanotube production, post-treatment, and bulk applications for composites and energy storage. *Small* **2013**, *9*, 1237–1265. [CrossRef]
235. Fang, X.L.; Li, Y.; Chen, C.; Kuang, Q.; Gao, X.Z.; Xie, Z.X.; Xie, S.Y.; Huang, R.B.; Zheng, L.S. pH-induced simultaneous synthesis and self-assembly of 3D layered beta-FeOOH nanorods. *Langmuir* **2010**, *26*, 2745–2750. [CrossRef]
236. Danaei, M.; Dehghankhold, M.; Ataei, S.; Hasanzadeh Davarani, F.; Javanmard, R.; Dokhani, A.; Khorasani, S.; Mozafari, M.R. Impact of Particle Size and Polydispersity Index on the Clinical Applications of Lipidic Nanocarrier Systems. *Pharmaceutics* **2018**, *10*, 57. [CrossRef]
237. Montes Ruiz-Cabello, F.J.; Trefalt, G.; Maroni, P.; Borkovec, M. Electric double-layer potentials and surface regulation properties measured by colloidal-probe atomic force microscopy. *Phys. Rev. E* **2014**, *90*, 012301. [CrossRef]
238. Jalil, A.H.; Pyell, U. Quantification of Zeta-Potential and Electrokinetic Surface Charge Density for Colloidal Silica Nanoparticles Dependent on Type and Concentration of the Counterion: Probing the Outer Helmholtz Plane. *J. Phys. Chem. C* **2018**, *122*, 4437–4453. [CrossRef]
239. Vidal-Iglesias, F.J.; Solla-Gullón, J.; Rodes, A.; Herrero, E.; Aldaz, A. Understanding the Nernst Equation and Other Electrochemical Concepts: An Easy Experimental Approach for Students. *J. Chem. Educ.* **2012**, *89*, 936–939. [CrossRef]
240. Schwegmann, H.; Feitz, A.J.; Frimmel, F.H. Influence of the zeta potential on the sorption and toxicity of iron oxide nanoparticles on *S. cerevisiae* and *E. coli*. *J. Colloid. Interface Sci.* **2010**, *347*, 43–48. [CrossRef] [PubMed]
241. Sharma, A.; Cornejo, C.; Mihalic, J.; Geyh, A.; Bordelon, D.E.; Korangath, P.; Westphal, F.; Gruettner, C.; Ivkov, R. Physical characterization and in vivo organ distribution of coated iron oxide nanoparticles. *Sci. Rep.* **2018**, *8*, 4916. [CrossRef] [PubMed]
242. Taupitz, M.; Wagner, S.; Schnorr, J.r.; Kravec, I.; Pilgrimm, H.; Bergmann-Fritsch, H.; Hamm, B. Phase I Clinical Evaluation of Citrate-coated Monocrystalline Very Small Superparamagnetic Iron Oxide Particles as a New Contrast Medium for Magnetic Resonance Imaging. *Investig. Radiol.* **2004**, *39*, 394–405. [CrossRef] [PubMed]

243. Moghimi, S.M.; Symonds, P.; Murray, J.C.; Hunter, A.C.; Debska, G.; Szewczyk, A. A two-stage poly(ethylenimine)-mediated cytotoxicity: Implications for gene transfer/therapy. *Mol. Ther.* **2005**, *11*, 990–995. [CrossRef] [PubMed]
244. Moore, T.L.; Rodriguez-Lorenzo, L.; Hirsch, V.; Balog, S.; Urban, D.; Jud, C.; Rothen-Rutishauser, B.; Lattuada, M.; Petri-Fink, A. Nanoparticle colloidal stability in cell culture media and impact on cellular interactions. *Chem. Soc. Rev.* **2015**, *44*, 6287–6305. [CrossRef]
245. Yallapu, M.M.; Chauhan, N.; Othman, S.F.; Khalilzad-Sharghi, V.; Ebeling, M.C.; Khan, S.; Jaggi, M.; Chauhan, S.C. Implications of protein corona on physico-chemical and biological properties of magnetic nanoparticles. *Biomaterials* **2015**, *46*, 1–12. [CrossRef]
246. Park, Y.; Whitaker, R.D.; Nap, R.J.; Paulsen, J.L.; Mathiyazhagan, V.; Doerrer, L.H.; Song, Y.Q.; Hurlimann, M.D.; Szleifer, I.; Wong, J.Y. Stability of superparamagnetic iron oxide nanoparticles at different pH values: Experimental and theoretical analysis. *Langmuir* **2012**, *28*, 6246–6255. [CrossRef]
247. Fleming, M.S.; Walt, D.R. Stability and Exchange Studies of Alkanethiol Monolayers on Gold-Nanoparticle-Coated Silica Microspheres. *Langmuir* **2001**, *17*, 4836–4843. [CrossRef]
248. Lazzari, S.; Moscatelli, D.; Codari, F.; Salmona, M.; Morbidelli, M.; Diomede, L. Colloidal stability of polymeric nanoparticles in biological fluids. *J. Nanopart. Res.* **2012**, *14*, 920. [CrossRef] [PubMed]
249. Khan, S.; Gupta, A.; Verma, N.C.; Nandi, C.K. Kinetics of protein adsorption on gold nanoparticle with variable protein structure and nanoparticle size. *J. Chem. Phys.* **2015**, *143*, 164709. [CrossRef] [PubMed]
250. Salvati, A.; Pitek, A.S.; Monopoli, M.P.; Prapainop, K.; Bombelli, F.B.; Hristov, D.R.; Kelly, P.M.; Aberg, C.; Mahon, E.; Dawson, K.A. Transferrin-functionalized nanoparticles lose their targeting capabilities when a biomolecule corona adsorbs on the surface. *Nature Nanotechnol.* **2013**, *8*, 137–143. [CrossRef]
251. Barakat, N.S. Magnetically modulated nanosystems: A unique drug-delivery platform. *Nanomedicine* **2009**, *4*, 799–812. [CrossRef] [PubMed]
252. Kurgan, E.; Gas, P. Simulation of the electromagnetic field and temperature distribution in human tissue in RF hyperthermia. *Przeglad Elektrotechniczny* **2015**, *91*, 169–172. [CrossRef]
253. Enriquez-Navas, P.M.; Garcia-Martin, M.L. Application of Inorganic Nanoparticles for Diagnosis Based on MRI. In *Frontiers of Nanoscience*; de la Fuente JM, G.V., Ed.; Elsevier: North Holland, The Netherlands, 2012; Volume 4, pp. 233–245.
254. Merbach, A.S.; Helm, L.; Tóth, É. *The Chemistry of Contrast Agents in Medical Magnetic Resonance Imaging*, 2nd ed.; John Wiley & Sons: Hoboken, NY, USA, 2013. [CrossRef]
255. Bao, Y.; Sherwood, J.A.; Sun, Z. Magnetic iron oxide nanoparticles as T1 contrast agents for magnetic resonance imaging. *J. Mater. Chem. C* **2018**, *6*, 1280–1290. [CrossRef]
256. Wei, H.; Bruns, O.T.; Kaul, M.G.; Hansen, E.C.; Barch, M.; Wiśniowska, A.; Chen, O.; Chen, Y.; Li, N.; Okada, S.; et al. Exceedingly small iron oxide nanoparticles as positive MRI contrast agents. *Proc. Natl. Acad. Sci. USA* **2017**, *114*, 2325–2330. [CrossRef]
257. Di Gregorio, E.; Ferrauto, G.; Furlan, C.; Lanzardo, S.; Nuzzi, R.; Gianolio, E.; Aime, S. The Issue of Gadolinium Retained in Tissues: Insights on the Role of Metal Complex Stability by Comparing Metal Uptake in Murine Tissues Upon the Concomitant Administration of Lanthanum- and Gadolinium-Diethylentriamminopentaacetate. *Investig. Radiol.* **2018**, *53*, 167–172. [CrossRef]
258. McDonald, R.J.; McDonald, J.S.; Kallmes, D.F.; Jentoft, M.E.; Murray, D.L.; Thielen, K.R.; Williamson, E.E.; Eckel, L.J. Intracranial Gadolinium Deposition after Contrast-enhanced MR Imaging. *Radiology* **2015**, *275*, 772–782. [CrossRef]
259. Amiri, H.; Bordonali, L.; Lascialfari, A.; Wan, S.; Monopoli, M.P.; Lynch, I.; Laurent, S.; Mahmoudi, M. Protein corona affects the relaxivity and MRI contrast efficiency of magnetic nanoparticles. *Nanoscale* **2013**, *5*, 8656–8665. [CrossRef]
260. LaConte, L.E.; Nitin, N.; Zurkiya, O.; Caruntu, D.; O'Connor, C.J.; Hu, X.; Bao, G. Coating thickness of magnetic iron oxide nanoparticles affects R_2 relaxivity. *J. Magn. Reson. Imaging* **2007**, *26*, 1634–1641. [CrossRef] [PubMed]
261. Roch, A.; Muller, R.N.; Gillis, P. Theory of proton relaxation induced by superparamagnetic particles. *J. Chem. Phys.* **1999**, *110*, 5403–5411. [CrossRef]
262. Gossuin, Y.; Orlando, T.; Basini, M.; Henrard, D.; Lascialfari, A.; Mattea, C.; Stapf, S.; Vuong, Q.L. NMR relaxation induced by iron oxide particles: Testing theoretical models. *Nanotechnology* **2016**, *27*, 155706. [CrossRef] [PubMed]

263. Bordonali, L.; Kalaivani, T.; Sabareesh, K.P.; Innocenti, C.; Fantechi, E.; Sangregorio, C.; Casula, M.F.; Lartigue, L.; Larionova, J.; Guari, Y.; et al. NMR-D study of the local spin dynamics and magnetic anisotropy in different nearly monodispersed ferrite nanoparticles. *J. Phys. Cond. Matter* **2013**, *25*, 066008. [CrossRef]
264. Ruggiero, M.R.; Crich, S.G.; Sieni, E.; Sgarbossa, P.; Forzan, M.; Cavallari, E.; Stefania, R.; Dughiero, F.; Aime, S. Magnetic hyperthermia efficiency and ^{1}H-NMR relaxation properties of iron oxide/paclitaxel-loaded PLGA nanoparticles. *Nanotechnology* **2016**, *27*, 285104. [CrossRef]
265. Gilchrist, R.K.; Medal, R.; Shorey, W.D.; Hanselman, R.C.; Parrott, J.C.; Taylor, C.B. Selective Inductive Heating of Lymph Nodes. *Ann. Surg.* **1957**, *146*, 596–606. [CrossRef]
266. Laurent, S.; Dutz, S.; Hafeli, U.O.; Mahmoudi, M. Magnetic fluid hyperthermia: Focus on superparamagnetic iron oxide nanoparticles. *Adv. Colloid. Interface Sci.* **2011**, *166*, 8–23. [CrossRef]
267. Del Bianco, L.; Spizzo, F.; Barucca, G.; Ruggiero, M.R.; Geninatti Crich, S.; Forzan, M.; Sieni, E.; Sgarbossa, P. Mechanism of magnetic heating in Mn-doped magnetite nanoparticles and the role of intertwined structural and magnetic properties. *Nanoscale* **2019**, *11*, 10896–10910. [CrossRef]
268. Gas, P.; Miaskowski, A. Specifying the ferrofluid parameters important from the viewpoint of magnetic fluid hyperthermia. In Proceedings of the 2015 Selected Problems of Electrical Engineering and Electronics (WZEE), Kielce, Poland, 17–19 September 2015; pp. 1–6. [CrossRef]
269. Bunaciu, A.A.; Udriştioiu, E.G.; Aboul-Enein, H.Y. X-ray diffraction: Instrumentation and applications. *Crit. Rev. Anal. Chem.* **2015**, *45*, 289–299. [CrossRef]
270. Giron, D. Thermal analysis and calorimetric methods in the characterisation of polymorphs and solvates. *Therm. Acta* **1995**, *248*, 1–59. [CrossRef]
271. Vyazovkin, S. Thermal analysis. *Anal. Chem.* **2010**, *82*, 4936–4949. [CrossRef] [PubMed]
272. Garcia Casillas, P.E.; Rodriguez, C.; Martinez Perez, C. Infrared Spectroscopy of Functionalized Magnetic Nanoparticles. In *Infrared Spectroscopy—Materials Science, Engineering and Technology, Theophile Theophanides*; IntechOpen: London, UK, 2012. [CrossRef]
273. Skoglund, S.; Hedberg, J.; Yunda, E.; Godymchuk, A.; Blomberg, E.; Odnevall Wallinder, I. Difficulties and flaws in performing accurate determinations of zeta potentials of metal nanoparticles in complex solutions-Four case studies. *PLoS ONE* **2017**, *12*, e0181735. [CrossRef] [PubMed]
274. Robson, A.L.; Dastoor, P.C.; Flynn, J.; Palmer, W.; Martin, A.; Smith, D.W.; Woldu, A.; Hua, S. Advantages and Limitations of Current Imaging Techniques for Characterizing Liposome Morphology. *Front. Pharmacol.* **2018**, *9*, 80. [CrossRef] [PubMed]
275. Sylvester, P.W. Optimization of the tetrazolium dye (MTT) colorimetric assay for cellular growth and viability. *Methods Mol. Biol.* **2011**, *716*, 157–168. [PubMed]
276. Rampersad, S.N. Multiple applications of Alamar Blue as an indicator of metabolic function and cellular health in cell viability bioassays. *Sensors* **2012**, *12*, 12347–12360. [CrossRef] [PubMed]
277. Riss, T.L.; Moravec, R.A.; Niles, A.L.; Duellman, S.; Benink, H.A.; Worzella, T.J.; Minor, L. Cell Viability Assays. In *Assay Guidance Manual [Internet]*; Sittampalam, G.S., Grossman, A., Brimacombe, K., Eds.; Eli Lilly & Company and the National Center for Advancing Translational Sciences: Bethesda, MD, USA, 2004. Available online: https://www.ncbi.nlm.nih.gov/books/NBK144065/ (accessed on 1 June 2019).
278. Patil, U.S.; Adireddy, S.; Jaiswal, A.; Mandava, S.; Lee, B.R.; Chrisey, D.B. In Vitro/In Vivo Toxicity Evaluation and Quantification of Iron Oxide Nanoparticles. *Int. J. Mol. Sci.* **2015**, *16*, 24417–24450. [CrossRef] [PubMed]
279. Oancea, M.; Mazumder, S.; Crosby, M.E.; Almasan, A. Apoptosis assays. *Methods Mol. Med.* **2006**, *129*, 279–290.
280. Macías-Martínez, B.I.; Cortés-Hernández, D.A.; Zugasti-Cruz, A.; Cruz-Ortíz, B.R.; Múzquiz-Ramos, E.M. Heating ability and hemolysis test of magnetite nanoparticles obtained by a simple co-precipitation method. *J. Appl. Res. Technol.* **2016**, *14*, 239–244. [CrossRef]
281. Chen, D.; Tang, Q.; Li, X.; Zhou, X.; Zang, J.; Xue, W.Q.; Xiang, J.Y.; Guo, C.Q. Biocompatibility of magnetic Fe_3O_4 nanoparticles and their cytotoxic effect on MCF-7 cells. *Int. J. Nanomed.* **2012**, *7*, 4973–4982. [CrossRef]

© 2020 by the authors. Licensee MDPI, Basel, Switzerland. This article is an open access article distributed under the terms and conditions of the Creative Commons Attribution (CC BY) license (http://creativecommons.org/licenses/by/4.0/).

Review

Smart Ligands for Efficient 3d-, 4d- and 5d-Metal Single-Molecule Magnets and Single-Ion Magnets [†]

Panagiota S. Perlepe [1], Diamantoula Maniaki [1], Evangelos Pilichos [1], Eugenia Katsoulakou [1,*] and Spyros P. Perlepes [1,2,*]

1 Department of Chemistry, University of Patras, 265 04 Patras, Greece; pennyperlepes@gmail.com (P.S.P.); dia.maniaki@gmail.com (D.M.); pilvag@gmail.com (E.P.)
2 Institute of Chemical Engineering Sciences, Foundation for Research and Technology-Hellas (FORTH/ICE-HT), Platani, P.O.Box 1414, 265 04 Patras, Greece
* Correspondence: eugeniachem@gmail.com (E.K.); perlepes@patreas.upatras.gr (S.P.P.); Tel.: +30-2610-996019 (E.K.); +30-2610-996730 (S.P.P.)
† This work is devoted to Research Director Vassilis Psycharis on the occasion of his 60th birthday. Dr. Psycharis is an excellent structural scientist, a fantastic collaborator and a precious friend.

Received: 16 April 2020; Accepted: 21 May 2020; Published: 29 May 2020

Abstract: There has been a renaissance in the interdisciplinary field of Molecular Magnetism since ~2000, due to the discovery of the impressive properties and potential applications of d- and f-metal Single-Molecule Magnets (SMMs) and Single-Ion Magnets (SIMs) or Monometallic Single-Molecule Magnets. One of the consequences of this discovery has been an explosive growth in synthetic molecular inorganic and organometallic chemistry. In SMM and SIM chemistry, inorganic and organic ligands play a decisive role, sometimes equally important to that of the magnetic metal ion(s). In SMM chemistry, bridging ligands that propagate strong ferromagnetic exchange interactions between the metal ions resulting in large spin ground states, well isolated from excited states, are preferable; however, antiferromagnetic coupling can also lead to SMM behavior. In SIM chemistry, ligands that create a strong axial crystal field are highly desirable for metal ions with oblate electron density, e.g., Tb^{III} and Dy^{III}, whereas equatorial crystal fields lead to SMM behavior in complexes based on metal ions with prolate electron density, e.g., Er^{III}. In this review, we have attempted to highlight the use of few, efficient ligands in the chemistry of transition-metal SMMs and SIMs, through selected examples. The content of the review is purely chemical and it is assumed that the reader has a good knowledge of synthetic, structural and physical inorganic chemistry, as well as of the properties of SIMs and SMMs and the techniques of their study. The ligands that will be discussed are the azide ion, the cyanido group, the tris(trimethylsilyl)methanide, the cyclopentanienido group, soft (based on the Hard-Soft Acid-Base model) ligands, metallacrowns combined with click chemistry, deprotonated aliphatic diols, and the family of 2-pyridyl ketoximes, including some of its elaborate derivatives. The rationale behind the selection of the ligands will be emphasized.

Keywords: ligands; molecular magnetism; single-ion magnets (SIMs) or monometallic single-molecule magnets; single-molecule magnets (SMMs); synthetic strategies; 3d-, 4d- and 5d-metal complexes as SIMs and SMMs

1. The History of Molecular Magnetism in Brief—The Era of Single-Molecule Magnets (SMMs) and Single-Ion Magnets (SIMs)

Molecular magnetism [1] is currently a "hot" interdisciplinary research field which started almost 35 years ago. It was a rather natural extension of the field of Magnetochemistry which was developed in the

1955–1985 period. In the latter field, inorganic chemists were using experimental values of magnetic moment and their variation with temperature to draw chemical and structural conclusions for molecular complexes. Well known examples [2,3] were: (i) The distinction between several stereochemistries (tetrahedral, square planar, octahedral) in complexes of some 3d-metal ions, e.g., Ni(II) and Co(II); (ii) the recognition of high-spin and low-spin configurations in octahedral complexes of $3d^4$–$3d^7$ metal ions; and (iii) the elegant determination of the singlet-triplet gap in copper(II) acetate hydrate by Bleaney and Bowers, before the X-ray solution of its dinuclear $[Cu_2(O_2CMe)_4(H_2O)_2]$ structure.

Before proceeding to a brief history of molecular magnetism, we can mention two advantages that molecular systems offer over conventional atom- or ion-based magnetic materials [4]: (a) The structures that can be formed are more complicated and more diverse that the structural types we normally meet in conventional inorganic materials. Molecular crystals can provide scientists with ensembles of identical structures and iso-orientated magnetic objects which permit the in-depth study of their physical behavior. Such molecular materials offer model systems to test existing theories on many-body problems and discover, e.g., new quantum, behavior; and (b) Molecular systems are ideal for incorporation of other useful functionalities, e.g., optical properties, conductivity, etc. This may be a synergistic property not coupled with magnetism, or it may be a coupling of different physical properties, e.g., light-induced magnetic ordering in spin-crossover (SCO) Prussian blue phases. This type of materials may find novel, highly specific applications.

In the field of molecular magnetism, chemists, chemical engineers, physicists and material scientists, both experimentalists and theoreticians, closely collaborate trying to design, synthesize, fully characterize and model the magnetic properties of molecule-based materials. In the initial period of its development, the focus of research was on simple model systems (homo- and heterometallic dinuclear complexes and coordination clusters); the goal was to test theories in solids about exchange interactions and electron delocalization at the molecular scale [5]. The field later expanded towards the study of 1D systems [1,6,7], e.g., homometallic chains, heterometallic chains and homometallic chains in which the metal centers are bridged by an organic radical. In the late 1980s and during the 1990s, chemists prepared a great variety of 3D complexes exhibiting spontaneous magnetization below a critical temperature (T_c) [8–10]. The Miller and Epstein groups broke the critical temperature barrier record ($T_c > 350$ K) with the ferrimagnet $\{[V^{III}(TCNE)_2]\cdot xCH_2Cl_2\}_n$ [11], where $TCNE^-$ is the radical anion of tetracyanoethylene, while other groups came close by using Prussian blue derivatives [9,10].

Simultaneously with the above developments, the molecular magnetism community paid more attention on octahedral $3d^4$–$3d^7$ systems (mainly on $3d^6$ iron(II) complexes) with a SCO behavior [12,13]. This phenomenon, discovered by Cambi in 1931, continues to attract the intense interest of scientists even after 90 years [14,15]. The goals are to construct systems which undergo spin transition near 300 K and to study the possibility of tuning this molecular bistability through application of external stimuli (temperature, pressure, light). The SCO phenomenon has already led to applications, e.g., sensors [16], actuators [17] and thermal displays [18].

Thus, the initial growth of molecular magnetism was mainly based on the deep understanding of the factors that could be used to design and synthesize novel crystalline molecule-based materials exhibiting useful magnetic properties, e.g., ferromagnetism and ferrimagnetism, similar to those observed in inorganic atom-/ion-based materials. More importantly, those molecular materials had special features such as low density, insulating nature and optical transparency; these distinctive physical properties provide material scientists with a number of fabrication advantages, as the materials are most often prepared using solution methods.

After the period of initial growth of molecular magnetism, its community turned the attention to three large classes of molecular compounds. The first class comprises the Single-Molecule Magnets (SMMs) and the Single-Ion Magnets (SIMs), the second the Single-Chain Magnets (SCMs) and the third involves the multifunctional molecular magnetic materials [19,20]. The focus of this review is on transition-metal SMMs and SIMs.

In the presence of axial magnetic anisotropy (D), the M_S levels of a transition-metal complex with total spin S will split under zero magnetic field according to the Hamiltonian $\hat{H} = D\hat{S}_z^2$. If the value of D is negative, the two $\pm M_S$ levels of maximal projection along the z axis form a bistable ground state because they are degenerate. If we reverse the magnetic moment by converting $-M_S$ to $+M_S$, this requires traversal of a spin-inversion barrier. This barrier is $U = S^2|D|$ for integer S or $U = (S^2 - 1/4)|D|$ for non-integer S, and the system passes through the $M_S = 0$ or the $M_S = 1/2$ levels, respectively, at the height of the energy barrier. The existence of such a barrier can lead to the slow relaxation of the magnetic moment at very low temperatures upon removal of the external dc field [21,22]. The presence of this barrier is often proven by the appearance of magnetic hysteresis of molecular origin as first observed for the iconic [$Mn_{12}O_{12}(O_2CMe)_{16}(H_2O)_4$] (**1**, Figure 1) SMM [21–23]. Clusters containing polynuclear molecules that exhibit such behavior have been named Single-Molecule Magnets. The magnetic behavior of each of these clusters can be described as a giant anisotropic spin as a result of the exchange coupling between the spins of neighboring metal ions. Because of the magnetic bistability, these polynuclear molecules were proposed for use in magnetic memory devices since they can remain magnetized in one of the two spin states, thus giving rise to a "bit" of memory. The aim during the first decade of SMM research (1993–2003) was to prepare SMMs with memory effects at higher temperatures [19]. Although synthetic inorganic chemists made many efforts to achieve this goal, the progress was little and the energy barriers that stabilize the magnetic bits against thermal fluctuations remained small. Another tremendously important consequence of the discovery of SMMs was the observation of quantum effects in mesoscopic magnets. At that time, physicists were looking for small magnetic particles, all identical to each other, to investigate if quantum effects could be observed in ensembles of such identical particles; however, the preparation of these collections proved difficult. Chemists solved the problem using a molecular approach to prepare identical cluster molecules in crystalline SMMs. A few years after the characterization of **1**, scientists revealed that its crystals exhibit quantum tunneling of magnetization (QTM) [24,25]; this phenomenon is considered one of the milestones in the study of spin during the 20th century [19]. Synthetic efforts were followed by advanced theoretical studies, and the latter provided strong evidence that the magnitude of D decreases as S increases; this implied that the construction of efficient SMMs with a large U cannot be achieved by only maximizing S and that control of D is equally important [21,22].

In the last 15 years or so, another exciting subclass of SMMs was discovered, the so-called Single-Ion Magnets (SIMs). These represent the smallest molecular nanomagnets we can imagine [21,26–34]. In the literature, SIMs are often referred to as mononuclear SMMs. However, neither SIMs nor mononuclear SMMs are perfect descriptors [26]. Monometallic SMMs is probably better, but herein we have chosen to use the SIM acronym rather, e.g., MSMM, which is probably more awkward. SIMs are mononuclear complexes containing a single magnetic d- or f-metal ion. The motivation behind the SIM research evolution was the belief of scientists that incorporating many paramagnetic centers into a cluster molecule may be a disadvantage in terms of generating complexes with large cluster anisotropies (D or better $D_{cluster}$). Many metal ions can, in principle, lead to a large S, but as the nuclearity of the molecule increases it is becoming very difficult (practically impossible) to control the mutual alignment of the anisotropy axes of the individual metal ions; this gives small $D_{cluster}$ values [27]. These considerations, combined with the observation of slow magnetization relaxation in complexes (Bu^n_4N)[$Ln(pc)_2$] (**2**), where Ln = Tb, Dy and pc = the dianion of phthalocyanine [35], by Ishikawa and co-workers, turned the attention of researchers to single-ion systems [19,21]. If we compare 4f-(Ln) and d-metal ions for use in SIMs, the former appear better because they possess: (i) larger magnetic moments, (ii) higher spin-orbit coupling constants, and (iii) weak coupling of the f orbitals to the ligand field which cannot quench first-orbital contributions to the magnetic moment. As a consequence, magnetic hysteresis has been observed at temperatures as high as 80 K for mononuclear organometallic Dy(III) complexes [36]. However, d-metal ion SIMs are important and cannot be considered simply as academic research curiosities. As scientists understand deeper the physics of d single ions in a ligand field, they can begin to design strategies to couple the anisotropy of individual transition metal ions together in order to create cluster SMMs in a rational way; such clusters could be ideal molecular analogues for magnetic nanoparticles. Additionally, the utilization of well-studied and understood SIM building blocks in the modular synthesis of 1D SCMs with high spin and uniaxial anisotropies (a concept which is active in the 4f-metal chemistry), and Metal–Organic Frameworks (MOFs) with SIM units as nodes (e.g., for the design of porous magnetic materials) is a currently "hot" research area, see information provided in reference [27].

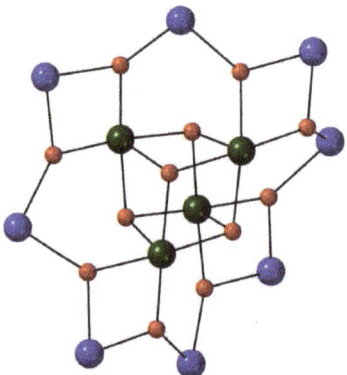

Figure 1. The $\{Mn^{III}_8Mn^{IV}_4(\mu_3\text{-O})_{12}\}^{16+}$ core of the archetype Single-Molecule Magnet (SMM) **1**. The Mn^{III} centers are shown with blue color, the Mn^{IV} centers with green color and the oxido ions are shown with red color. The four central Mn^{IV} atoms are weakly ferromagnetically coupled, and the remaining $Mn^{III}\cdots Mn^{IV}$ and $Mn^{III}\cdots Mn^{III}$ exchange interactions are all antiferromagnetic, with the former much stronger than the latter. As a result, the stronger $Mn^{III}\cdots Mn^{IV}$ interactions overcome the weaker $Mn^{III}\cdots Mn^{III}$ ones within each triangular $Mn^{III}_2Mn^{IV}$ subunit of the core, aligning the spins of the outer Mn^{III} atoms all parallel, and thus antiparallel to the central Mn^{IV} atoms; this gives an $S = 16 - 6 = 10$ ground state.

Despite high energy barriers for magnetization reversal, often SMMs and SIMs suffer from fast relaxation processes, not just via QTM, but also via interactions between the spin states and lattice phonons [27,30]. The complex behavior of relaxation dynamics is a specialized topic beyond the scope of this review. However, it is helpful to consider the SMM and SIM systems as being composed of two parts, the spin part and the lattice part. The interaction between lattice vibrations (phonons) and spin provide the system with additional relaxation processes, which "shortcut" the desired thermal pathway. The spin-lattice relaxation mechanisms are of three types (direct, Orbach and Raman processes). Since the tunneling pathways are very sensitive to changes of molecular symmetry, synthetic chemists try to control the molecular symmetry of SMMs/SIMs through careful design of the ligands used. Additional tools for minimizing QTM through the ground state are magnetic dilution and utilization of a Kramers metal ion; for the latter, breaking of the M_S degeneracy and therefore QTM is formally forbidden in strictly zero magnetic field. However, the fast quantum tunneling offers an advantage because these molecules are candidates to realize quantum bits (qubits), the basic units of quantum computers [37].

A second class of molecular compounds which has captured the intense interest of the molecular magnetism community consists of slow-relaxing 1D magnets, or SCMs [20,21,38,39]. These consist of chains, isolated from each other, presenting a slow relaxation of the magnetization; they cannot present a long-range magnetic order. However, they exhibit a short-range order caused by the occurrence of domains where N spins are oriented in the same direction, interrupted by a reversed spin or by chain defects. A finite magnetization can thus be frozen in the absence of an applied magnetic field at low temperatures. Like in SMMs, the system should have an Ising-type magnetic anisotropy, i.e., the spins must preferentially orient in one direction. The main concept of dynamics is the probability of a spin to flip within the chain, taking into account only the nearest neighbors' interactions, with an Hamiltonian of the type $\hat{H} = -J\Sigma_{i=1,N-1}\hat{S}_i\hat{S}_{i+1}$. The prototype compound is the 1D complex $\{[Co^{II}(hfac)_2(NITPhOMe)]\}_n$ (3) [40], where hfac is the hexafluoroacetylacetonato(−1) ligand and NITPhOMe is the neutral nitronyl nitroxide, bridging radical 4'-methoxyphenyl-4,4,5,5-tetramethylimidazoline-1-oxyl-3-oxide. The Ising nature of the chain has been attributed to the presence of Co(II) as this center yields significant anisotropic effects when it is 6-coordinate.

The third class of molecular compounds, of great interest to scientists, consists of multifunctional materials whose study is beyond molecular magnetism. These molecular materials combine two, technologically interesting, electronic properties, e.g., ferromagnetism and superconductivity, or SMM properties and photoluminescence; such properties are very difficult to be included in purely inorganic solids. Examples of crystalline, multifunctional molecular materials include porous magnets [41], chiral magnets [42,43], conducting magnets [44] and luminescent SMMs [45,46], among others.

In the last few years, the interdisciplinary field of molecular magnetism is rapidly shifting to magnetic molecules and materials in physics- and nanotechnology-related fields [19], such as molecular spintronics, quantum technologies, 2D materials and MOFs [37,47–51]. For example, in the field of quantum technologies, the achievements in the design of molecular spin qubits with long quantum coherence times and in the implementation of quantum operations give hopes for the use of molecular spin qubits in quantum computation.

Closing this introductory section, we would like to emphasize some advantages of the use of nd metal ions (compared to nf ones) in the SMM/SIM research. The magnetic anisotropy of a transition metal center can be more rationally tuned by chemical design. For a given d^n configuration, both the sign and the magnitude of D are controlled, mainly by the geometry of the coordination sphere of the nd metal ion, and thus they can be tuned more easily than for a nf metal ion by the chemist. Very large $|D|$ values are found for complexes with low coordination numbers, which are more accessible for the nd metal ions (at least with $n = 3$) than for the nf metal centers. Such chemical control on the magnetic anisotropy of monometallic SMMs can help scientists to develop more efficient polynuclear molecular magnetic species.

2. Ligands in SMM and SIM Chemistry

Alfred Stock, a pioneer in borane and silane chemistry, was the first scientist to introduce the term "ligand" in the second decade of the 20th century; the word has its roots in the Latin language where "ligare" means "to bind" [52]. The term came into use among English-speaking inorganic chemists after the 2nd world war, mainly through the wide popularity of the PhD Thesis of Jannik Bjerrum [52]. The proper utilization of known ligands and the design of new ones is behind many spectacular developments in coordination and metallosupramolecular chemistry [53]. Theoretical concepts related to ligands are the chelate effect, the macrocycle effect, the cryptate effect, the isoelectronic and isolobal relationships, the conformation of chelating rings and the reactivity of coordinated ligands [54].

In SMM and SIM chemistry, the ligands play a crucial role [55,56], sometimes equally important to that of the magnetic metal ions. An impressive example comes from the 4f-metal SIM chemistry. By combining two different organometallic ligands of sufficient bulk (but not too bulky to avoid close approach of the ligands), the groups of Tong, Mansikkamäki and Layfield synthesized the mononuclear complex $[(Cp^{iPr5})Dy^{III}(Cp^*)][B(C_6F_5)_4]$ (4), Equations (1) and (2), Figure 2; Cp^{iPr5} is the penta-iso-propylcyclopentadienyl(−1) ligand and Cp* is the pentamethylcyclopentadienyl(−1) ligand. This design of ligand framework allowed the two key structural parameters—that is, the Dy–Cp$_{cent}$ distances (cent refers to the centroids of the cyclopentadienyl ligands) and the Cp–Dy–Cp bend angle-to be short and wide, respectively, thus achieving an axial crystal field of sufficient strength to give a SIM that shows magnetic hysteresis above 77 K [36]; this is the first step for the development of nanomagnet devices that function at relatively practical temperatures.

$$[(Cp^{iPr5})Dy\{HB(\mu\text{-}H)_3\}_2(THF)] + K(Cp*)] \rightarrow [(Cp^{iPr5})Dy(Cp*)\{H_2B(\mu\text{-}H)_2\}] + K(BH_4) + THF \quad (1)$$

$$[(Cp^{iPr5})Dy(Cp*)\{H_2B(\mu\text{-}H)_2\}] + [(Et_3Si)_2(\mu\text{-}H)][B(C_6F_5)_4] \rightarrow [(Cp^{iPr5})Dy(Cp*)][B(C_6F_5)_4] + 2\ Et_3SiH + 0.5\ B_2H_6 \quad (2)$$

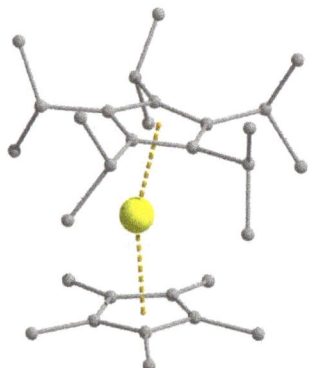

Figure 2. The molecular structure of the cation [(Cp^{iPr5})Dy^{III}(Cp*)]$^+$ that is present in the Single-Ion Magnets (SIM) **4**, which holds the world record for the magnetic blocking temperature (T_B = 80 K); one description of this parameter refers to the maximum temperature at which it is possible to observe hysteresis in the field dependence of the magnetization, subject to the field sweep rate. The Dy^{III} center is shown in yellow color. Modified from reference [36].

3. Scope and Organization of this Review

As briefly mentioned in Section 1, in the last 25 years there has been a renaissance in the field of molecular magnetism. The main reason for this was the discovery of the exciting properties and the efforts for improvement of SMMs and SIMs. One of the consequences of these discoveries was an explosive growth of synthetic molecular inorganic and organometallic chemistry. Many groups around the world have been working on the synthesis of SMMs and SIMs with higher effective energy barriers for magnetization reversal (U_{eff}) and T_B values, with the Holy Grail in this area being their technological applications. Novel structural types of metal complexes and new, smart synthetic methods have been reported to realize this general goal. *Inorganic and organic ligands are central "players" in this game.* Concerning transition-metal ion polynuclear SMMs, the ligands are of various types. The inorganic ligands are mainly the hydroxido, oxido, azido, dicyanamido, cyanato, cyanido and halogenido groups. Simple carboxylato, azolato, deprotonated diol and triol, and thiolato organic ligands are frequently used, but the most employed ones are polydentate ligands involving a combination of functional groups, e.g., alkoxido, phenoxido, carboxylato, oximato, etc. Concerning transition-metal ion monometallic SMMs (SIMs), the ligands should be terminal and the most popular ones are simple heterocycles, phosphines and thioureas, thiolates, phenolates, bis(trimethylsilyl)amide, 2,9-dialkylcarboxylate-1,10-phenantrholines,

6,6′-dialkylcarboxylate-2,2′-bipyridines, chelating Schiff bases and various organometallic-type ligands, including tris(trimethylsilyl)methanide. We should mention here that the U_{eff} term is used to indicate thermally induced reversal of the magnetization, the rate of which is dependent on the energy barrier in which the system must surmount to reverse the spin [30]. Due to tunneling effects, U_{eff} is always lower than U defined in Section 1.

The general aim of this review is to highlight the use of few (we emphasize the word "few") inorganic and organic ligands in the chemistry of 3d-, 4d-, and 5d-metal SMMs and SIMs, through selected examples. According to the authors' opinion (a subjective opinion!), these ligands have contributed much into the development of the transition-metal SMM/SIM area, but simultaneously they are promising and have the potential for exciting achievements in the future. Ligands that have been used in f-metal-containing (both homometallic and heterometallic) SMM and SIM chemistry will not be described.

The review's content is shaped by a few specific features. First, it is important to specify *what this review is not*. It is *not* a comprehensive review on the chemistry of transition-metal SMMs and SIMs; there are excellent reviews and books covering this wide topic [21–23,26,27,29,57–60]. We never considered the idea of being exhaustive; each of the sections could have been on its own a subject of another review. It is *not* a survey of recent interesting results; such results can be found in the current literature. Thus, we apologize to the outstanding researchers whose excellent work will not be cited.

Second, the content of the review is chemical and we assume that the readers have a good knowledge of synthetic, structural and physical coordination and organometallic chemistry, as well as of the properties of SMMs/SIMs and the techniques of their study. Structural and magnetic information will be confined to the absolute minimum. To avoid long synthetic discussions, balanced chemical equations (written using molecular and not-ionic formulae) will be used. This, of course, implies that only one reaction occurs in solution (which is certainly not the case, at least in SMM chemistry that involves polynuclear species); however, we do believe that writing chemical equations offers a great help to the reader to understand the processes, better than presenting a long text. We shall try to explain the synthetic philosophy behind the reactions with emphasis on the choice of ligands and metal ion sources. Many of the references for the general information are reviews and book chapters.

Third, the method that will be used to describe the coordination of ligands to transition-metal ions is mainly the "Harris Notation"; occasionally, and in simple cases, the traditional η/μ rotation will be also used. The "Harris Notation" [61] is an already widely accepted method for the description of the ligands' binding to metal ions. It is written in the form $X.Y_1Y_2Y_3 \ldots Y_i$, where X is the total number of metal ions bound by the whole ligand, and each Y value refers to the number of metal centers attached to the different donor atoms. The order of Y atoms follows the Cahn–Ingold–Prelog priority rules; hence, for most of the ligands reported in this review, O comes before N. For clarity, the coordination modes of the ligands in most examples of this work will be presented schematically.

A fourth, more personal point, is that this review is a distillation from life-long experience of our group as a synthetic one, preparing molecules and molecular assemblies specially designed to exhibit given physical properties. Few of the ligands that will be discussed reflect our poor knowledge of and minor contributions into the area of the chemistry of SMMs and SIMs with 3d- and 4f-metal ions. Thus, few examples (but not the majority of them) will be work from our group.

Fifth, it is important to realize that the discipline of designing appropriate ligands for efficient transition-metal SMMs and SIMs has reached such a state of maturity that the present attempt cannot give (we are afraid) innovative ideas, but it will be a trip to the great achievements of selected (and not all) groups in the area.

As far as we are aware, this is the first attempt to highlight in a review the great influence of ligands on SMM and SIM properties. The topic has been partly covered in many of the books [21,22], chapters in books [55] and reviews [23,26,27,57–60] available dealing with the chemistry, physics, properties and potential applications of molecular nanomagnets.

4. The Azide Ion: An Evergreen "Tree" in the Chemistry of Transition-Metal SMMs

The azide ion (N_3^-) is the conjugate base of hydrazoic acid. Aqueous solutions of HN_3 were first prepared by T. Curties in 1890, Equation (3). Such solutions are weakly acidic (pK_a = 4.75).

$$N_2H_5^+ + HNO_2 \rightarrow HN_3 + H^+ + 2\,H_2O \qquad (3)$$

The azido ligand is very popular in transition-metal chemistry and especially for the synthesis of coordination clusters and coordination polymers [62–64]. In Coordination Chemistry, it is found either as a terminal ligand (η^1) or as a bridging one. The to-date crystallographically established coordination modes of the bridging azido ligand are illustrated in Figure 3.

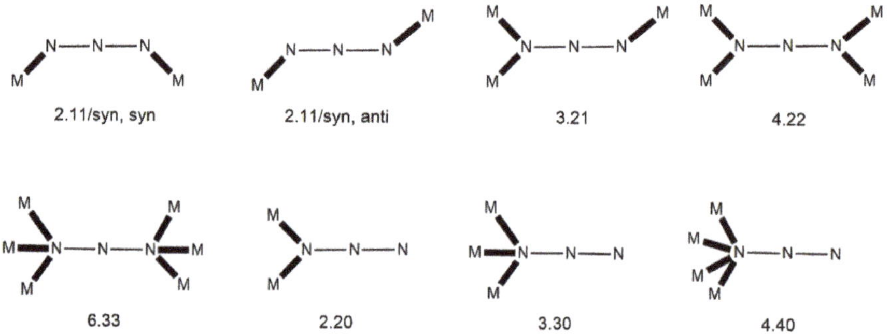

Figure 3. The to-date crystallographically established bridging coordination modes of the azido ligand and the Harris notation [61] that describes these modes. Coordination bonds are drawn with bold lines.

The bridging azido ligand has been one of the most investigated ligands in Magnetochemistry and Molecular Magnetism [63]. The variety of its bridging modes and its ability to propagate exchange interactions has led to compounds with several kinds of magnetic behaviors, such as antiferromagnetism, ferrimagnetism, ferromagnetism, canted weak ferromagnetism, spin-flop, and SMM and SCM properties [63]. End-to-end (EE) azido groups generally propagate antiferromagnetic interactions between paramagnetic metal ions, whereas end-on (EO) N_3^- ions are generally ferromagnetic couplers; exceptions of this general rules are known [63]. In addition to the coordination mode, other structural and electronic parameters (bridging and dihedral angles, bond lengths, orthogonality of magnetic orbitals, spin polarization, delocalization of unpaired electrons, et al.) play an important role in determining the sign and strength of the magnetic coupling. The ability of N_3^- to promote ferromagnetic exchange interaction has been utilized in the SMM chemistry of 3d-metal ions [62–64].

We should mention here that an excellent experimental and theoretical study by Sarkar, Neese, Meyer and coworkers [65] has opened the doors for the use of the terminal azido group in 3d-metal SIM chemistry. Contrary to previous suggestions, it was shown that the N_3^- ligand behaves as a strong σ and π donor. Magnetostructural correlations have revealed a remarkable increase in the negative D value with shortening of the axial Co^{II}–N_{azido} bond lengths, i.e., with increasing Lewis basicity.

Returning to SMM chemistry, our group, in collaboration with the group of Escuer at Barcelona, have developed a general synthetic strategy for the remarkable increase of the ground-state spin in coordination clusters, which often "switches on" SMM behavior [66–69]. The strategy is based on the substitution of bridging hydroxido or alkoxido groups, which most often propagate antiferromagnetic exchange interactions, in pre-formed coordination clusters by EO azido groups which propagate ferromagnetic interactions. The core changes, but the nuclearity does not. The incoming azido groups (always EO) introduce ferromagnetic components in the superexchange scheme of the molecule and, as a consequence, the ground-state spin increases significantly, sometimes inducing SMM behavior.

The reaction between the pre-formed cluster $[Fe^{II}_9(OH)_2(O_2CMe)_8\{(py)_2C(O)_2\}_4]$ (**5**) and a slight excess of NaN$_3$ in refluxing MeCN under N$_2$ gives the corresponding azido cluster $[Fe^{II}_9(N_3)_2(O_2CMe)_8\{(py)_2C(O)_2\}_4]$ (**6**) in ~40% yield, Equation (4); (py)$_2$C(O)$_2^{2-}$ is the dianion of the *gem*-diol form of di-2-pyridyl ketone, (py)$_2$CO [68,69]. The reactant and the product have similar molecular structures, the only difference being the presence of two 4.40 azido groups in the latter instead of two 4.4 hydroxido groups in the former. The nine Fe^{II} atoms in **6** adopt the topology of two square pyramids sharing a common apex. Bridging within each square base is achieved through four syn, syn 2.11 MeCO$_2^-$ ligands, four alkoxido oxygen atoms from the four 5.3311 (py)$_2$C(O)$_2^{2-}$ ligands (**A** in Figure 4) and one extremely rare 4.40 azido group (Figure 3). Each alkoxido oxygen is μ_3 bridging connecting two metal centers from a square base to the central Fe^{II} atom; the latter is thus 8-coordinate. The core is $\{Fe^{II}_9(4.40\text{-}N_3)_2(\mu_3\text{-}OR)_8\}^{8+}$, Figure 5. The two square bases have a slightly staggered conformation which results in a square antiprismatic coordination geometry about the central Fe^{II} atom, the chromophore being $\{Fe^{II}O_8\}$. Dc magnetic susceptibility studies in the 2–300 K temperature range indicate that the substitution of the 4.4 OH$^-$s in **5** by the 4.40 N$_3^-$s in **6** induces ferromagnetic coupling in the latter; its ground spin state is not well isolated from low-lying excited states and it cannot be accurately determined. Compound **6** is EPR silent at the X-band frequency at 4.2 K, but it is an SMM with a U_{eff} value of 29(1) cm^{-1}. The slow magnetic relaxation is also evident using zero-field ^{57}Fe Mössbauer spectroscopy.

$$[Fe^{II}_9(OH)_2(O_2CMe)_8\{(py)_2C(O)_2\}_4] + 2\ NaN_3 \xrightarrow[\Delta]{MeCN,\ N_2} [Fe^{II}_9(N_3)_2(O_2CMe)_8\{(py)_2C(O)_2\}_4] + 2\ NaOH \quad (4)$$

Almost all the azido-bridged transition-metal clusters and SMMs contain chelating or bridging organic ligands. The presence of bridging organic ligands is usually a disadvantage because it introduces antiferromagnetic components in the superexchange scheme, thus decreasing the effect of the ferromagnetic EO azido groups. The group of Stamatatos have recently developed [70–72] a novel strategy for the synthesis of 3d-metal, EO azido-bridged coordination clusters with high-spin *S* values and SMM properties by avoiding the presence of any organic chelating/bridging ligand. The azido groups in the cores of the clusters are exclusively EO, which ensures the presence of ferromagnetic exchange interactions between the metal spin carriers and therefore the attainment of the maximum *S* value. In the presence of magnetic anisotropy, induced by the choice of the appropriate metal ion, some of the clusters also exhibit remarkable SMM properties. The peripheral ligation around the metallic core is completed by terminal, volatile MeCN ligands, whose solvation/de-solvation effects sometimes lead to interesting magnetic phenomena. The key reagent for this chemistry is Me$_3$SiN$_3$. This has some remarkable differences compared to its all-inorganic analogue, which is almost 100% used in metal–azide chemistry. Firstly, Me$_3$SiN$_3$ is more soluble in organic solvents than NaN$_3$, and this allows reactions to be performed in a variety of such solvents. Secondly, Me$_3$SiN$_3$ can abstract OH$^-$ ions from the reaction media and thus bridging hydroxide ligands are not incorporated in the products; this allows the dominance of the N$_3^-$ ions in solution which can act as ligands without competition from the OH$^-$ species. And third, Me$_3$Si$^+$ cannot coordinate and Na$^+$ ions are not incorporated in the clusters, which is often the case when NaN$_3$ is used as the azide source. Representative examples of this strategy are briefly mentioned below.

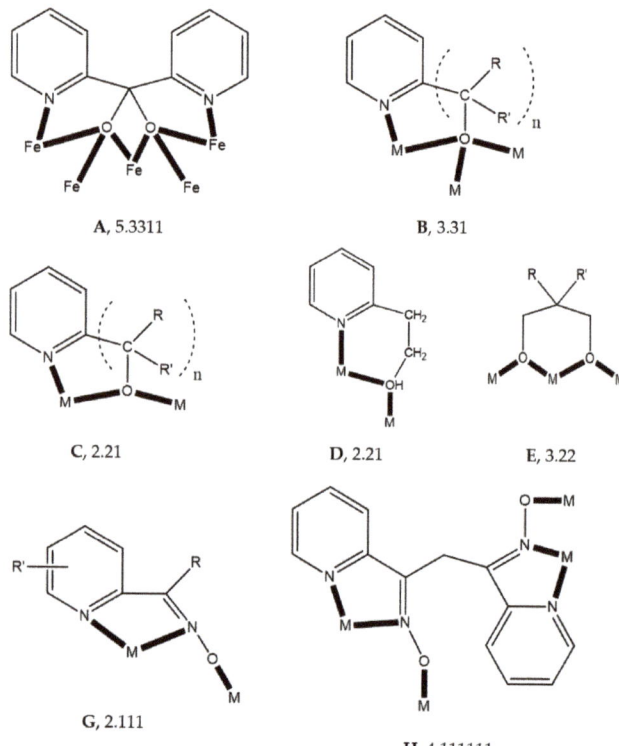

Figure 4. The coordination modes of some of the ligands discussed in the text and the Harris notation that describes these modes. M = metal ion (this is specified in the cited examples); n = 1, 2.

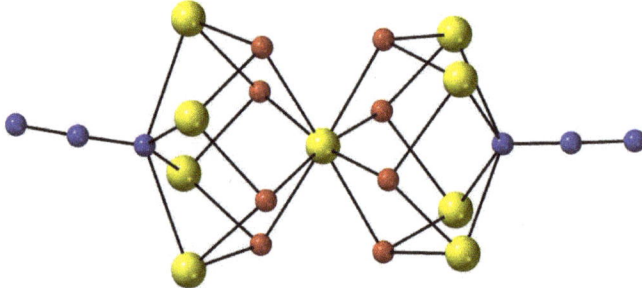

Figure 5. The $\{Fe^{II}_9(4.40\text{-}N_3)_2(\mu_3\text{-}OR)_8\}^{8+}$ core of cluster **6**. The Fe^{II} centers are shown in dark yellow, the nitrogen atoms in blue and the oxygen atoms are shown in red. Modified from reference [69].

The 1:1:4 reaction between $Fe^{II}(ClO_4)_2\cdot 6H_2O$, Et_3N and Me_3SiN_3 in MeCN under N_2 gives a dark red solution from which orange crystals of $[Fe^{II}_7(N_3)_{12}(MeCN)_{12}](ClO_4)_2\cdot 4MeCN$ (**7**·4MeCN) can be isolated in a ~55% yield, Equation (5). The crystals were treated in two ways for magnetic and spectroscopic studies [70,71]. A portion of the crystalline material was immediately transferred and sealed in an NMR tube representing the structurally characterized sample. The other portion was collected by filtration and dried under N_2 for 3 h; its analytical data corresponded to the formula {Fe^{II}$_7$(N$_3$)$_{12}$(MeCN)$_2$(ClO$_4$)$_2$} (**7a**). The IR spectra of **7·4MeCN** and **7a** indicate the presence of the EO azido group, with the band due to the v_{as}(N$_3$) mode appearing at ~2100 cm^{-1}. The almost complete de-solvation of **7·4MeCN** to give **7a** is evidenced by the nearly complete disappearance of the IR bands due to the v(C≡N) mode of the MeCN molecules from the spectrum of the latter (at 2310 and 2279 cm^{-1} in the spectrum of the former).

$$7\ Fe^{II}(ClO_4)_2 \cdot 6\ H_2O + 12\ Me_3SiN_3 + 16\ MeCN + 12\ Et_3N \xrightarrow[N_2]{MeCN}$$
$$[Fe^{II}_7(N_3)_{12}(MeCN)_{12}](ClO_4)_2 \cdot 4\ MeCN + 12\ Me_3SiOH + 12\ (Et_3NH)(ClO_4) + 30\ H_2O \quad (5)$$

The heptanuclear cation that is present in the crystal structure of **7·4MeCN** (Figure 6) contains a nearly ideal planar hexagon of metal ions centered on the seventh, central Fe^{II} atom [71]. The central Fe^{II} atom is a crystallographic inversion center; the {Fe^{II}$_7$} disk-like cation possesses virtual S_6 symmetry. The seven octahedral Fe^{II} atoms are held together through six 2.20 and six 3.30 EO azido groups. The six 3.30 azides connect the outer {Fe^{II}$_6$} hexagon with the central metal ion, while the six 2.20 azides bridge exclusively the outer Fe^{II} centers. Peripheral ligation is completed by twelve MeCN molecules, two on each of the outer metal ions. The {Fe^{II}$_7$(3.30-N$_3$)$_6$(2.20-N$_3$)$_6$}$^{2+}$ core can also be described as consisting of six {Fe^{II}$_3$(N$_3$)$_4$} defective cubane units, each double face-sharing; a vertex of each cubane unit is shared with the common vertex of the six cubanes which is the central Fe^{II} atom. The metal–nitrogen bond lengths are indicative of high-spin $(t_{2g})^4(e_g)^2$ Fe^{II} atoms with N-ligation. The intramolecular Fe^{II}···Fe^{II} distances are ~3.35 Å, while the Fe^{II}–N–Fe^{II} angles span the range 95.5–105°. The intermolecular Fe^{II}···Fe^{II} distances are large (>8 Å) due to the packing of the heptanuclear cations and to the presence of the coordinated MeCN molecules.

Both the as-synthesized (**7·4MeCN**) and dried (**7a**) forms of the complex were magnetically studied [70,71]. Dc magnetic susceptibility studies reveal the maximum possible ground-state spin ($S = 14$). Both forms are SMMs; however, their ac magnetic dynamics are different, revealing a "Janus"-faced SMM behavior for the pristine and dried samples which have been attributed to solvation/de-solvation effects from the coordinated solvent molecules. Sample **7a** exhibits two individual relaxation processes, which are both thermally assisted; the 2.7–5.0 K process is characterized by a U_{eff} value of 30.5(1) cm^{-1}, and the 1.8–2.6 K process by a U_{eff} value of 15.3(2) cm^{-1}. Data are shown in Figure 7. In contrast, the as-synthesized (pristine) sample **7·4MeCN** exhibits only one relaxation process below ~3 K with a U_{eff} value of 10.0(2) cm^{-1}. The different number of relaxation processes and the different values of effective energy barriers for magnetization reversal can be rationalized in terms of the differences in intermolecular, i.e., intercationic, interactions and the different molecular anisotropies arising from different crystal fields around the peripheral Fe^{II} atoms [71].

Complexes [Co$_7$(N$_3$)$_{12}$(MeCN)$_{12}$](ClO$_4$)$_2$ (**8**) and [Ni$_7$(N$_3$)$_{12}$(MeCN)$_{12}$](ClO$_4$)$_2$ (**9**) were prepared by a reaction similar to that used for **7·4MeCN**, simply replacing the metal perchlorate starting material [72]. The crystal structures of **8** and **9** (these complexes are isomorphous) do not contain solvent molecules in the lattice. The clusters have a similar molecular structure to its {Fe^{II}$_7$} analogue. The crystals of **8** and **9** are stable at room temperature and no degradation is observed after 24 h exposure to the normal laboratory atmosphere; the static and dynamic properties of their wet- and "dried"-forms are identical for each complex. As expected, both clusters are strongly ferromagnetically coupled. The Ni(II) cluster ($S = 7$) has a negligible magnetoanisotropy and, consequently, it does not exhibit out-of-phase ac magnetic susceptibility signals, i.e., it is not a SMM, in either the absence or the presence of an external dc field. The Co(II) cluster exhibits SMM properties under the application of a weak external dc field of 0.1 T with a U_{eff} value of 19.6(1) cm^{-1} [72]. The application of weak dc fields during the dynamic susceptibility studies helps to suppress QTM, which is otherwise strong for systems in low-symmetry crystal environments. This experimental practice is very helpful in elucidating the mechanisms operating in magnetization relaxation processes. However, the researchers should be cautious when interpreting ac susceptibility results in the presence of external dc fields; an excellent description of such a cautionary note is provided in reference [27].

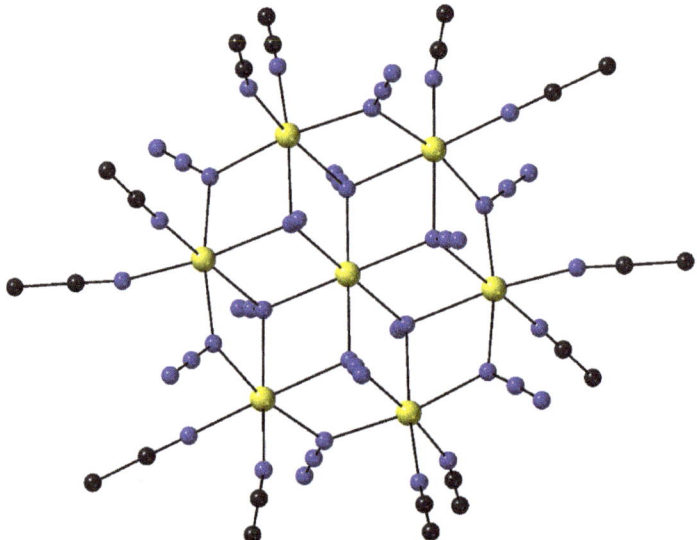

Figure 6. The molecular structure of the heptanuclear cation of cluster 7·4MeCN. The Fe^{II} centers are shown with dark yellow color, the nitrogen atoms with blue color and the carbon atoms with black color. Modified from reference [31].

The great utility of the above mentioned synthetic approach, which can potentially stimulate the research in transition-metal azido chemistry to meet new directions, is demonstrated by the 2-nm-sized spherical cluster $[Mn_{29}O_{24}(N_3)_{10}(dma)_{28}]$ (**10**) [73], where dma is the 3,3-dimethylacrylate(−1) ligand. The cluster contains a $\{Mn^{II}Mn^{III}_{28}(4.4\text{-}O)_8(3.3\text{-}O)_{16}(4.40\text{-}N_3)_2(2.20\text{-}N_3)_8\}^{28+}$ core, and has an $S = 9/2$ ground state. Despite the appreciable number of EO azido groups, the complex is not an SMM due to the simultaneous presence of bridging oxido and carboxylato groups, which promote antiferromagnetic exchange interactions and "force" the Mn–N_{azido}–Mn angles to be rather large (average 111.4°) presaging antiferromagnetic interactions between the respective azido-bridged metal ions.

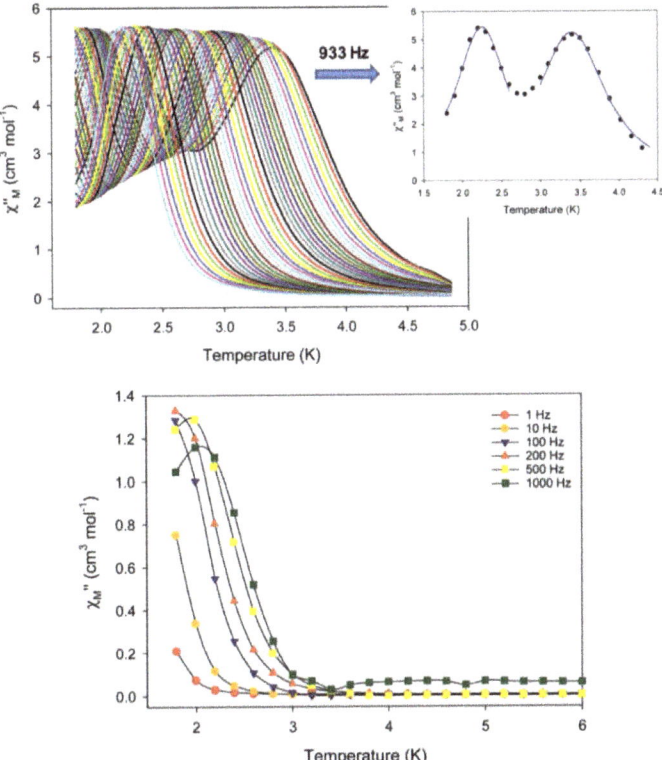

Figure 7. (Top) Out-of-phase (χ''_M) ac molar magnetic susceptibility signals for **7a**. The inset shows the χ''_M vs. T plot at the representative ac frequency of 933 Hz revealing the two separation relaxation processes in two different T regions for the sample; the solid blue lines are the fits of the data. (Bottom) Frequency dependent χ''_M signals of **7**·4MeCN. Reproduced from reference [70] with permission from the Royal Society of Chemistry.

5. Cyanido-Directed Assembly of Transition-Metal SMMs

Many transition-metal SMMs contain hydroxido (OH$^-$), alkoxido-type (RO$^-$) or oxido (O^{2-}) bridging groups. The problem is that the oxygen atom can bridge two or more (up to six with the O^{2-} ligand) metal centers, with a wide variety of M–O–M angles. Since the pairwise magnetic exchange interactions (which should be ferromagnetic and strong for an efficient SMM) are very sensitive to the bridging angles (and more generally to local geometry), chemists cannot often predict the magnetic properties of a complex structure. As a consequence, the search for new metal–hydroxido/alkoxido/oxido SMMs remains a rather serendipitous effort. The "good news" is that there is an alternative small-sized inorganic bridging ligand which might overcome the aforementioned difficulty. This is the cyanide (CN$^-$) group. Given that Prussian Blue, Fe$^{III}_4$[Fe^{II}(CN)$_6$]$_3$·xH$_2$O (x = 14–16), the first coordination complex and the first molecule-based magnetic solid was discovered in 1704 by the painter Diesbach in Berlin, it is amazing that after more than 300 years the CN$^-$ ion is in the forefront of coordination chemistry and molecular magnetism [74]. Perhaps, Andreas Ludi, who called Prussian Blue "An Inorganic Evergreen" in an article written in 1981, was more visionary when he gave this nickname [75].

The preference of CN⁻ for binding just two metal sites, one at each end, leading to a linear bridging arrangement is well established in inorganic chemistry. Thus, solution assembly reactions can be designed with the expectation that the product will possess linear M–CN–M' groups, thus providing synthetic chemists with a degree of synthetic and structural control [58]. In addition, because of the linear bridging arrangement, there is a satisfactory level of predictability in the expected nature of the magnetic exchange coupling between octahedral M and M' spin carriers. Unpaired spin density from orthogonal metal-based orbitals ($t_{2g} + e_g$) will leak over into orthogonal CN⁻-based orbitals, leading to ferromagnetic exchange via Hund's rule. On the contrary, the unpaired spin density from metal-based orbitals of compatible symmetry ($t_{2g} + t_{2g}$ or $e_g + e_g$) will leak over into the same cyanide-based orbitals, leading to antiferromagnetic exchange via Pauli exclusion principle. The antiferromagnetic exchange interactions are generally stronger and tend to dominate the superexchange in a competitive coupling scheme. Such predictions are useful in the design of cyanido-bridged clusters with high values of the ground-state spin [21,55,58].

Metal–cyanido chemistry can rather easily lead to molecules with high-spin ground states [21,58]. The difficult problem is to introduce axial magnetic anisotropy in these systems, i.e., negative D. The incorporation of metal centers that have a large single-ion anisotropy is a good first step in obtaining large D values in a cluster. The anisotropy of an individual metal ion is determined by the coupling between its spin angular momentum and its orbital angular momentum. Thus, metal ions with orbitally degenerate ground states are expected to have a large zero-field splitting and would be appropriate for incorporation in SMMs. This is, for example, the case in **1** (Figure 1) where the outer Mn^{III} atoms, with a $t_{2g}^3 e_g^1$ configuration contribute much to the anisotropy barrier. Thus, there are two approaches for instilling magnetic anisotropy in metal–cyanido clusters. The first one is to utilize metal ions already known to give hydroxido-, alkoxido- and/or oxido-bridged 3d-metal SMMs, e.g., V^{III}, high-spin Mn^{III}, high-spin Co^{II}, etc. Such high-spin configurations are obtained easily with O-based donors; however, low-spin configurations often occur in metal–cyanido clusters and certain of these (with orbital degeneracy) may be particularly effective. The second approach is to utilize 4d- and 5d-metal ions. Because of the relativistic nature of the spin-orbit coupling phenomenon (which is a source of magnetic anisotropy), this will generally result in significant increase of single-ion anisotropy for heavier transition-metal ions. For example, replacement the Cr^{III} atoms in known cyanide clusters with Mo^{III} atoms can lead to clusters with enhanced magnetic anisotropy while preserving the ground-state spin [58]. Of particular importance in this context is that theory has predicted that single-ion anisotropy originating from spin-orbit coupling will give SMMs with higher T_B values [76].

Based on the facts mentioned above, the main method for synthesizing cyanido-based homo- and, especially, heterometallic transition-metal SMMs is a building-block approach, often called modular process or "complexes as ligands and complexes as metal ions" strategy. A typical cluster preparation involves two building units. One bears one or more coordination sites occupied by labile (loosely coordinated) solvent molecules (the "metal ion") and the other is a complex that possesses one or more terminal cyanido ligands (the "metalloligand") [21,55,58,77–85]. In solution, the nucleophilic "free" nitrogen atoms of the terminal cyanido ligands displace the labile solvent molecules, leading to dinuclear or polynuclear assemblies. An important synthetic parameter is the nature and denticity of the capping (chelating) ligands for either of the two building units (precursors); these characteristics often control the nuclearity of the product. The "metal ion" in this building-block approach can also be a simple metal "salt" without a capping ligand and/or labile solvent molecules. In addition to this structural control, there are available qualitative rules for predicting the nature and the strength of the magnetic exchange interaction between the cyanido-bridged metal spin carriers established via detailed studies on hundreds of Prussian

Blue analogues [15,21,58,86]. This fruitful combination of structural and magnetic predictability has led to the rational design, synthesis and study of many cyanido-bridged SMMs.

A significant advantage of transition metal–cyanido cluster chemistry is that once a stable new structural motif is identified, the researcher can be confident that she/he can replace some metal ions with certain other metal ions in the structure [58].

After this, rather long, introduction we proceed with few representative examples. The metalloligand $(Et_4N)[Re^{II}(CN)_3(triphos)]$, where triphos is the bulky tridentate phosphine 1,1,1-tris(diphenylphosphinomethyl)ethane (Figure 8a), was prepared by Dunbar's group. Direct reaction of $[(triphos)Re^{II}(\mu-Cl)_3Re^{II}(triphos)]Cl$ with 6 equivs of $(Et_4N)CN$ in MeCN leads to homolytic scission of the $\{Re^{II}{\equiv}Re^{II}\}^{4+}$ unit to give the bright yellow precursor metalloligand, Equation (6), in ~35% yield [87]. The coordination geometry of Re^{II} in the mononuclear anion is *fac* pseudo-octahedral; three coordination sites are occupied by the C-bonded cyanido ligands, while the other three sites are filled by the P atoms of the tridentate chelating triphos ligand. Mononuclear 17-electron complexes of Re(II) are attractive in molecular magnetism due to strong spin-orbit coupling effects (λ = 2000–3000 cm^{-1}) arising from the low-spin 5d^5 configuration. The 1:1 reaction between $(Et_4N)[Re^{II}(CN)_3(triphos)]$ and anhydrous $MnCl_2$ in $Me_2CO/MeCN$ under reflux gives the orange-red cluster $[Mn^{II}_4Re^{II}_4Cl_4(CN)_{12}(triphos)_4]$ (**11**), Equation (7). The octanuclear molecule has an approximate cubic topology (Figure 8b) [78,79] with alternating octahedral Re^{II} and tetrahedral Mn^{II} concerns. The edges of the cube are spanned by the 12 bridging cyanido groups (Re^{II}–CN–Mn^{II}) that link the metal ions. The coordination geometry around the Re^{II} atoms is similar to that of the mononuclear starting metalloligand, i.e., distorted octahedral with the triphos ligand behaving as a facially capping, tridentate chelating ligand with the carbon end of the three cyanido ligands completing the coordination sphere. The Mn^{II} centers adopt a distorted tetrahedral geometry, the donor atoms for each metal ion being three coordinated nitrogen atoms from bridging cyanido groups and a fourth terminal chlorido ligand extending out of the cube. It is obvious that the steric bulk of the triphos ligands enforces the distorted tetrahedral coordination environments at each of the 3d-metal sites. The structurally similar cubic complexes $[Mn^{II}_4Re^{II}_4I_4(CN)_{12}(triphos)_4]$ (**12a**) and $[M^{II}_4Re^{II}_4Cl_4(CN)_{12}(triphos)_4]$ (M = Fe, **12b**; M = Co, **12c**; M = Ni, **12d**; M = Zn, **12e**) have also been prepared through analogous reactions [79]. Variable-temperature dc magnetic susceptibility data for **11** indicate antiferromagnetic coupling between the "S = 1/2" Re^{II} and S = 5/2 Mn^{II} atoms. The complex is a weak SMM with a U_{eff} value of ~9 cm^{-1}. Micro-SQUID temperature-dependent scans reveal hysteretic behavior for the cluster. The data also show a prominent step at zero field, resulting from fast QTM. The step becomes temperature-independent below 0.2 K but remains sweep rate-dependent, suggesting a ground-state resonant tunneling process at H = 0 [79].

$$[(triphos)Re^{II}(\mu\text{-}Cl)_3Re^{II}(triphos)]Cl + 6\ (Et_4N)CN \xrightarrow[N_2]{MeCN} 2\ (Et_4N)[Re^{II}(CN)_3(triphos)] + 4\ (Et_4N)Cl \quad (6)$$

$$4\ (Et_4N)[Re^{II}(CN)_3(triphos)] + 4\ MnCl_2 \xrightarrow[N_2]{Me_2CO/MeCN,\ T} [Mn^{II}_4Re^{II}_4Cl_4(CN)_{12}(triphos)_4] + 4\ (Et_4N)Cl \quad (7)$$

Staying at Re, an interesting metalloligand is $[Re^{IV}(CN)_7]^{3-}$, which was prepared by Long's group through a simple ligand metathesis reaction [88], Equation (8). This anion has a low-spin 5d^3 pentagonal bipyramidal geometry. The complex has a strong magnetic anisotropy which can be attributed to a combination of its large spin-orbit coupling associated with the heavy rhenium ion and the unquenched orbital angular momentum of its $^2E_1{}'$ electronic ground state. Thus, the incorporation of $[Re^{IV}(CN)_7]^{3-}$ into a high-spin coordination cluster would "transfer" magnetic anisotropy to the product, possibly leading to slow magnetic relaxation. Complex $[Mn^{II}(PY5Me_2)(MeCN)](PF_6)_2$ is an ideal "metal ion" for a building-block reaction for three reasons: (i) The Mn^{II} ion has 5 unpaired electrons; (ii) the six-coordinate cation possesses a labile MeCN molecule; and (iii) the potentially pentadentate chelating ligand 2,6-bis(1,1-bis(2-pyridyl)ethyl)pyridine ligand (PY5Me$_2$, Figure 9a) is an efficient capping moiety which can ensure the formation of star-like clusters that are magnetically isolated. Reaction of (Bu$^n_4$N)$_3$[Re^{IV}(CN)$_7$] with 4 equivs of [Mn^{II}(PY5Me$_2$)(MeCN)](PF$_6$)$_2$ in MeCN at −40 °C produces a blue solution from which the blue, temperature-sensitive cluster [Mn$^{II}_4$Re^{IV}(CN)$_7$(PY5Me$_2$)$_4$](PF$_6$)$_5$ (**13**) is isolated in a good yield, Equation (9). If the same reaction is performed at room temperature, an immediate color change from blue to green and then to yellow is observed affording complex [Mn$^{II}_4$Re^{III}(CN)$_7$(PY5Me$_2$)$_4$](PF$_6$)$_4$ (**13a**); this complex is formed via a spontaneous reduction of Re^{IV} ($S = 1/2$) to Re^{III} ($S = 0$) within the cluster [80].

$$(Bu^n_4N)_2[Re^{IV}Cl_6] + 7\,(Bu^n_4N)CN \xrightarrow[3\,\text{days},\,N_2]{DMF,\,85\,°C} (Bu^n_4N)_3[Re^{IV}(CN)_7] + 6\,(Bu^n_4N)Cl \tag{8}$$

$$(Bu^n_4N)_3[Re^{IV}(CN)_7] + 4\,[Mn^{II}(PY5Me_2)(MeCN)](PF_6)_2 \xrightarrow[N_2]{MeCN,\,-40\,°C}$$
$$[Mn^{II}_4Re^{IV}(CN)_7((PY5Me_2)_4](PF_6)_5 + 3\,(Bu^n_4N)(PF_6) + 4\,MeCN \tag{9}$$

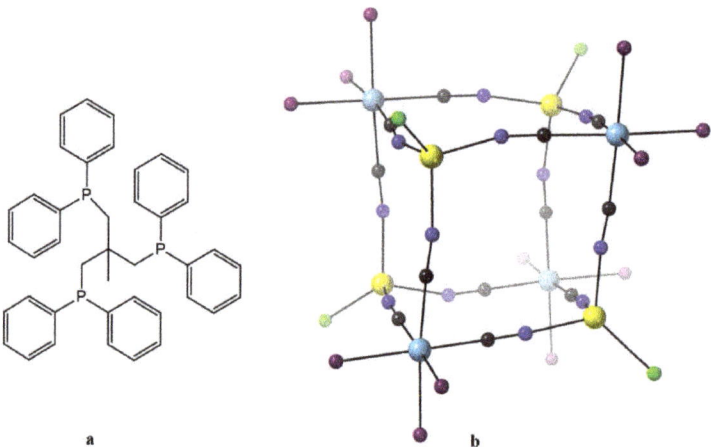

Figure 8. (a) The structural formula of triphos. (b) An aspect of the molecular structure of the cubic cluster **11**; only the P atoms of the tridentate chelating phosphine triphos have been drawn. The Re^{II}, Mn^{II}, Cl$^-$, N, C and P atoms/ions are shown with turquoise, yellow, green, blue, gray and mauve colors, respectively. Modified from reference [78].

X-ray analysis on single crystals of **13** reveals the presence of a four-point star-like topology for the cation of the complex, as shown in Figure 9. The [Re^{IV}(CN)$_7$]$^{3-}$ unit is at the center of the star and is bridged through cyanido groups to four {Mn^{II}(PY5Me$_2$)}$^{2+}$ pendant units; three cyanido ligands remain terminal at the central Re^{IV} atom. The coordination polyhedron of the 5d-metal ion is close to that of an ideal pentagonal bipyramid, with an essential liner axial C$_{ax}$–Re^{IV}–C$_{ax}$ angle of 179.9°. The arrangement of the four 3d-metal ions can be described as a square, with two of the Mn^{II} atoms binding axial cyanido groups of Re^{IV} and the other two binding non-neighboring equatorial cyanido groups of Re^{IV}. The magnetic exchange interactions between the central Re^{IV} atom ($S = 1/2$) and the surrounding Mn^{II} centers ($S = 5/2$ each) are ferromagnetic, resulting in a high-spin ground state (most probably 21/2). The high-spin ground state of the cluster, combined with the negative D value of −0.44 cm^{-1} gives SMM properties in **13** with an effective relaxation barrier of U_{eff} = 33 cm^{-1} [80]. Analogous reaction schemes lead to structurally similar clusters [M$^{II}_4$Re^{IV}(CN)$_7$(PY5Me$_2$)$_4$](PF$_6$)$_5$ (M = Ni, **13b**; M = Cu, **13c**), which are also SMMs with U_{eff} values of ~17 and ~8 cm^{-1}, respectively [81]; the D values are −0.93 cm^{-1} (**13b**) and −1.33 cm^{-1} (**13c**), the corresponding S values being 9/2 and 5/2.

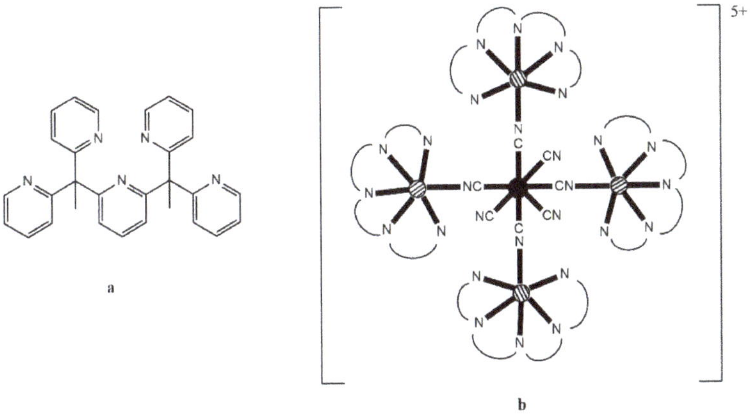

Figure 9. (**a**) The structural formula of the "open" pentadentate ligand PY5Me$_2$. (**b**) Schematic drawing of the molecular structure of the pentanuclear {Mn$^{II}_4$Re^{IV}} cation that is present in **13**; the solid circle represents the Re^{IV} atom, the semi-dashed circles represent the Mn^{II} atoms and N^N^N^N^N represents the pentadentate chelating ligand PY5Me$_2$, while the coordination bonds are drawn with bold lines.

Today the cyanido ligand is one of the most popular ligands in SMM chemistry [21,85], because of the variety of metalloligands that are available. For example, octacyanidometallates are unique building blocks that are useful in the construction of various types of magnetic clusters with topologies ranging from square, trigonal bipyramidal, octahedral, to pentadecanuclear six-capped-body-centered cubes and even larger molecules, some of which are SMMs [85].

6. Tris(trimethylsilyl)methanide, an Old Organometallic "Friend" Joins the Chemistry of 3d-Metal SIMs

Several groups around the world re-investigate the magnetic anisotropies of complexes based on transition-metal ions in a general effort to find magnetic alternatives to the f-block metal ions. As it has already been mentioned in Section 1, mononuclear transition metal–ion complexes are not good candidates for large U_{eff} values because of their smaller magnetic moments and, for the 3d-metal ions, lower spin-orbit coupling constants. Also, the larger ligand-field splitting energies of the d orbitals lower the orbital contributions to the magnetic properties required to develop significant magnetic anisotropy. There are two manifestations of this effect: (i) The first-order orbital angular momentum can be quenched as the result of a Jahn–Teller distortion; and (ii) the second-order contribution to the magnetic anisotropy, i.e., the zero-field splitting, is becoming very small due to the large energy separation between ground and excited electronic states, which reduces the degree of mixing. These ligand-field effects can become much less important by synthesizing transition metal–ion complexes with very low coordination numbers [27]. In this way, the ligand-field splitting energies of the d orbitals fall within a narrow range, similar to the energies of the 4f orbitals in Ln(III) complexes. A linear two-coordinate geometry at the metal center is ideal for suppressing ligand-field effects in 3d-metal complexes and generating a large anisotropy barrier. For a strictly linear coordination geometry, with a local D$_{\infty h}$ symmetry at the metal ion, the energies of the d orbitals are split as $(d_{xy}, d_{x^2-y^2}) < (d_{xz}, d_{yz}) < d_{z^2}$ (these relative energies do not consider the possibility of s-d_{z^2} mixing, which is known to lower the energy of d_{z^2} below that of the d_{xy}, d_{yz} pair), with the d_{z^2} and (d_{xz}, d_{yz}) orbital energies being destabilized by σ- and π-metal–ligand interactions, respectively; on the contrary, the $(d_{xy}, d_{x^2-y^2})$ orbitals have δ symmetry and are thus not participating in bonding with the ligands' orbitals. A high-spin 3d^6 electron configuration is expected to give a large first-order contribution to orbital angular momentum that will not be quenched through a Jahn–Teller distortion; this maximizes the magnetic moment and the two donor atoms define an axis for its easy alignment [89].

One of the ligands that has played an important role in the chemistry of two-coordinate 3d-metal SIMs is the tris(trimethylsilyl)methanide, $^-$C(SiMe$_3$)$_3$. The neutral compound HC(SiMe$_3$)$_3$ and its conjugate base have their own history in organic [90,91] and organometallic [92,93] chemistry, respectively. The tris(trimethylsilyl)methanide ('trisyl') anion has been used as a very bulky ligand in main group and Ln(III) chemistry. The advantages of this group are numerous: (a) Due to its bulkiness, it can lead to very low coordination numbers and stabilize coordinatively unsaturated transition-metal complexes which often exhibit unusual structural features or exciting reactivity; (b) its steric bulk-which is estimated to be similar to that of Cp* anion or P(^{t}Bu)$_3$-confers kinetic stability to complexes; and (c) it lacks α-and β-hydrogens prohibiting undesired reactivity. In a sense, it may be viewed as combining the electronic features of a methyl anion and the steric requirements of the Cp* anion [93].

Complex [Fe^{II}{C(SiMe$_3$)$_3$}$_2$] (**14**) was synthesized [93,94] by the reaction illustrated in Equation (10). Li{C(SiMe$_3$)$_3$}·2THF can be prepared by the metallation of tris(trimethylsilyl)methane with MeLi in THF/Et$_2$O under reflux [92]. The two-coordinate Fe^{II} center sits on a crystallographic inversion center, resulting in a strictly linear geometry (C–Fe^{II}–C = 180.0°). Variable-temperature dc magnetic susceptibility data indicate a high-spin 3d^6 Fe^{II} center (t$_{2g}^4$e$_g^2$). Low-temperature magnetization data show a saturation value of 3.24 B.M.; this value is lower than 4 B.M. that would be expected for a spin-only $S = 2$ center. The results from the dc susceptibility and magnetization data indicate that **14** has a highly anisotropic magnetic moment. Both data sets could not be fit, suggesting that the magnetic anisotropy of the molecule is not due to spin-only phenomena [94]. Ac susceptibility data reveal slow magnetic relaxation under an external dc field. Arrhenius plots for the linear complex were fit by employing a sum of tunneling, direct, Raman and Orbach relaxations, resulting in a U_{eff} value of ~146 cm^{-1}. Theoretical calculations (CASSCF/NEVPT2 on the crystal structure) were performed to explore the influence of deviation from rigorous D$_{\infty h}$ geometry on the splittings of the 3d orbitals and the electronic state energies. The calculations suggest that the ligand field asymmetry quenches the orbital angular momentum of **14**, but finally spin-orbit coupling is strong enough to compensate and regenerate the orbital moment. The non-observation of a single Arrhenius behavior has been attributed to a combination of the asymmetry of the ligand field and the influence of vibronic coupling.

$$\text{Fe}^{II}\text{Cl}_2 + 2\ \text{Li}\{\text{C}(\text{SiMe}_3)_3\}\cdot 2\text{THF} \xrightarrow[\text{N}_2]{\text{THF}, -78\ °C} [\text{Fe}^{II}\{\text{C}(\text{SiMe}_3)_3\}_2] + 2\ \text{LiCl} + 4\ \text{THF} \quad (10)$$

The observation of slow magnetic relaxation in **14**, and related mononuclear two-coordinate Fe(II) complexes, requires the application of an applied dc field which suppress fast magnetization reversal through QTM; the latter is very efficient for such systems because of the small non-Kramers $S = 2$ ground state prohibiting slow relaxation in zero magnetic field [27]. An alternative way to minimize tunneling of the magnetic moment, caused by mixing of the ground-state ± M_S levels, is employment of half-integer spin systems, according to the Kramers' theorem [95]. Such systems should not require the application of a dc field to display slow magnetic relaxation. The group of Long reported the structurally interesting and magnetically impressive two-coordinate SIM [K(crypt-222)][Fe^I{C(SiMe$_3$)$_3$}$_2$] (**15**), which possesses a $S = 3/2$ ground state [96]; crypt is the ligand 4,7,13,16,21,24-hexaoxa-1,10-diazabicyclo[8,8,8]-hexacosane, which chelates the K$^+$ ion though its six ether oxygen atoms. The synthesis of **15** was achieved by the one-electron reduction of **14** using KC$_8$ as reductant, Equation (11). The possibility of the one-electron reduction of the Fe(II) complex had been recognized by cyclic voltammetry studies in difluorobenzene, which show a reversible process corresponding to the [Fe{C(SiMe$_3$)$_3$}$_2$]$^{0/-1}$ couple. The product is obtained in yields of ~50% and has a bright yellow-green color. Single-crystal X-ray crystallography reveals a linear geometry around Fe^I (Figure 10a), the C–Fe^I–C bond angle being 179.2°. The SiMe$_3$ groups are eclipsed, contrary to the structure of **14**, in which they are staggered. Complex **15** not only shows, as expected, one of the highest U_{eff} values [226(4) cm^{-1}] yet reported for SMMs and SIMs that contain transition-metal ions, magnetic hysteresis below 29 K and a T_B value of 4.5 K, but also has improved permanency in zero field [96,97]; the U_{eff} value starts to approach values that we meet in Ln(III)-based systems. Quantum-chemical and spectroscopic studies suggest an electronic structure (Figure 10b), which would not be expected by most coordination chemists. The theoreticians in the research team provided strong evidence that strong 4s-3d$_{z^2}$ mixing stabilizes the 3d$_{z^2}$ orbital. The result is supported by advanced spectroscopic characterization, and the calculated orbital degeneracies explain well the unquenched orbital moment. The U_{eff} value is close to the calculated energy gap between the ground M_J = 7/2 pair and the first excited M_J = 5/2 doublet, suggesting that the relaxation occurs via this latter state. Excellent ^{57}Fe-Mössbauer studies of **14** and **15** under zero applied dc field between 295 and 5 K yield their magnetization dynamics on a significantly faster time scale (the lifetime of the measurement is 10^{-8} to 10^{-9} s) than it is possible with ac magnetometry. From the modeling of the Mössbauer profiles, Arrhenius plots between 295 and 5 K were obtained for the two complexes [98]. The high-temperature regimes suggest Orbach relaxation with U_{eff} values of 178(9) (**14**) and 246(3) (**15**) cm^{-1}, in good agreement with the values obtained from magnetism. In **15**, two distinct high-temperature regimes of magnetic relaxation are observed with mechanisms corresponding to two distinct single-excitation Orbach processes within the ground-state spin–orbit coupled manifold of the Fe^I atom.

$$[\text{Fe}^{II}\{C(\text{SiMe}_3)_3\}_2] + \text{KC}_8 + \text{crypt-222} \xrightarrow[\text{N}_2]{\text{THF}} [\text{K}(\text{crypt-222})][\text{Fe}^I\{C(\text{SiMe}_3)_3\}_2] + 8\,C \quad (11)$$

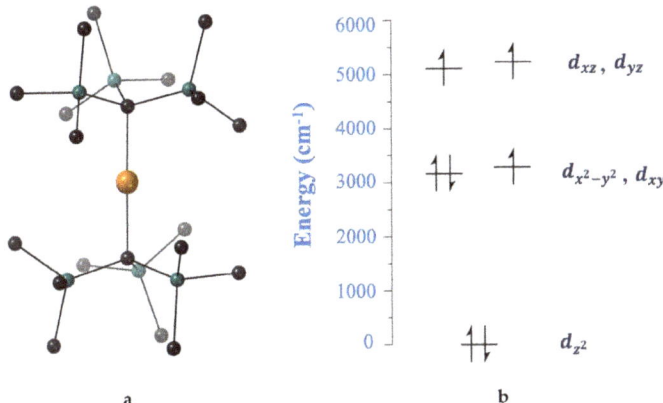

Figure 10. (a) The molecular structure of the anion that is present in **15**. (b) Energy splitting of the 3d orbitals of [Fe^I{C(SiMe$_3$)$_3$}$_2$]$^-$ derived from an ab initio computational analysis; this orbital scheme is the origin of a strong magnetic moment along the z axis. Color code: Fe dark orange, Si turquoise, C black. H atoms have been omitted for clarity. Modified from reference [96].

It has become clear from the preceding information that orbital angular momentum L gives rise to magnetic anisotropy, an essential property for efficient SIMs. Unquenched L arises from an odd number of electrons in degenerate orbitals; this is observed only for free ions and in f-element complexes. In transition-metal complexes, however, the ligand field removes any orbital degeneracy, leading to practically zero L. Any magnetic anisotropy in such complexes is a weak effect arising from mixing of the electronic ground state and excited states induced by spin–orbit coupling. The maximum value L for a transition metal is 3 and, at first glance, it seems impossible. An $L = 3$ ground state in a linear complex requires two sets of degenerate orbitals, $(d_{x^2-y^2}, d_{xy})$ with $m_l = \pm 2$ and (d_{xz}, d_{yz}) with $m_l = \pm 1$, and an odd number of electrons in each. Such a system would imply a non-Aufbau configuration, wherein the electrons do not fill the d orbitals in the usual manner from lowest to highest energy, and likely exhibit a large magnetic anisotropy. Having as scientific arsenal: (A) The previous characterization of **15** with unquenched orbital angular momentum, large mangetic anisotropy and non-influence by Jahn–Teller distortions that would otherwise remove orbital degeneracy. (B) Experiments which have shown that Co atoms, deposited on a MgO surface under vacuum (referred to as adatoms) and adopting a coordination number of 1, have a $J = 9/2$ ($S = 3/2$, $L = 3$) ground state giving rise to near-maximal magnetic anisotropy [99]; and (C) Calculations on the hypothetical complex $[Co^{II}\{C(SiMe_3)_3\}_2]$ which have shown an $L = 3$ ground state arising from a non-Aufbau $(d_{x^2-y^2}, d_{xy})^3$, $(d_{xz}, d_{yz})^3 (d_{z^2})^1$ filling of the 3d orbitals and further predicting a gap of ~455 cm^{-1} between ground and first excited M_J states [100], a mutli-national research team led by Long, Neese and van Slageren set out efforts to synthesize and characterize $[Co^{II}\{C(SiMe_3)_3\}_2]$, i.e., the 3d^7 analogue of **14**.

The strongly reducing nature of the carbanion ligand $^-C(SiMe_3)_3$ hinders isolation of the desired compound [101]. Metathesis reactions of $\{C(SiMe)_3\}^-$ salts and CoX_2 (X = Cl, Br, I) gave only amorphous solids that could not be characterized. However, lowering the basicity of the central carbanion through the introduction of electron-withdrawing aryloxide groups provided access to the dialkyl complex $[Co^{II}\{C(SiMe_2ONaph)_3\}_2]$ (**16**), where Naph is the naphthyl group. The synthetic process is illustrated in Figure 11. The Zn(II) congener, $[Zn\{C(SiMe_2ONaph)_3\}_2]$ (**16a**) was obtained in an analogous manner. Using the same reaction conditions with a mixture of $ZnBr_2$ and $CoBr_2\cdot THF$ (molar ratio ~900:1) enabled the preparation of the magnetically dilute sample $Co_{0.02}Zn_{0.98}\{C(SiMe_2ONaph)_3\}_2$ [101].

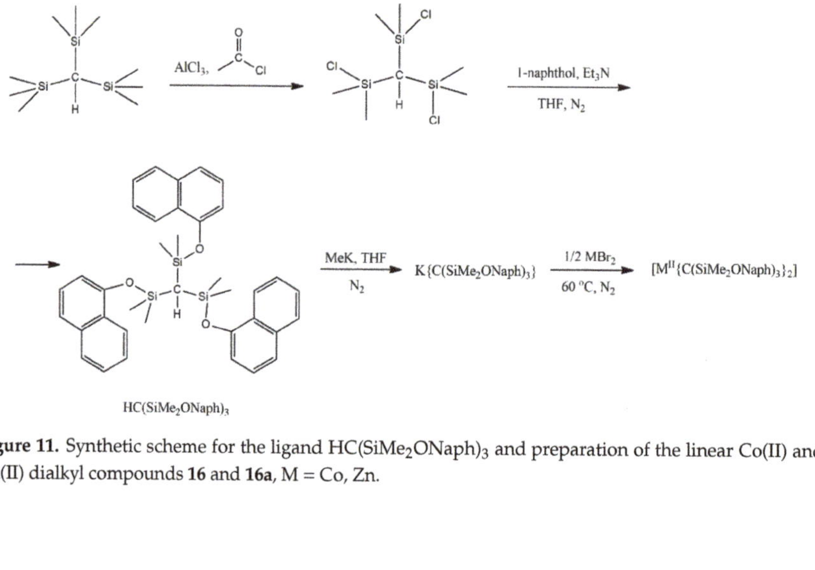

Figure 11. Synthetic scheme for the ligand HC(SiMe$_2$ONaph)$_3$ and preparation of the linear Co(II) and Zn(II) dialkyl compounds **16** and **16a**, M = Co, Zn.

Complexes **16** and **16a** are isomorphous and feature a linear C–M^{II}–C axis imposed by the S_6 site symmetry. The staggered orientation of the ligands facilitates close sp^3–CH···π and sp^2–CH···π contacts (2.69 and 2.82 Å, respectively). This indicates that interligand interactions contribute into the stabilization of the structures [101]. Ab initio calculations on **16** predict a ground state with $S = 3/2$, $L = 3$ and $J = 9/2$ which arises from the non-Aufbau electron configuration $(d_{x^2-y^2}, d_{xy})^3 (d_{xz}, d_{yz})^3 (d_{z^2})^1$, Figure 12; this deviates from the expected Aufbau filling of $(d_{x^2-y^2}, d_{xy})^4 (d_{xz}, d_{yz})^2 (d_{z^2})^1$. This deviation can be explained in terms of the competing effects of ligand-field stabilization and interelectron repulsion. As for Ln(III) complexes, the ligand field is weak so that interelectron repulsion and spin–orbit coupling determine the electronic ground state. Dc magnetic susceptibility results reveal a well-isolated $M_J = \pm 9/2$ ground state. Variable-field far-infrared spectra suggest a magnetically active excited state at ~450 cm^{-1} that, together with variable-temperature ac susceptometry and theoretical calculations, has assigned to the $M_J = \pm 7/2$ state. A d-orbital filling scheme with equally occupied of $(d_{x^2-y^2}, d_{xy})$ and (d_{xz}, d_{yz}) orbital sets is also indicated by modeling of experimental charge density maps. The U_{eff} barrier of ~450 cm^{-1} determined for **16** is the largest reported to date for a transition-metal SMM or SIM. As a consequence of its large orbital angular momentum, the magnetically dilute sample Co$_{0.02}$Zn$_{0.98}${C(SiMe$_2$ONaph)$_3$}$_2$ exhibits a coercive field of 600 Oe at 1.8 K. Although its magnetic properties mainly pertain at very low temperature, the synthesis, structure and properties of **16** provide scientists with a valuable, general design principle [101].

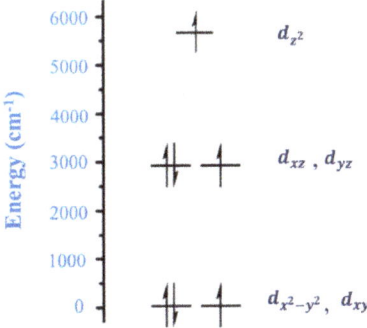

Figure 12. Energy diagram and occupations of the 3d orbitals by electrons in the linear complex **16**. Modified from reference [101].

7. Softer Ligands for More Efficient SIMs

The main advantage of transition-metal SIM chemistry is that the ground-state spin of the mononuclear molecule is fixed and D is the only parameter that affects U_{eff}. As we have seen, the orbital angular momentum is the main factor that dictates the magnitude of D; the former depends on the oxidation state and the coordination number/geometry of the 3d-metal ion (for first-row transition metals). Thus, scientists can fine-tune the electronic structure of the complex by playing with the oxidation state and the geometry of the metal ion. The classical approach involves the lowering of the coordination number which provides the system with a large orbital angular momentum resulting in high U_{eff} values [94,96,101,102]. In addition to this approach, several efforts have been performed to alter the spin Hamiltonian parameters of various metal ions by other methods [103]. Slow magnetization relaxation is normally not observed in integer-spin systems, even under application of external dc magnetic fields, because of underbarrier tunneling mechanisms. As a consequence of this, the interest has focused on non-integer spin systems for a better design of SIMs. Cobalt is a good candidate for efficient SIMs due to strong first-order spin-orbit coupling displayed by the metal in its high-spin +II oxidation state [26,27]; this is particularly so in pseudotetrahedral symmetry. Several parameters can potentially be employed to stabilize easy-axis (or Ising-type) magnetic anisotropy including the softness (HSAB) of the donor atoms (and hence the Co^{II}–ligand covalency), the influence of the other commonly used coligands (e.g., halides) and the variation of the counter cation in anionic complexes. Representative examples are briefly described below.

The first transition-metal SIM without requiring the application of an external dc field to suppress quantum relaxation processes was the tetrahedral complex $(Ph_4P)_2[Co(SPh)_4]$ (**17**) [104], prepared as illustrated in Equation (12). The complex has an $S = 3/2$ spin ground state, with a large, negative, axial zero-field splitting ($D = -70$ cm^{-1}) and a low rhombicity ($E/D < 0.9$). The large magnetic anisotropy of the anion can be explained by examining Figure 13, derived from angular overlap model calculations. The near degeneracy of the filled $3d_{x^2-y^2}$ orbital and the singly occupied $3d_{xy}$ orbital leads to a low-lying electronic excited state that can couple to the ground electronic state through spin-orbit coupling, thus affording a large D value and resulting in a U_{eff} value of 21(1) cm^{-1}. Dilution of $[Co(SPh)_4]^{2-}$ within the isostructural $[Zn(SPh)_4]^{2-}$ matrix eliminates quantum tunneling pathways, indicating that they occur via intermolecular dipolar interactions [104].

$$CoCl_2 + 4\ K(SPh) + 2\ (Ph_4P)Br \xrightarrow[N_2]{MeCN} (Ph_4P)_2[Co(SPh)_4] + 2\ KCl + 2\ KBr \qquad (12)$$

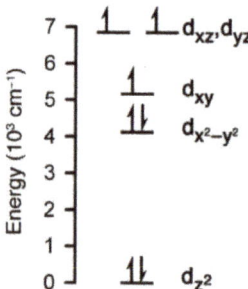

Figure 13. Electronic configuration and splitting of the 3d-orbital energy levels for the complex anion in **17**, derived using the angular overlap model. The diagram was reproduced from reference [104] with the permission of the American Chemical Society.

Continuing the above mentioned work, the group of Long prepared the salts of $[Co(EPh)_4]^{2-}$ (E = O, Se) $(Ph_4P)[Co(OPh)_4]$ (**18**), $K(Ph_4P)[Co(OPh)_4]$ (**18a**), $(Ph_4P)_2[Co(SePh)_4]$ (**19**), from reactions of $CoCl_2$ or CoI_2 with excess amounts of $K(EPh)$ and $(Ph_4P)Br$ in MeCN [105]. All anions possess pseudotetrahedral coordination environments with tetragonal distortions to give exactly or approximately D_{2d} symmetry. One of the goals was to correlate D and U_{eff}. The values of the former are ~ -11, -24 and -83 cm^{-1} for **18**, **18a** and **19**, respectively. Arrhenius plots of the ac data indicate U_{eff} values of 21(1) and 19(1) cm^{-1} for **18** and **19**, respectively, whereas the plot for **18a** shows substantial curvature indicating strong intermolecular interactions. Dilution experiments with $[Zn(OPh)_4]^{2-}$ allowed the observation of thermally-activated magnetic relaxation for **18a** with a U_{eff} value of 34.0(5) cm^{-1}. The trend in U_{eff} for **17, 18, 18a** and **19** does not follow the trend in D values; this possibly indicates that magnetization relaxation in **17, 18a** and **19** is not fully thermally activated (the relation between D and U for an $S = 3/2$ system is $U = 2D$). An analysis of the four complexes within the framework of ligand field theory shows that the increase in $|D|$ takes place in concert with a decrease in the Racah parameter B, i.e., increased covalency. This suggest the importance of soft donor ligands in the efforts to obtain systems with a large magnetic anisotropy.

The story does not end here. Inspired from the above described studies, the groups of Neese and Atanasov reported a systematic theoretical study developing magnetostructural correlations in the anions $[Co^{II}(EPh)_4]^{2-}$ (E = O, S, Se, Te; the Te derivative is hypothetical) based on multireference quantum chemical methods and ab initio-based ligand field theory; they also discussed the correlation of D with softness of the ligands, relativistic nephelauxetic effects and covalency [106]. The $|D|$ value increases when the ligand field decreases across the series from O to Te. It has been shown that due to the π-anisotropy of the S and Se donor atoms, magnetostructural correlations in $[Co(OPh)_4]^{2-}$ and $[Co(EPh)_4]^{2-}$ (E = S, Se) differ. In the case of the isotropic PhO$^-$ ligand, only variations within the first coordination sphere of Co^{II} affect magnetic properties; in the case of the PhE$^-$ (E = S, Se) ligands, variations in the first and second coordination sphere affect equally the magnetic properties. The influence of the counter cations on the spin Hamitonian parameters was also studied in the two salts $(Ph_4P)_2[Co(SPh)_4]$ (**17**; D_{2d} symmetry) and $(Et_4N)_2[Co(SPh)_4]$ (**17a**; S_4 symmetry). The characterization techniques employed were high-field/high-frequency EPR, multifield SQUID magnetometry, frequency domain Fourier-transform THz-EPR and variable-field variable-temperature magnetic circular dichroism [107]. The $[Co(SPh)_4]^{2-}$ anion in **17** shows strong axial magnetic anisotropy as already was known [104,105], whereas the anion in **17a** shows rhombic anisotropy with D = +11 cm^{-1} and E/D = ~0.20 [107]. It has been verified, also with the help of multireference ab initio calculations, that the differences observed in the two complexes are associated with slight changes of the S–Co–S bond angles and C–S–Co–S torsion angles around the {Co^{II}S$_4$} unit.

Another excellent experimental study on mononuclear Co(II) complexes is in agreement with the fact that ligands with heavier and softer main group donor atoms increase the magnetic anisotropy of the complexes, as evidenced by the increased $|D|$ values [108]. The reactions of CoI$_2$ and the monodentate ligands quinoline (qun) and Ph$_3$P in anhydrous EtOH, and Ph$_3$As in MeNO$_2$, all under refluxing conditions, give complexes $[CoI_2(qun)_2]$ (**20**), $[CoI_2(Ph_3P)_2]$ (**21**) and $[CoI_2(Ph_3As)_2]$ (**22**) in good yields. The crystal structures of the complexes reveal a pseudo-tetrahedral local coordination environment around the central Co^{II} atom. The D value was found to vary from +9.2 cm^{-1} in **20** to −37 cm^{-1} in **21** and −75 cm^{-1} in **22**. However, the dynamic properties reveal only a minor effect on the U_{eff} value. Compound **20** does not show an out-of-phase ac magnetic susceptibility signal under a zero or applied dc field; complexes **21** and **22** exhibit slow magnetization relaxation below 4 K under an applied dc field of 1000 Oe with U_{eff} values of ~21 and ~23 cm^{-1}, respectively. It is obvious that the observed increase in the energy barrier for the As-based complex **22** (~−21 cm^{-1}→~−23 cm^{-1}) is much smaller than the corresponding increase in the D value (~−37 cm^{-1}→~−75 cm^{-1}).

Analogous studies have been performed by the groups of Rajaraman and Shanmugam [109]. They used the exocyclic mesoionic ligands 2,3-diphenyl-1,2,3,4-tetrazolium-5-olate (L$_1$) and 2,3-diphenyl-1,2,3,4-tetrazolium-5-thiolate (L$_2$), whose general structural formula is shown in Figure 14. Mesoionic ligands contain dipolar 5- or 6-membered rings whose canonical resonance structures cannot be represented without any additional charges in them. Their coordination chemistry is almost completely unexplored. Such ligands offer flexibility which allows researchers to selectively change the coordinating substituents and, thus, to investigate the influence of the donor atoms on the magnetic anisotropy. The 1:1 reactions of CoX$_2$·2H$_2$O (X = Cl, Br) with L$_1$ or L$_2$ in MeOH provide access to blue complexes $[CoX_2(L_1)(MeCN)]$ (X = Cl, **23a**; X = Br, **23b**) or green $[CoX_2(L_2)(MeCN)]$ (X = Cl, **24a**; X = Br, **24b**) when crystallized from MeCN; the preparation of **23a** and **23b** is illustrated in Equation (13). The complexes are pseudotetrahedral, the donor atoms of monodentate L$_1$ and L$_2$ being the exocyclic oxygen and sulfur atoms, respectively. The D values, deduced from magnetization data, are +15.6 cm^{-1} (**23a**), +11.2 cm^{-1} (**23b**), −11.3 cm^{-1} (**24a**) and −10.3 cm^{-1} (**24b**). Thus, simple substitution of L$_1$ (O-donor in **23a** and **23b**) by L_2 (S-donor in **24a** and **24b**) switches the single-ion magnetic anisotropy parameter from positive to negative. All four complexes are field-induced SIMs with U_{eff} values of 10.3 cm^{-1} (**23a**), 8.2 cm^{-1} (**23b**), 20.2 cm^{-1} (**24a**) and 13.8 cm^{-1} (**24b**).

$$CoX_2 \cdot 6H_2O + L_1 + MeCN \xrightarrow[\Delta]{MeOH/MeCN} [CoX_2(L_1)(MeCN)] \tag{13}$$

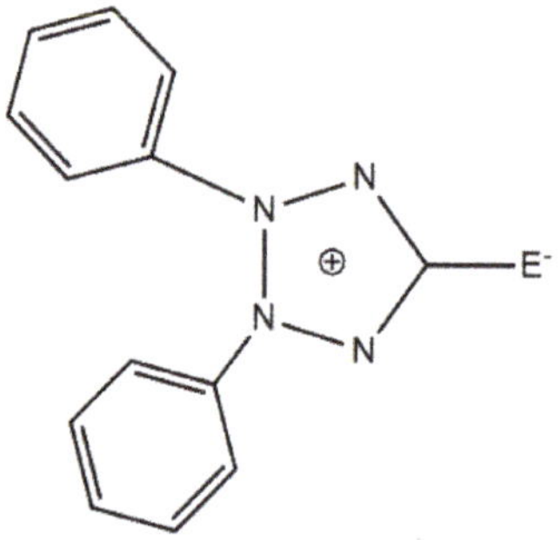

Figure 14. The general structural formula of the exocyclic mesoionic ligands L_1 (E = O) and L_2 (E = S).

8. Combined Metallacrown and Click Chemistry as a Tool for SMM Research

Metallacrowns (MCs) are analogues of crown ethers. They consist of a repeat unit of –{M'–N–O}$_n$– in a cyclic arrangement where the ring metal ion and the nitrogen atom replace the methylene carbon atoms of a crown ether. MCs are named and abbreviated on the basis of the ring size and the number of oxygen atoms that act as donors. For example, in the abbreviation 12-MC-4, MC represents a metallacrown that is a 12-membered ring comprising 4 repeating –{M'–N–O}– units with 4 donating oxygen atoms. The nomenclature/abbreviation also includes the bound central metal ion M, the ligand, and any bound or unbound ions. Thus, in the typical representation [MX{ring size-MC$_{M'Z(L)}$-ring oxygens}]Y, M is the bound central metal with its oxidation state, X is any bound anion, M' is the ring metal with its oxidation state, Z is the third heteroatom of the ring (usually N), L is the organic ligand of the complex, and Y is any unbound anion. Sometimes there are unbound cations, which are placed before the bound central metal M. An example of the above naming scheme is [Gd^{III}(NO$_3$)$_2${15-MC$_{Cu}{}^{II}{}_{N(picha)}$-5}](NO$_3$), where H$_2$picha is picoline hydroxamic acid. The molecular structure of the hexanuclear cation is shown in Figure 15. The –{M'–N–O}– repeat unit is now a quite general motif in inorganic chemistry and the connectivity has significantly grown to include a variety of bridges such as –{N–N}–, –{O–P}–, –{N–C–O}–, –{N–C–N}–, –{O–C–O}– and –{X}– (X is a nonmetal). After the initial great efforts to synthesize many MCs and to discover a great variety of structural types, the research activity in the past 15 years or so has shifted towards the use of these unique compounds for applications, e.g., for selective binding of cations or anions, as MRI contrast agents, in catalysis, as mimics of surface science, as building blocks for 1-, 2- and 3-dimensional solids, in liquid crystals and in various aspects of Molecular Magnetism [110]. Metallacrown chemistry has a brilliant potential for growth.

In addition to its use in inorganic chemistry (Section 4) as an anion, the azide functional group was widely used after the second world war in the synthesis of N-containing natural products and medicinal formulae. For example, the azide group is one of the most efficient amine precursors, and its ability to undergo 1,3-dipolar cycloadditions and diazo-transfer reactions is a valuable tool in organic synthesis. The [3 + 2] cycloadditions between azides and alkynes were first observed by Michael in 1893 and examined later by Huisgen [111]. The reaction has a high activation energy barrier and elevated temperatures or pressures are required to facilitate it. Almost 20 years ago, it was discovered that Cu(I) catalysis accelerates the rate of formal cycloaddition between azides and terminal alkynes, which proceeds at ambient temperatures and pressures affording 1,4-disubstitued 1,2,3-triazoles exclusively. This variant of the Huisgen cycloaddition has been termed CuAAC for Cu-catalyzed azide alkyne cycloaddition. CuAAC is an exciting example of "click chemistry", a term used by Sharpless and coworkers to describe a category of chemical reactions that link two components in high yields and with minimal byproducts [112].

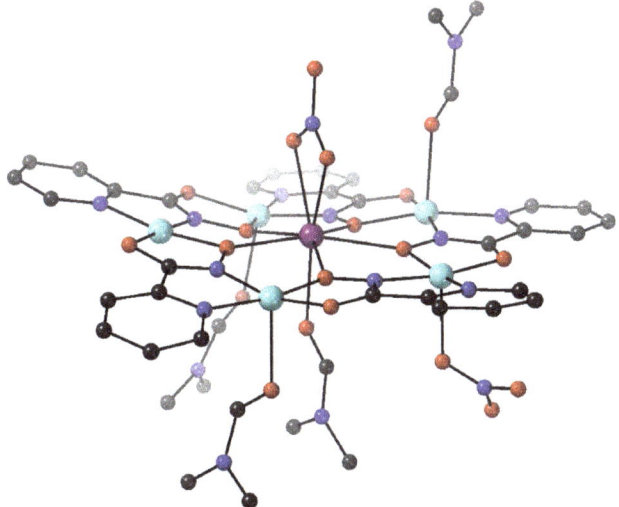

Figure 15. The molecular structure of the hexanuclear cation $[Gd^{III}Cu^{II}_5(NO_3)_2(picha)_5(DMF)_4]^+$ that is present in $[Gd^{III}(NO_3)_2\{15\text{-}MC_{Cu^{II}N(picha)}\text{-}5\}](NO_3)$. Color code: Gd^{III} dark mauve, Cu^{II} turquoise, O red, N dark blue, C, gray. Modified from reference [110].

The group of Rentschler has developed an approach which allows for the rational decoration of Cu(II) MCs with SIMs or SMMs using click chemistry [113]. The CuAAC is suitable for covalently linking magnetic building blocks which bear azide and terminal alkyne functional groups. Under mild conditions, the process leads selectively to a 1,4-disubstituted 1,2,3-triazole as a conjugated bridge which ensures communication between two or more magnetic building blocks. In order for the approach to be successful, several conditions must be satisfied: (i) To isolate thermodynamically stable complexes, polydentate chelating ligands are required. (ii) The number of introduced modified ligands must be limited to prevent an overloading of the complex. (iii) A short distance between the metal ions and the periphery of the ligands, as well as a conjugated π system are important to enable strong magnetic communication between the building blocks; and (iv) The latter should preserve their structural integrity in the reaction solvent. It is worth mentioning that this click concept can be further used in other challenging research areas of Molecular Magnetism, e.g., covalent anchoring of SMMs on surfaces, development of new cluster-based inorganic-organic hybrid materials, etc. We give an example of this strategy.

The two functionalized complexes are $(Me_4N)_2[Cu^{II}\{12\text{-}MC_{Cu(II)N(eshi)}\text{-}4\}]$ (**25**) [114] and $[Co^{II}(oda)(aterpy)]$ (**26**) [115]. Compound **25** was prepared by the high-yield reaction illustrated in Equation (14), where H_3eshi is 4-ethynylsalicylhydroxamic acid. In the centrosymmetric pentanuclear dianion

(Figure 16), each of the Cu^{II} atoms has an almost square planar coordination geometry. The peripheral four metal ions are bridged by N–O groups of the eshi^{3-} ligands forming a 12-membered metallacrown planar ring, which encapsulates the central Cu^{II} site through coordination of the μ_2 oxygen atoms of the N–O groups. The other two, terminally ligated oxygen atoms of each eshi^{3-} ligand complete coordination at neighboring Cu^{II} atoms. The four alkynyl groups protrude from the planar cluster dianion and each of them is orthogonal to its two neighbors. The complex exhibits an isolated $S = 1/2$ ground state with $J_1 = -158$ cm^{-1}, $J_2 = -65$ cm^{-1} and $g = 2.13$, adopting the Heisenberg Hamiltonian $\Sigma_{ij} (-2J_{ij} \cdot \hat{S}_i \cdot \hat{S}_j)$; J_1 and J_2 are the coupling constants corresponding to the radial magnetic interaction between the central Cu^{II} atom and the ring Cu^{II} atoms, and the tangential exchange interaction between two neighboring ring metal ions, respectively. The solution integrity of the pentanuclear dianion of **25** is proven by paramagnetic 1D and 2D ^{1}H NMR spectroscopy in d$_6$-DMSO and ESI-MS studies in a DMSO/MeOH solvent matrix. These results suggest that the 12-MC-4 Cu(II) complex is highly stable and thus suitable to perform CuAAC click reactions.

$$5\,[Cu_2(O_2CMe)_4(H_2O)_2] + 8\,H_3eshi + 4\,(Me_4N)(O_2CMe) \xrightarrow{MeOH}$$
$$2\,(MeN)_2[Cu_5(eshi)_4] + 24\,MeCO_2H + 10\,H_2O \tag{14}$$

Compound **26** was prepared [115] by the reaction illustrated in Equation (15) in ~30% yield; the neutral ligand aterpy is 4'-azido-2,2':6',2''-terpyridine and oda is the oxodiacetate(−2) ligand. Complex [Zn(oda)(terpy)] (**26a**) was prepared in a similar manner (yield ~50%). The two complexes are isomorphous. The metal ion is in an octahedral coordination environment with rhombic (C_2) distortion because of the rigidity of the planar, tridentate chelating, tripodal aterpy and oda^{2-} ligands (Figure 16). The chelate effect of the two ligands provides the complexes with high thermodynamic stability in DMSO, as proven by ESI-MS and ^{1}H NMR (for the diamagnetic complex **26a**) studies. Thus, both complexes are suitable for CuAAC click reactions. Ac susceptibility data under a static dc field of 0.15 T reveal that the mononuclear Co(II) complex is a weak SIM.

$$[Co^{II}(oda)(H_2O)_2] + aterpy \xrightarrow[\Delta]{CHCl_3/H_2O} [Co^{II}(oda)(aterpy)] + 2\,H_2O \tag{15}$$

The click reactions of **25** with **26** or **26a** in d$_6$-DMSO at 80 °C with copper(I) iodide as catalyst lead to complexes (Me$_4$N)$_2$[Cu^{II}{12-MC$_{Cu(II)N([M(II)(oda)(ttshi)]}$-4}] (**27**, M = Co; **27a**, M = Zn) [113,114], Equation (16) and Figure 15; H$_3$ttshi is 4-(2,2':6',2''-terpyridyl-1H-[1–3]-triazol-4-yl)salicylhydroxamic acid. The nature and structural type of the complexes were verified by IR and UV/VIS spectroscopic techniques, ESI-MS studies, as well as by paramagnetic ^{1}H NMR methods for the {Cu$_5$Zn$_4$} cluster **27a**. Magnetic studies of **27a** indicate fairly strong antiferromagnetic Cu^{II}···Cu^{II} exchange interactions and an $S = 1/2$ ground state, similar to the behavior of the precursor complex **25**. Orbital contributions and spin-orbit coupling effects from the additional Co^{II} atoms make difficult the analysis of the magnetic properties of **27**. The important fact, however, is that ac susceptibility measurements of this complex reveal a weak out-of-phase signal at low temperatures and low-field oscillation frequencies. This is the first time (and a very promising approach) where the CuAAC click chemistry has been used to upgrade the magnetic properties of a simple copper(II) metallacrown with remarkable magnetic features from the attached Co(II) SIM units, opening new doors for the creation of interesting molecular magnetic materials and providing a contribution toward the future development of SMM-based quantum computers [113]; the latter require linking of SMMs/SIMs in a deliberate manner, like the click chemistry approach presented here. Whatever the future form of quantum technology will be, it is likely that the role of chemists will be the design and optimization of molecules that couple to an external stimulus in the adequate energy rate, while offering some elementar functionality [37]. SMMs/SIMs, being much more versatile than magnetic atoms, and yet microscopic are among the quantum objects with the highest capacity to form non-trivial ordered states at the nanoscale and to be replicated in large numbers by means of chemical methods.

$$(Me_4N)_2[Cu_5(eshi)_4] + 4\,[Co^{II}(oda)(aterpy)] \xrightarrow[CuI]{DMSO,\,80\,°C} (Me_4N)_2[Cu_5Co^{II}_4(oda)_4(ttshi)_4] \quad (16)$$

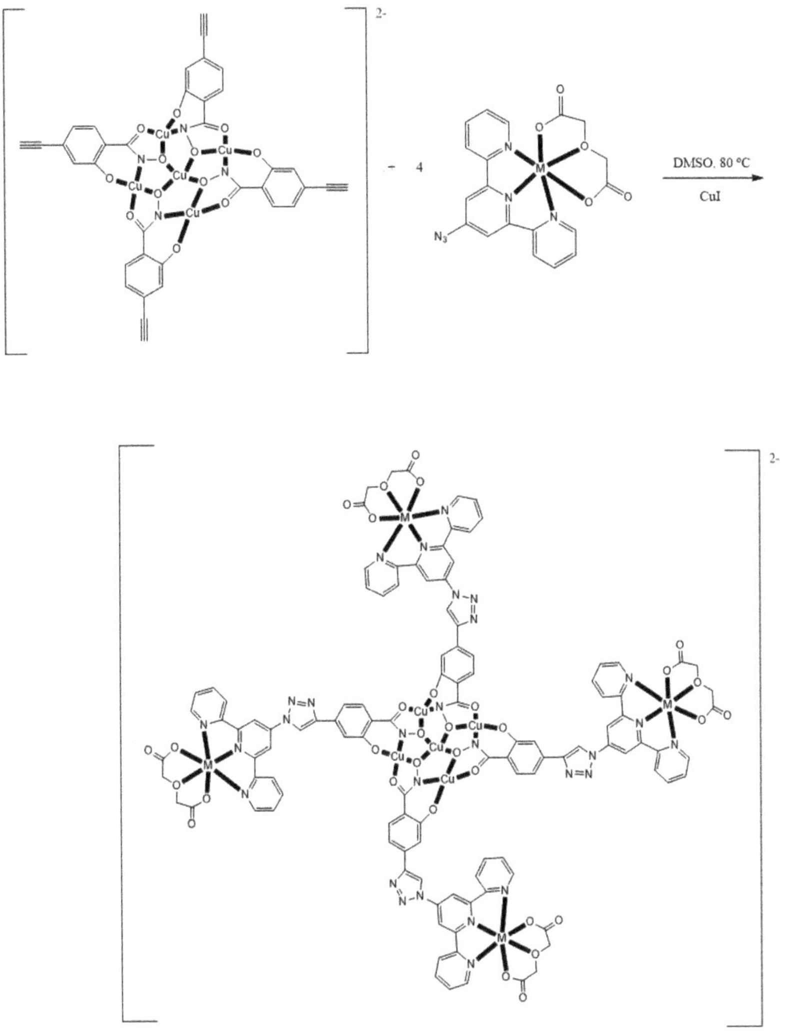

Figure 16. Representation of the CuAAC click reaction between **25** and **26/26a** that leads to the enneanuclear 3d/3d' heterometallic clusters **27/27a**. The cations Me$_4$N$^+$ are not shown. The coordination bonds are drawn with bold lines. M = Co, Zn.

9. Deprotonated 2-Pyridyl Alcohols: Central "Players" in the Chemistry of Transition-Metal SMMs

When developing routes and strategies for the synthesis of transition-metal SMMs, the design of bridging ligands is of paramount importance. In addition to bridging ability, the propagation of strong magnetic exchange interaction between the metal spin carriers and the simultaneous formation of chelating rings are highly desirable. The anions of 2-pyridyl alcohols (Figure 17) have proven to be versatile chelating and bridging groups that have yielded a number of 3d-metal, especially Mn, and 3d/4f-metal clusters with various structural motifs [116–133]. Moreover, the bridging deprotonated oxygen atom, i.e., the alkoxido group, often supports ferromagnetic coupling between the metal ions and has thus yielded polynuclear complexes with large S values and SMM properties. We give below examples from the use of such ligands in 3d-metal SMM chemistry.

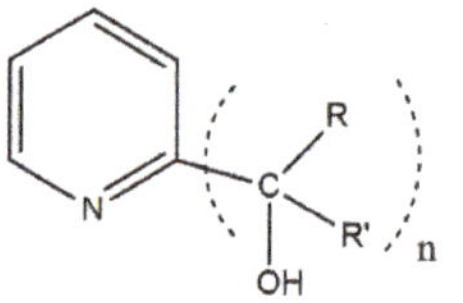

Figure 17. The general formula of the neutral 2-pyridyl alcohols; n can be 1 or 2, while R, R' are H and various non-donor organic groups (Me, Ph, ...).

The 1:1:1 NiCl$_2$·6H$_2$O/Hhmp/NaOR reaction mixtures in alcohols (MeOH, EtOH) give clusters [Ni$_4$(hmp)$_4$Cl$_4$(ROH)$_4$] (**28**, R = Me; **28a**, R = Et), Equation (17); hmp is the anion of 2-(hydroxymethyl)pyridine (R = R' = H and n = 1 in Figure 17). The tetranuclear cluster molecules possess a distorted cubane {Ni$_4$(μ_3-OR")$_4$}$^{4+}$ core, where R"– = (2-pyridyl)CH$_2$–, with the Ni^{II} atoms and the oxygen atoms from the 3.31 hmp$^-$ ligands (**B** with R = R' = H and n = 1 in Figure 4) occupying alternate vertices of the cube. A terminal chlorido ligand, an alcohol molecule and a 2-pyridyl nitrogen atom complete the octahedral coordination sphere of each metal ion [121]. Variable-temperature dc magnetic susceptibility studies indicate ferromagnetic Ni^{II}···Ni^{II} exchange interactions and a S = 4 ground state. Single-crystal high-frequency EPR spectra clearly suggest that each of the complexes has a total spin of 4 in the ground state with a negative D value. Magnetization vs. magnetic field measurements performed on single crystals with a micro-SQUID magnetometer show that these {Ni$_4$} clusters are SMMs. An appreciable exchange bias is evident in the hysteresis loops. The first resonant tunneling step is shifted considerably from zero field by virtue of intermolecular antiferromagnetic exchange interactions between the tetranuclear molecules of **28** and **28a** in their crystals [121].

$$4\ NiCl_2·6H_2O + 4\ Hhmp + 4\ NaOR \xrightarrow{ROH} [Ni_4(hmp)_4Cl_4(ROH)_4] + 4\ NaCl + 24\ H_2O \quad (17)$$

Deprotonated 2-pyridyl alcohols have contributed a lot into the development of Mn SMMs. For example, the reaction of Hhmp with Mn(O$_2$CPh)$_2$ in the presence of Et$_3$N in CH$_2$Cl$_2$/MeOH leads to the isolation of [Mn^{II}$_6$Mn^{III}$_{10}$O$_8$(OH)$_2$(O$_2$CPh)$_{12}$(hmp)$_{10}$(H$_2$O)$_2$](O$_2$CPh)$_2$ (**29**) in 30% yield, Equation (18). The core structure consists of a linked pair of complete cubanes {Mn^{II}$_2$Mn^{III}$_2$(μ_4-O)$_2$(μ_3-OR")$_2$}$^{4+}$, on either side of which is attached a tetrahedral {Mn^{II}Mn^{III}$_3$(μ_4-O)}$^{9+}$ unit. Among the ten hmp$^-$ groups, four are bridging within the central cubanes in a 3.31 mode (**B** with R = R' = H and n = 1 in Figure 4) and the other six are bridging within the outer tetrahedral units in a 2.21 mod (**C** with R = R' = H and n = 1 in Figure 4). The Mn ions are all six-coordinate with distorted octahedral geometry, except for the Mn^{II} atoms of the tetrahedral units which are 7-coordinate. Solid-state dc and ac magnetic susceptibility measurements on **29** establish that it possesses a S = 8 ground state. The complex displays frequency-dependent out-of-phase

(χ''_M) ac susceptibility signals below 3 K suggestive of SMM behavior. The D and U_{eff} value are −0.11 cm^{-1} and 7 cm^{-1}, respectively. Magnetization vs. applied dc field sweeps on single crystals of the complex down to 0.04 K exhibit hysteresis, confirming **29** to be a SMM, albeit weak [131]. Comparison of the structure of **29** ({Mn$_{16}$}) with {Mn$_{12}$} and {Mn$_6$} clusters obtained under the same reaction conditions but with two Me (R = R' = Me and n = 1 in Figure 17) or two Ph (R = R' = Ph and n = 1 in Figure 17) groups, respectively, added next to the alkoxido O atom of hmp$^-$ indicate their influence on the nuclearity and structure of the products [130,131] as being due to the overall bulk of the ligand plus the decreased ability of the deprotonated oxygen atom to bridge.

$$16\ Mn^{II}(O_2CPh)_2 + 10\ Hhmp + 5/2\ O_2 + 18\ Et_3N + 7\ H_2O \xrightarrow{CH_2Cl_2/MeOH}$$
$$[Mn_6^{II}Mn_{10}^{III}O_8(OH)_2(O_2CPh)_{12}(hmp)_{10}(H_2O)_2](O_2CPh)_2 + 18\ (Et_3NH)(O_2CPh) \quad (18)$$

The ligand Hhep (R = R' = H and n = 2 in Figure 17) also gives structurally and magnetically interesting clusters. The reaction of a 2:1 mixture of [Mn$^{III}_3$O(O$_2$CMe)$_6$(py)$_3$](ClO$_4$) and [Mn$^{II,III,III}_3$O(O$_2$CMe)$_6$(py)$_3$] with 4.5 equivs. of Hhep in MeCN affords cluster [Mn$^{II}_2$Mn$^{III}_{16}$O$_{14}$(O$_2$CMe)$_{18}$(hep)$_4$(Hhep)$_2$(H$_2$O)$_2$](ClO$_4$)$_2$ (**30**) in 20% yield. The core appears to be {Mn$_{18}$(μ_4-O)$_4$(μ_3-O)$_{10}$(μ_3-O$_{acetato}$)$_2$(μ_2-acetato)$_2$(μ_2-O$_{hep^-/Hhep}$)$_6$}$^{14+}$ [119,125]. It can be described as a central {Mn$_4$O$_6$} unit (containing a linear Mn$_4$ chain) linked by its μ_3-O^{2-} ions to two {Mn$_7$O$_9$} units, one on each side. Each of the heptanuclear units comprises a face-sharing set of one {Mn$_4$O$_4$} cubane and two {Mn$_3$O$_4$} partial cubanes. All the metal ions are six-coordinate. The hep$^-$ and Hhep groups behave as 2.21 (**C** with R = R' = H and n = 2 for hep$^-$ and **D** for Hhep in Figure 4) ligands. Fitting of magnetization data establish that **30** possesses a total spin of 13 in the ground state and a D value of -13 cm^{-1}. The complex is SMM (U_{eff} = 15 cm^{-1}), and this is confirmed by the appearance of hysteresis loops in magnetization vs. dc field sweeps on a single crystal. Below 0.2 K, the relaxation becomes temperature-independent, consistent with relaxation only by QTM through the anisotropy barrier via the lowest-energy M_S = ±13 levels of the S = 13 spin manifold. Although the high nuclearity of the complex is mostly due to the presence of 4.4- and 3.3-O^{2-} groups, as well as 4.31, 2.21 and 2.11 MeCO$_2^-$ ligands, the 2.21 hep$^-$ and Hhep moieties certainly contribute into its interesting magnetic properties.

2-pyridyl alcohols can be chiral leading to 3d-metal clusters with unprecedented structural motifs and interesting magnetic properties. The ligand α-methyl-2-pyridine-methanol (Hmpm; R = H, R' = Me and n = 1 in Figure 17) presents similar coordination features to Hhmp, but offers slightly different steric and electronic effects. The Hmpm ligand can be prepared via the reduction of 2-acetylpyridine by NaBH$_4$ [133]. The 2:1:2 reaction between Mn(O$_2$CPh)$_2$·2H$_2$O, rac-Hmpm and Et$_3$N in MeOH gives a deep red solution, which upon slow solvent evaporation at room temperature affords cluster [Mn$^{II}_2$Mn$^{III}_{28}$Mn$^{IV}_4$(OH)$_2$(OMe)$_{24}$O$_{24}$(O$_2$CPh)$_{16}$(rac-mpm)$_2$] (**31**) in ~30% yield, Equation (19). The core (Figure 18, left) can be described as a consecutive array of edge-sharing {Mn$_4$(μ_4-O)} tetrahedra and {Mn$_3$(μ_3-O)} triangles that are linked to each other via bridging MeO$^-$ and O^{2-} groups [133]. An alternative description of the core is as consisting of seven parallel layers (Figure 18, right) of four types (A, B, C, D) with an ABCDCBA arrangement. Layers A and B are simple Mn^{III} monomeric and {Mn$^{III}_4$} butterfly subunits, respectively, attached to each other through a 4.4 oxido group. Layer C is a {Mn^{II}Mn$^{III}_3$} cluster moiety containing three edge-sharing {Mn$_3$} triangles. Layer D comprises a {Mn$^{III}_8$Mn$^{IV}_4$} rod-like cluster unit that can be further seen as a central, planar disk-like {Mn$^{III}_6$Mn$^{IV}_4$} moiety with two additional Mn^{III} atoms above and below the disk. The layers are held together by a combination of bridging methoxido and oxido groups. The rac-mpm$^-$ groups behave as 2.21 ligands (**C** with R = H, R' = Me and n = 1 in Figure 4). The molecule is spherical with a diameter of ~2.5 nm. The cluster is: (i) One of the largest 3d-metal clusters, (ii) the second highest-nuclearity Mn complex containing an odd number of metal centers and the second cluster with a nuclearity of 31, and (iii) the largest SMM (S = 23/2, U_{eff} = ~40 cm^{-1}) that possesses entirely resolved out-of-phase ac peaks and magnetization hysteresis loops below 5 K.

$$31\ Mn^{II}(O_2CPh)_2 \cdot 2H_2O + 2\ rac\text{-}Hmpm + 46\ Et_3N + 15/2 O_2 + 24\ MeOH \xrightarrow{MeOH}$$
$$[Mn_2^{II}Mn_{28}^{III}Mn^{IV}(OH)_2(OMe)_{24}O_{24}(O_2CPh)_{16}(rac\text{-}mmp)_2] + 46\ (Et_3NH)(O_2CPh) + 51\ H_2O \quad (19)$$

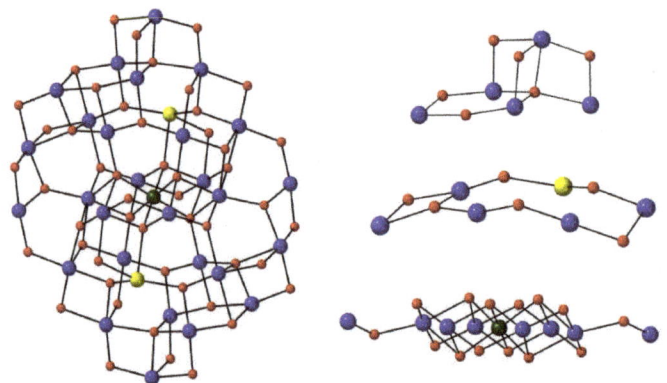

Figure 18. The metal–oxygen core (left) and the four types of constituent layers (right) along the crystallographic c axis. H atoms have been omitted for clarity. Color code: Mn^{II} yellow, Mn^{III} blue, Mn^{IV} olive green, O red. Modified from reference [133].

10. Deprotonated Aliphatic Diols: Simple and Efficient Ligands in the Chemistry of 3d-Metal SMMs

Molecules with two hydroxyl groups on different aliphatic carbon atoms have some interesting chemical characteristics [134,135] which are more prominent for vicinal diols (*vic*-diols). In the last 15 years or so, there has been an intense interest in the use of diols in the chemistry of transition-metal clusters and SMMs [136]. The simplistic rationale here is that each deprotonated alkoxido oxygen atom can bridge two or three metal ions and the latter can be linked together into polymetallic arrays with hopefully large ground-state spins, anisotropies and SMM properties. The structures of the products depend on (i) the level of deprotonation of the ligand (singly or doubly deprotonated), (ii) the existence of other donor groups on the diol (e.g., a pyridyl group, –NH$_2$, –COOH, ...), and (iii) the presence of other bridging co-ligands in the reaction system. A family of simple aliphatic diols, with no other donor groups, consist of 1,3-propanediol and its derivatives (Figure 19). We give below one notable example from the use of a member of this family in Mn SMM chemistry. The cited complexes belong to the class of giant homometallic 3d clusters (nuclearities higher than 30) with the metals in moderate oxidation states. Such giant clusters are of great interest not only for their impressive structures with nano dimensions, but also because they often display interesting magnetic properties (including SMM behavior). In addition, they possess the properties of both classical and quantum world, thus giving the opportunity for the discovery of new physical phenomena and the deeper understanding of the existing ones [137].

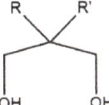

Figure 19. 1,3-propanediol (H$_2$pd; R = R' = H) and its derivatives.

The 1:4:1 reaction of [Mn$^{II,III,III}_3$O(O$_2$CMe)$_6$(py)$_3$] with H$_2$pd and NaN$_3$ in MeCN leads to cluster [{Mn$^{II}_2$Mn$^{III}_8$NaO$_2$(O$_2$CMe)$_{13}$(pd)$_6$(py)$_2$}$_4$] (**32**) in 35% yield [138,139], Equation (20). Since the giant cluster contains Na$^+$ but not N$_3^-$, it has been assumed that NaN$_3$ is important as the source of Na$^+$ and perhaps to provide additional weak base (N$_3^-$) for ligand deprotonation and further oxide ion formation; however, the main sources for ligand deprotonation are the MeCO$_2^-$ ions, as shown in Equation (20), and/or the pyridine molecules. This is supported by the same reactions, but using NaCN, NaOCN, NaSCN or Na{N(CN)$_2$} instead of NaN$_3$, which all give **32** in slightly lower yields (18–26%). The molecule of **32** consists of four {Mn$^{II}_2$Mn$^{III}_8$} loops linked through four Na$^+$ ions to give a supramolecular aggregate with a saddle-like topology; one loop with one Na$^+$ ion is shown in Figure 20. The Mn ions are all in a distorted octahedral coordination environment; the eight high-spin 3d^4 Mn^{III} centers display the Jahn–Teller axial elongation, but the Jahn–Teller axes are not co-parallel. Each loop consists of two triangular {Mn$^{III}_3$(μ$_3$-O)}$^{7+}$ and two dinuclear {Mn^{II}Mn^{III}} subunits linked by the oxygen atoms of the six 3.22 pd^{2-} groups (**E** with R = R' = H in Figure 4), as well as by 2.11 and 4.22 MeCO$_2^-$ ligands. The triangular subunits are connected by two pd^{2-} oxygen atoms, whereas the dinuclear subunits are connected by two pd^{2-} oxygen atoms and a 2.11 MeCO$_2^-$ ligand. The Mn^{II} and Mn^{III} atoms within each dinuclear subunit are bridged by a 2.11 acetato group, a pd^{2-} oxygen atom and an oxygen atom from a 4.22 MeCO$_2^-$ ligand. The Mn^{III} atoms of each triangular subunit are bridged by a 3.3 O^{2-} group, two oxygen atoms from two different pd^{2-} groups, one oxygen atom from a 4.22 MeCO$_2^-$ ligand and two 3.21 MeCO$_2^-$ ligands. The latter and an additional acetato group link each triangular subunit to a Na$^+$ ion; the two alkali metal ions attached to the decanuclear {Mn$^{II}_2$Mn$^{III}_8$} loop connect it in an equivalent manner to a neighboring loop giving a giant loop-of-loops aggregate. Four of the pd^{2-} groups bridge one Mn^{II} and two Mn^{III} centers, and the remaining two bridge three Mn^{III} ions. The crystal structure shows that the {Mn$^{II}_8$Mn$^{III}_{32}$Na$_4$} aggregates pack as tail-to-tail {Mn$^{II}_8$Mn$^{III}_{32}$Na$_4$}$_2$ dimers, thus leading to egg-shaped stacks. A detailed magnetic study of **32** reveals that each decanuclear loop has a $S = 4$ ground-state spin, and displays frequency-dependent in-phase and out-of-phase ac magnetic susceptibility signals. The aggregate also exhibits hysteresis loops. The hysteresis loops are not typical of SMM behavior because of the presence of interloop exchange interactions through the diamagnetic Na$^+$ ions, and also intermolecular interactions between different {Mn$^{II}_8$Mn$^{III}_{32}$Na$_4$} aggregates [137,139].

$$40\ [\text{Mn}_3^{II,III,III}\text{O}(\text{O}_2\text{CMe})_6(\text{py})_3] + 72\ \text{H}_2\text{pd} + 12\ \text{NaN}_3 + 4\ \text{O}_2 \xrightarrow{\text{MeCN}}$$
$$3\ [\{\text{Mn}_2^{II}\text{Mn}_8^{III}\text{NaO}_2(\text{O}_2\text{CMe})_{13}(\text{pd})_6(\text{py})_2\}_4] + 84\ \text{MeCO}_2\text{H} + 12\ \text{HN}_3 + 96\ \text{py} + 24\ \text{H}_2\text{O} \quad (20)$$

Once the heterometallic (Na/Mn) character of **32** and its influence on the magnetic properties had been studied, the synthesis of the magnetically discrete, homometallic {Mn$_{44}$} analogue of **32** was sought, as a means of strengthening the interloop exchange interactions and appearance of better SMM properties. The desired product [Mn$^{II}_{12}$Mn$^{III}_{32}$O$_8$(O$_2$CMe)$_{52}$(pd)$_{24}$(py)$_8$]$^{4+}$ was isolated [139] from the 1:5:1 [Mn$^{II,III,III}_3$O(py)$_3$]/H$_2$pd/Mn^{II}(ClO$_4$)$_2$·6H$_2$O reaction mixture in CH$_2$Cl$_2$, Equation (21), i.e., with the use of Mn^{II}(ClO$_4$)$_2$·6H$_2$O instead of NaN$_3$. The product is, as expected, cationic with the formula [Mn$^{II}_{12}$Mn$^{III}_{32}$O$_8$(O$_2$CMe)$_{52}$(pd)$_{24}$(py)$_8$](OH)(ClO$_4$)$_3$ (**33**) and its yield is ~25%. The molecular structure of the cation of **33** is very similar to that of the molecule of **32**, the main difference being the fact that the

{Mn^{II}$_2$Mn^{III}$_8$} loops in the former are linked through Mn^{II} ions, whereas those of the latter by Na$^+$ ions; as a result the {Mn$_{44}$} cluster is positively charged, whereas the {Mn$_{40}$Na$_4$} aggregate is neutral. The packing of **33** is similar to that in **32**, and thus supramolecular {Mn$_{44}$}$_2$ dimers are formed in the crystal. In accord with the strongest magnetic exchange interactions between the four {Mn^{II}$_2$Mn^{III}$_8$} loops mediated through the connecting Mn^{II} centers, magnetic susceptibility studies reveal that **33** has a $S = 6$ ground-state spin and displays frequency-dependent in-phase and out-of-phase signals. Magnetization vs. dc magnetic field sweeps on single crystals of **33** display hysteresis loops below 0.7 K whose coercivities increase with decreasing temperature and with increasing sweep rate, confirming that this giant cluster is one of the largest Mn SMMs.

$$40\ [\text{Mn}_3^{\text{II,III,III}}\text{O}(\text{O}_2\text{CMe})_6(\text{py})_3] + 72\ \text{H}_2\text{pd} + 12\ \text{Mn}^{II}(\text{ClO}_4)_2\cdot 6\text{H}_2\text{O} + 4\ \text{O}_2 \xrightarrow{\text{CH}_2\text{Cl}_2}$$
$$3\ [\text{Mn}_{12}^{II}\text{Mn}_{32}^{III}\text{O}_8(\text{O}_2\text{CMe})_{52}(\text{pd})_{24}(\text{py})_8](\text{OH})(\text{ClO}_4)_3 + 84\ \text{MeCO}_2\text{H} + 15\ (\text{pyH})(\text{ClO}_4) + 81\ \text{py} + 93\ \text{H}_2\text{O} \tag{21}$$

The ligand 2-methyl-1,3-propanediol (H$_2$mpd; R = H and R' = Me in Figure 19) gives the heterometallic aggregate [{Mn^{II}$_2$Mn^{III}$_8$NaO$_2$(O$_2$CMe)$_{13}$(mpd)$_6$(py)$_2$}$_4$] (**34**) [139], which is structurally and magnetically similar to **32**. The ligand pd^{2-} has been also used by Tasiopoulos' group to construct giant Mn/Co^{II} and Mn/Ni^{II} clusters with exciting structures and large ground-state spins, but with no SMM properties [140,141]. For example, cluster [Mn^{II}$_8$Mn^{III}$_{28}$Ni^{II}$_4$O$_{12}$Cl$_{10}$(O$_2$CMe)$_{26}$(pd)$_{24}$(py)$_4$(H$_2$O)$_2$] (**35**) possesses an unprecedented "loop-of loops-and-supertetrahedra" structural topology and displays a high ground-state spin state value of 26 ± 1 [141].

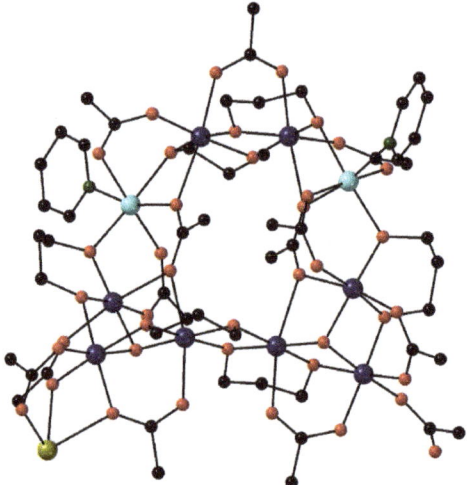

Figure 20. The structure of one {Mn^{II}$_2$Mn^{III}$_8$NaO$_2$(O$_2$CMe)$_{13}$(pd)$_6$(py)$_2$} loop that is present in **32**. Color code: Mn^{II} cyan, Mn^{III} dark blue, Na$^+$ dark yellow, O red, N green. Modified from reference [138].

11. Molecular and Supramolecular Approaches in the Chemistry of Manganese SMMs Using Simple and Elaborate 2-Pyridyl Oximes

The deprotonated oxime (oximate, R$_2$CNO$^-$) group has played an important role in the chemistry of 3d-metal clusters and SMMs. This diatomic group can bridge up to three metal ions and often propagates ferromagnetic exchange interactions; the latter property may lead to SMM properties. In most cases, the oxime group is part of an organic ligand that possesses one or more other donor groups, often in a position that enables the formation of a stable chelating ring with the participation of the oximate nitrogen atom. The two most studied families of oxime groups in Molecular Magnetism are the salicyl aldo(keto)ximes and the 2-pyridyl aldo(keto)ximes, Figure 21. The former [142–155] have led to a variety of clusters with exciting molecular structures and magnetic properties; they have been used, among others, in the development of a synthetic process widely known as "ground-state spin switching and enhancing SMM properties via targeted structural distortion" strategy [56,142–145,147–149]. The latter [156–174], in addition to their involvement in the synthesis of 3d-metal SIMs [158], have provided access to interesting 3d-metal clusters and SMMs. They have also been used in the realization of a synthetic scheme best known as "'switching on' SMM properties upon conversion of low-spin complexes into high-spin ones without changing the core" strategy [168,169] and, when derivatized by design, in the development of an innovative approach which allows the covalent linking of SMMs by applying principles of supramolecular chemistry [172–174]. Below we describe briefly the molecular and supramolecular approaches in the chemistry of Mn SMMs by using 2-pyridyl ketoximes and its derivatives as key synthetic tools.

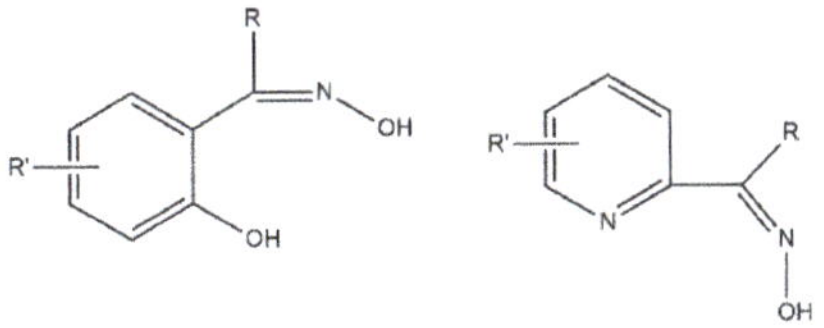

Figure 21. Salicyl aldo(keto)ximes (left) and 2-pyridyl aldo(keto)ximes (right), two families of ligands that have widely been used in the chemistry of transition-metal SMMs; R and R' are various non-donor groups. When R = H the ligands are aldoximes and when R ≠ H the products are ketoximes.

An in-depth studied family of Mn(III) carboxylate clusters consists of the triply oxido-bridged, triangular complexes [Mn$_3$O(O$_2$CR")$_6$L$_3$]X (R" = Me, Et, Ph, ... ; X = monoanionic counterions; L = neutral monodentate ligands) [175]. These complexes possess the {Mn$^{III}_3$(μ_3-O)}$^{7+}$ core with peripheral ligation provided by 2.11 carboxylato groups and terminal L ligands. The triangular cations are characterized by antiferromagnetic Mn^{III}···Mn^{III} exchange interactions and have low S values in the ground state; they are not thus SMMs. It was a general belief that this common triangular topology could never give complexes with SMM properties. Using 2-pyridyl ketoximes, however, these complexes can be converted into triangular clusters with the same {Mn$_3$(μ_3-O)}$^{7+}$ core, but with ferromagnetic Mn^{III}···Mn^{III} interactions. The strategy is illustrated in Figure 22. The 1:3 reaction between [Mn$_3$O(O$_2$CR")$_6$(py)$_3$](ClO$_4$) (R" = Me, **36a**; R" = Et, **36b**; R" = Ph, **36c**, see Figure 22, left) and methyl 2-pyridyl ketoxime (Hmpko; R = Me and R' = H in Figure 21) in MeCN/MeOH give dark brown solutions; evaporation of the reaction solutions to dryness and crystallization of the residues from CH$_2$Cl$_2$/n-hexane give dark red crystals of [Mn$_3$O(O$_2$CR")$_3$(mpko)$_3$](ClO$_4$) (R" = Me, **37a**; R" = Et, **37b**; R" = Ph, **37c**) in high yields (>80%), Equation (22) [168]. The 1:3 molar ratio of the reactants was chosen to allow for the incorporation of one mpko$^-$ ligand onto each edge of the {Mn$_3$(μ_3-O)}$^{7+}$ core. The reaction can thus be described as a simple ligand substitution with the replacement of three R"CO$_2^-$ groups and three py ligands by three mpko$^-$ ones, without change of the Mn oxidation level. This reaction scheme is quite general and can be extended to other carboxylate groups and 2-pyridyl ketoximes (Figure 21), where R is a non-donor group.

$$[Mn_3O(O_2CR")_6(py)_3](ClO_4) + 3\,Hmpko \xrightarrow{MeCN/MeOH}$$
$$[Mn_3O(O_2CR")_3(mpko)_3](ClO_4) + 3\,R"CO_2H + 3\,py \quad (22)$$

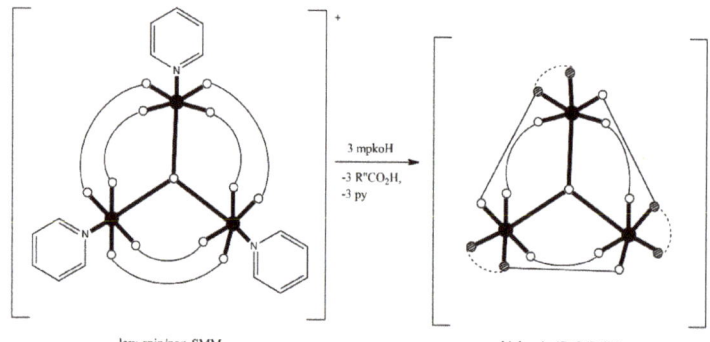

Figure 22. Schematic drawing of the conversion of the low-spin cations of **36a**, **36b**, **36c** to the high-spin ($S = 6$) SMM cations **37a**, **37b**, **37c**. The large solid (•) and the single small open (o) circles represent the Mn^{III} centers and oxido (O^{2-}) groups, respectively. The curved solid lines represent triatomic carboxylate groups. The mixed dashed/solid lines represent the N,N (small dashed circles), O (small open circle) donor sets of the 2.111 mpko$^-$ ligands. The coordination bonds are drawn with bold lines.

As in **36a**, **36b**, **36c**, the cations of **37a**, **37b**, **37c** (Figure 22, right) possess a {Mn$^{III}_3$(μ_3-O)}$^{7+}$ triangular core, but each Mn$_2$ edge is now bridged by a 2.11 R″CO$_2^-$ and a 2.111 mpko$^-$ (**G** with R = Me and R′ = H in Figure 4) group. The three R″CO$_2^-$ groups lie on one side of the plane defined by the Mn^{III} atoms and the three oximato groups on the other. The tridentate ligation mode of mpko$^-$ causes a buckling of the formerly planar {Mn$^{III}_3$(μ_3-O)}$^{7+}$ unit, giving rise to a relative twisting of the octahedra of the metal ions, a slight non-planarity of the Mn^{III}–N–O–Mn^{III} units and a displacement of the central oxido group which is ~0.3 Å above the {Mn$^{III}_3$} plane on the same side as the carboxylate groups. These structural distortions lead to ferromagnetic Mn^{III}···Mn^{III} interactions resulting in a $S = 6$ ground state. Fits of dc magnetization data collected in the 10–70 KG and 1.8–10.0 K ranges confirm the ground-state spin and give a D value of ~−0.35 cm^{-1}. Complexes **37** exhibit frequency-dependent out-of-phase (χ''_M) ac magnetic susceptibility signals suggesting a possible SMM behavior. Relaxation rate vs. T data down to 1.8 K obtained from the χ''_M vs. T studies were supplemented with rate vs. T data measured to 0.04 K via magnetization vs. time decay studies, and these were used to construct Arrhenius plots from which U_{eff} values of ~8 cm^{-1} were derived. Magnetization vs. dc field sweeps on single crystals of **37a**·3CH$_2$Cl$_2$ show hysteresis loops which exhibit steps due to QTM. The loops are temperature-independent below 0.3 K and this indicates only ground-state QTM between the M_S = ±6 levels. Complexes **37** were the first confirmed triangular SMMs of any transition metal. High-frequency EPR spectra of single crystals of **37a**·3CH$_2$Cl$_2$ give $D = -0.3$ cm^{-1} and provide evidence of a rather significant transverse anisotropy ($|E| \geq 0.015$ cm^{-1}). DFT calculations provide strong evidence [169] that the unusual ferromagnetic exchange interactions in complexes **37** originate from a combination of several factors including the non-planarity (with respect to the {Mn$^{III}_3$} plane) of the bridging oximato groups, the non-parallel alignment of the Jahn–Teller axes and the shift of the triply-bridging oxido group out of the plane defined by the three Mn^{III} ions. The above results demonstrate that structural distortions of a magnetic core imposed by peripheral ligands (deprotonated 2-pyridyl oximes in this case) can "switch on" SMM properties [168].

Ligands containing two 2-pyridyl ketoxime moieties are of special interest in the realm of the linking of SMMs [172–174]. Since SMMs have been proposed as qubits for quantum information processes and as components in molecular spintronics, a great challenge is their quantum mechanical coupling to each other or to other components of a device, while maintaining the intrinsic properties of each SMM; this maintenance requires very weak coupling between the SMM precursors. Excluding utilization of hydrogen bonds, with which it is difficult to control oligomerization and to achieve retention of the supramolecular structure in solution, the best solution is the designed linking of SMMs via coordination bonds. The groups of Papaefstathiou, Escuer, Brechin and Christou [150–153,170–174], among others, have used building-block strategies to link SMMs together employing carefully chosen linkers that provide inter-SMM interactions and ensure the formation of discrete oligomeric (and not polymeric) species.

The SMM cations of **37a**, **37b** and **37c** (Figure 22, right) are excellent candidates to be used as building blocks for such a strategy. They have their R"CO$_2^-$ and mpko$^-$ ligands on opposite sites of the {Mn$^{III}_3$} plane. The tripodal arrangement of the carboxylato and oximato groups suggests that their replacement with dicarboxylato [170] or bis(2-pyridyl) dioximato groups [172–174], respectively, could lead to discrete aggregates (oligomers) rather than polymeric complexes. We give an example in which a bis(2-pyridyl) dioxime provides the inter-SMM linkage.

The ligand of choice was H$_2$dpd (Figure 23) and its selection was based on principles of the supramolecular chemistry field [174]. The molecule can be seen as a fusion of two Hmpko units (Figure 21, right, with R = Me and R' = H) at the Me group. The single sp^3 central carbon atom reduces the conformational flexibility, and according to the directional bonding approach of supramolecular chemistry, the combination of a tritopic {Mn$^{III}_3$} unit with a ~109° ditopic dioximate should give a {Mn$^{III}_3$}$_2$ "dimer" with three linkers and parallel {Mn$^{III}_3$} planes. It was also expected that the coordination by tridentate 2-pyridyloximate groups of three dpd^{2-} ligands would give a rigidity in the resulting dimeric product, a favorable fact for the retention of the structure in solution. All these design principles turned out to be successful, Equation (23). The I$_3^-$ counterions, which are present in the product [Mn$_6$O$_2$(O$_2$CMe)$_6$(dpd)$_3$](I$_3$)$_2$ (**38**), come from I$_2$ in the reaction mixture, probably from reductive agents, e.g., EtOH, through the 2e$^-$ + 3 I$_2$ → 2 I$_3^-$ process.

$$2\,[\text{Mn}_3^{III}\text{O}(\text{O}_2\text{CMe})_6(\text{py})_3](\text{ClO}_4) + 3\,\text{H}_2\text{dpd} + 2\,\text{H}^+ + 2\,\text{I}_3^- \xrightarrow{\text{CH}_2\text{Cl}_2/\text{EtOH}}$$
$$[\text{Mn}_6^{III}\text{O}_2(\text{O}_2\text{CMe})_6(\text{dpd})_3](\text{I}_3)_2 + 2\,(\text{pyH})(\text{ClO}_4) + 4\,(\text{pyH})(\text{O}_2\text{CMe}) + 2\,\text{MeCO}_2\text{H} \tag{23}$$

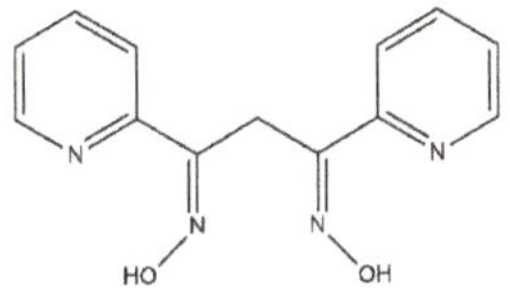

H$_2$dpd

Figure 23. The free ligand 1,3-di(pyridin-2-yl)propane-1,3-dione dioxime which has been used for the synthesis of a covalently linked dimer of {Mn$^{III}_3$} SMMs.

At targeted by the selection of H$_2$dpd, the structure of the hexanuclear dication consists of two {Mn$^{III}_3$(μ$_3$-O)}$^{7+}$ subunits connected by three 4.111111 dpd^{2-} ligands (**H** in Figure 5) to give a {Mn$^{III}_3$} "dimer" of D$_3$ symmetry; the two {Mn$^{III}_3$} planes are thus parallel. Each triangular subunit is structurally very similar to that of the "monomer" **37a**. Solid-state dc and ac magnetic susceptibility studies show that each {Mn$^{III}_3$} subunit of the "dimer" is a separate SMM with an S = 6 ground state and that the two subunits are very weakly ferromagnetically exchange-coupled. Single-crystal high-frequency EPR spectra on **38** display signal splittings suggesting quantum superposition/entanglement of the two SMM subunits. Remarkably, the same spectral behavior is observed in MeCN/toluene (1:1 v/v) frozen solutions, indicating that the structure of the "dimer" is retained in solution and the weak inter-SMM interaction persists. This work proves that the synthesis of covalently linked oligomers of exchange-coupled SMMs is feasible, with a careful ligand design, and the products can retain their oligomeric nature and inter-SMM quantum mechanical coupling in solution. These results provide scientists with a good background as efforts of using solution methods for deposition of SMMs on surfaces and other substrates continue.

12. Concluding Comments and Brief Prognosis for the Future

We hope that this review has provided the readers with a taste of the synthetic chemistry and reactivity studies of 3d-, 4d- and 5d-metal SIMs and SMMs, with emphasis on some ligands or families of ligands that have been used successfully in this area. Other authors could have selected other ligands from a plethora available. We do believe that some of the ligands discussed, e.g., the azido and cyanido groups, ligands containing soft donor atoms, aliphatic diols and oximate-based ligands, are promising for further developments.

Although the current interest in this field is shifted to f-elements, it is our opinion that the chemistry of transition-metal SMMs and SIMs has brilliant perspectives. In the SMM area, the major advantage of using d-block metal ions is their ability to create strongly coupled systems; this is in contrast to the situation with the lanthanoid ions (with the exception of radical-bridged 4f-metal SMMs [176]) where the core-like character of the 4f orbitals prohibits this. The d-block SIM chemistry appears to grow exponentially and it is striking that few of the compounds reported to date are SIMs in zero field. One approach to increase the number of zero-field d-metal SIMs is to design ligand-field environments, which can preserve strict axial symmetry around the metal ion. In both areas, and from a synthetic inorganic chemistry point of view, the synthesis of 4d- and 5d-metal SIM and SMM systems is expected to gain more and more attention. The spin-orbit coupling constants of these metal ions are larger than their first-row counterparts, and this can lead to improved SIM properties. Moreover, the increased radial extension of the 4d and 5d orbitals gives the possibility for stronger magnetic exchange interactions, a crucial consideration in the design of efficient SMMs.

Regardless of how the chemistry of SIMs and SMMs develops over the next few years, one thing is certain: the design and synthesis of new ligands, and the proper use of existing ones will remain to the fore.

Concluding, we hope that synthetic inorganic chemists active in the transition-metal SIM and SMM areas or scientists who just enter into this field will find this review useful. We shall be happy if the readers enjoy the review as much we enjoyed writing it.

Author Contributions: P.S.P., D.M. and E.P. studied the literature and proposed some of the cited examples. P.S.P. and E.K. proposed some ligand systems. D.M. prepared the figures and typed the manuscript. S.P.P. developed the concept of the review and wrote the manuscript jointly with E.K. All authors have read and agreed to the published version of the manuscript.

Funding: This research was funded by the research program THALES, grand number 377365, D. 533.

Acknowledgments: This work has been supported by the European Union (European Social Fund-ESF) and Greek National funds through the Operational Program "Educational and Lifelong Learning" of the National Strategic Reference Framework (NSRF)—Research Funding Programs: THALES: "Investigating in knowledge society through the European Social Fund (to S.P.P. and E.K.).

Conflicts of Interest: The authors declare no conflict of interest.

References

1. Kahn, O. *Molecular Magnetism*; Wiley-VCH: New York, NY, USA, 1993.
2. Housecroft, C.E.; Sharpe, A.G. *Inorganic Chemistry*, 5th ed.; Pearson: Harlow, UK, 2018; pp. 661–795.
3. Miessler, G.L.; Fischer, P.J.; Tarr, D.A. *Inorganic Chemistry*, 5th ed.; Pearson: Boston, MA, USA, 2014; pp. 313–470.

4. Murrie, M.; Price, D.J. Molecular Magnetism. *Annu. Rep. Prog. Chem. Sect.* **2007**, *103*, 20–38. [CrossRef]
5. Kahn, O. Dinuclear complexes with predictable magnetic properties. *Angew. Chem. Int. Ed. Engl.* **1985**, *24*, 834–850.
6. Coronado, E.; Drillon, M.; Nugteren, P.R.; de Jongh, L.J.; Beltran, D.; Georges, R. Low-Temperature investigation of the ferrimagnetic chains MnM′(EDTA)·6H$_2$O (M′ = Co, Ni, and Cu(II)): Thermal and magnetic properties. *J. Am. Chem. Soc.* **1989**, *111*, 3874–3880.
7. Caneschi, A.; Gatteschi, D.; Sessoli, R.; Rey, P. Toward molecular magnets: The metal-radical approach. *Acc. Chem. Res.* **1989**, *22*, 392–398. [CrossRef]
8. Stumpf, H.O.; Quahab, L.; Pei, Y.; Grandjean, D.; Kahn, O. A molecular-based magnet with a fully interlocked 3-dimensional structure. *Science* **1993**, *261*, 447–449. [CrossRef]
9. Ferlay, S.; Mallah, T.; Quahes, R.; Veillet, P.; Verdaguer, M. A room-temperature organometallic magnet based on Prussian blue. *Nature* **1995**, *378*, 701–703. [CrossRef]
10. Holmes, S.M.; Girolami, G.S. Sol-Gel Synthesis of KV^{II}[Cr^{III}(CN)$_6$]·2H$_2$O: A Crystalline molecule-based magnet with a magnetic ordering temperature above 100 °C. *J. Am. Chem. Soc.* **1999**, *121*, 5593–5594. [CrossRef]
11. Manriquez, J.M.; Yee, G.T.; McLean, R.S.; Epstein, A.J.; Miller, J.S. A Room-Temperature molecular/organic-based magnet. *Science* **1991**, *252*, 1415–1417. [CrossRef] [PubMed]
12. Gütlich, P.; Garcia, Y.; Goodwin, H.A. Spin crossover phenomena in Fe(II) complexes. *Chem. Soc. Rev.* **2000**, *29*, 419–427. [CrossRef]
13. Brooker, S. Spin crossover with thermal hysteresis: Practicalities and lessons learnt. *Chem. Soc. Rev.* **2015**, *44*, 2880–2892. [PubMed]
14. Lada, Z.G.; Andrikopoulos, K.S.; Chrissanthopoulos, A.; Perlepes, S.P.; Voyiatzis, G.A. A known Iron(II) Complex in Different Nanosized Particles: Variable-Temperature Raman Study of Its Spin-Crossover Behavior. *Inorg. Chem.* **2019**, *58*, 5183–5195. [CrossRef] [PubMed]
15. Aguilà, D.; Prado, Y.; Koumousi, E.S.; Mathonière, C.; Clérac, R. Switchable Fe/Co Prussian blue networks and molecular analogues. *Chem. Soc. Rev.* **2016**, *45*, 203–224. [CrossRef] [PubMed]
16. Linares, J.; Codjovi, E.; Garcia, Y. Pressure and temperature spin crossover sensors with optical detection. *Sensors* **2012**, *12*, 4479–4492. [CrossRef] [PubMed]
17. Molnár, G.; Rat, S.; Salmon, L.; Nicolazzi, W.; Bousseksou, A. Spin crossover nanomaterials: From fundamental concepts to devices. *Adv. Mater.* **2018**, *30*, 1703862. [CrossRef]
18. Kahn, O.; Jay Martínez, C. Spin-transition polymers: From molecular materials toward memory devices. *Science* **1998**, *279*, 44–48. [CrossRef]
19. Coronado, E. Molecular magnetism: From chemical design to spin control in molecules, materials and devices. *Nat. Rev. Mater.* **2020**, *5*, 87–104. [CrossRef]
20. Gatteschi, D.; Bogani, L.; Cornia, A.; Mannini, M.; Sorace, L.; Sessoli, R. Molecular magnetism, status and perspectives. *Solid State Sci.* **2008**, *10*, 1701–1709. [CrossRef]
21. Launay, J.-P.; Verdaguer, M. *Electrons in Molecules*, revised ed.; Oxford University Press: Oxford, UK, 2018; pp. 187–190, 207–223.
22. Gatteschi, D.; Sessoli, R.; Villain, J. *Molecular Nanomagnets*; Oxford University Press: Oxford, UK, 2006.
23. Bagai, R.; Christou, G. The Drosophila of single-molecule magnetism: [Mn$_{12}$O$_{12}$(O$_2$CR)$_{16}$(H$_2$O)$_4$]. *Chem. Soc. Rev.* **2009**, *38*, 1011–1026. [CrossRef]
24. Thomas, L.; Lionti, F.; Ballou, R.; Gatteschi, D.; Sessoli, R.; Barbara, B. Macroscopic quantum tunneling of magnetization in a single crystal of nanomagnets. *Nature* **1996**, *383*, 145–147. [CrossRef]
25. Friedman, J.R.; Sarachik, M.P.; Tejada, J.; Ziolo, R. Macroscopic measurement of resonant magnetization tunneling in high-spin molecules. *Phys. Rev. Lett.* **1996**, *76*, 3830–3833. [CrossRef]
26. Craig, G.A.; Murrie, M. 3d-single ion magnets. *Chem. Soc. Rev.* **2015**, *44*, 2135–2147. [CrossRef] [PubMed]
27. Frost, J.M.; Harriman, K.L.M.; Murugesu, M. The rise of 3d single-ion magnets in molecular magnetism: Towards materials from molecules? *Chem. Sci.* **2016**, *7*, 2470–2491. [CrossRef] [PubMed]

28. Martínez-Lillo, J.; Mastropierto, T.F.; Lhotel, E.; Paulsen, C.; Cano, J.; De Munno, G.; Faus, J.; Lloret, F.; Julve, M.; Nellutla, S.; et al. Highly anisotropic Rhenium(IV) complexes: New examples of mononuclear single-molecule magnets. *J. Am. Chem. Soc.* **2013**, *135*, 13737–13748. [CrossRef] [PubMed]
29. Feng, M.; Tong, M.-L. Single ion magnets from 3d to 5f: Developments and strategies. *Chem. Eur. J.* **2018**, *24*, 7574–7594. [CrossRef]
30. Harriman, K.L.M.; Errulat, D.; Murugesu, M. Magnetic axiality: Design principles from molecules to materials. *Trends Chem.* **2019**, *1*, 425–439. [CrossRef]
31. Woodruff, D.N.; Winpenny, R.E.P.; Layfield, R.A. Lanthanide single-molecule magnets. *Chem. Rev.* **2013**, *113*, 5110–5148. [CrossRef]
32. Pointillart, F.; Cador, O.; Le Guennic, B.; Quahab, L. Uncommon lanthanide ions in purely 4f Single Molecule Magnets. *Coord. Chem. Rev.* **2017**, *346*, 150–175. [CrossRef]
33. Gupta, S.K.; Murugavel, R. Enriching lanthanide single-ion magnetism through symmetry and axiality. *Chem. Commun.* **2018**, *54*, 3685–3696. [CrossRef]
34. Meihaus, K.R.; Long, J.R. Actinide-based single-molecule magnets. *Dalton Trans.* **2015**, *44*, 2517–2528. [CrossRef]
35. Ishikawa, N.; Sugita, M.; Ishikawa, T.; Koshihara, S.-U.; Kaizu, Y. Lanthanide double-decker complexes functioning as magnets at the single-molecule level. *J. Am. Chem. Soc.* **2003**, *125*, 8694–8695. [CrossRef]
36. Guo, F.-S.; Day, B.M.; Chen, Y.-C.; Tong, M.-L.; Mansikkamäki, A.; Layfield, R.A. Magnetic hysteresis up to 80 Kelvin in a Dysprosium metallocene single-molecule magnet. *Science* **2018**, *362*, 1400–1403. [CrossRef] [PubMed]
37. Gaita-Ariño, A.; Luis, F.; Hill, S.; Coronado, E. Molecular spins for quantum computation. *Nat. Chem.* **2019**, *11*, 301–309. [CrossRef] [PubMed]
38. Miyasaka, H.; Julve, M.; Yamashita, M.; Clérac, R. Slow Dynamics of the Magnetization in One-dimensional coordination polymers: Single-chain magnets. *Inorg. Chem.* **2009**, *48*, 3420–3437. [CrossRef] [PubMed]
39. Rams, M.; Jochim, A.; Böhme, M.; Lohmiller, T.; Ceglarska, M.; Rams, M.M.; Schnegg, A.; Plass, W.; Näther, C. Single-Chain magnet based on Cobalt(II) thiocyanate as XXZ spin chain. *Eur. J. Inorg. Chem.* **2020**, 2837–2851. [CrossRef]
40. Caneschi, A.; Gatteschi, D.; Lalioti, N.; Sangregorio, C.; Sessoli, R.; Venturi, G.; Vindigni, A.; Rettori, A.; Pini, M.G.; Novak, M.A. Cobalt(II)-Nitronyl nitroxide chains as molecular magnetic nanowires. *Angew. Chem. Int. Ed.* **2001**, *40*, 1760–1763. [CrossRef]
41. Maspoch, D.; Ruiz-Molina, D.; Wurst, K.; Domingo, N.; Cavallini, M.; Biscarini, F.; Tejada, J.; Rovira, C.; Veciana, J. A nanoporous molecular magnet with reversible solvent-induced mechanical and magnetic properties. *Nat. Mater.* **2003**, *2*, 190–195. [CrossRef]
42. Coronado, E.; Gómez-Garcia, C.J.; Nuez, A.; Romero, F.M.; Waerenborgh, J.C. Synthesis, chirality and magnetic properties of bimetallic Cyanide-bridged Two-dimensional ferromagnets. *Chem. Mater.* **2006**, *18*, 2670–2681.
43. Train, C.; Gheorge, R.; Krstic, V.; Chamoreau, L.-M.; Ovanesyan, N.S.; Rikken, G.L.J.A.; Gruselle, M.; Verdaguer, M. Strong magneto-chiral dichroism in enantiopure chiral ferromagnets. *Nat. Mater.* **2008**, *7*, 729–734.
44. Coronado, E.; Galán-Mascarós, J.R.; Gómez-Garcia, C.J.; Laukhin, V. Coexistence of ferromagnetism and metallic conductivity in a molecule-based layered compound. *Nature* **2000**, *408*, 447–449.
45. Errulat, D.; Marin, R.; Gálico, D.A.; Harriman, K.L.M.; Pialat, A.; Gabidullin, B.; Iikawa, F.; Couto, J.O.D.D.; Moilanen, J.O.; Hemmer, E.; et al. A Luminescent thermometer exhibiting slow relaxation of the magnetization: Toward Self-Monitored building blocks for Next-Generation optomagnetic devices. *ACS Cent. Sci.* **2019**, *5*, 1187–1198. [CrossRef]
46. Anastasiadis, N.C.; Granadeiro, C.M.; Mayans, J.; Raptopoulou, C.P.; Bekiari, V.; Cunha-Silva, L.; Psycharis, V.; Escuer, A.; Balula, S.S.; Konidaris, K.F.; et al. Multifunctionality in two families of dinuclear Lanthanide(III) complexes with a tridentate schiff-base ligand. *Inorg. Chem.* **2019**, *58*, 9581–9585. [CrossRef] [PubMed]
47. Aromí, G.; Aguilà, D.; Gamez, P.; Luis, F.; Roubeau, O. Design of magnetic coordination complexes for quantum computing. *Chem. Soc. Rev.* **2012**, *41*, 537–546. [CrossRef] [PubMed]
48. Katoh, K.; Isshiki, H.; Komeda, T.; Yamashita, M. Molecular spintronics based on single-molecule magnets composed of mutliple-decker phthalocyaninato Terbium(III) complex. *Chem. Asian J.* **2012**, *7*, 1154–1169. [CrossRef] [PubMed]

49. Fernandez, A.; Ferrando-Soria, J.; Moreno Pineda, E.; Tuna, F.; Vitorica-Yrezabal, I.J.; Knappke, C.; Ujma, J.; Muryn, C.A.; Timco, G.A.; Barran, P.E.; et al. Making hybrid [n]-rotaxanes as supramolecular arrays of molecular electron spin qubits. *Nat. Commun.* **2016**, *7*, 10240.
50. Pedersen, K.S.; Perlepe, P.S.; Aubey, M.L.; Woodruff, D.N.; Reyes-Lillo, S.E.; Reinholdt, A.; Voigt, L.; Li, Z.; Borup, K.; Rouzières, M.; et al. Formation of the layered conductive magnet $CrCl_2(pyrazine)_2$ through redox-active coordination chemistry. *Nat. Chem.* **2018**, *10*, 1056–1061.
51. Minguez Espallargas, G.; Coronado, E. Magnetic functionalities in MOFs: From the framework to the pore. *Chem. Soc. Rev.* **2018**, *47*, 533–557.
52. Ribas Gispert, J. *Coordination Chemistry*; Wiley-VCH: Weinheim, Germany, 2008; pp. XXXIX, XL, 31–57.
53. Cotton, F.A.; Wilkinson, G.; Murillo, C.A.; Bochmann, M. *Advanced Inorganic Chemistry*, 6th ed.; Wiley: New York, NY, USA, 1999.
54. Busch, D.H. The Compleat coordination chemistry-one practioner's perspective. *Chem. Rev.* **1993**, *93*, 847–860. [CrossRef]
55. Lada, Z.G.; Katsoulakou, E.; Perlepes, S.P. Synthesis and Chemistry of Single-molecule Magnets. In *Single-Molecule Magnets: Molecular Architectures and Building Blocks for Spintronics*; Holynska, M., Ed.; Wiley-VCH: Weinheim, Germany, 2019; pp. 245–313.
56. Maniaki, D.; Pilichos, E.; Perlepes, S.P. Coordination clusters of 3d-metals that behave as Single-Molecule Magnets (SMMs): Synthetic routes and strategies. *Front. Chem.* **2018**, *6*, 461. [CrossRef]
57. Bar, A.K.; Pichon, C.; Sutter, J.-P. Magnetic anisotropy in two- to eight-coordinated transition-metal complexes: Recent developments in molecular magnetism. *Coord. Chem. Rev.* **2016**, *308*, 346–380. [CrossRef]
58. Beltran, L.M.; Long, J.R. Directed assembly of metal cyanide cluster magnets. *Acc. Chem. Res.* **2005**, *38*, 325–334. [CrossRef]
59. Aromi, G.; Brechin, E.K. Synthesis of 3d metallic single-molecule magnets. *Struct. Bond.* **2006**, *122*, 1–67.
60. Milios, C.J.; Winpenny, R.E.P. Cluster-based single-molecule magnets. *Struct. Bond.* **2015**, *164*, 1–109.
61. Coxall, R.A.; Harris, S.G.; Henderson, D.K.; Parsons, S.; Tasker, P.A.; Winpenny, R.E.P. Inter-ligand reactions: In situ formation of new polydentate ligands. *J. Chem. Soc. Dalton Trans.* **2000**, 2349–2356. [CrossRef]
62. Escuer, A.; Aromi, G. Azide as a bridging ligand and magnetic coupler in transition metal clusters. *Eur. J. Inorg. Chem.* **2006**, *2006*, 4721–4736. [CrossRef]
63. Escuer, A.; Esteban, J.; Perlepes, S.P.; Stamatatos, T.C. The bridging azido ligand as a central "player" in high-nuclearity 3d-metal cluster chemistry. *Coord. Chem. Rev.* **2014**, *275*, 87–129. [CrossRef]
64. Stamatatos, T.C.; Christou, G. Azide groups in higher oxidation state manganese cluster chemistry. *Inorg. Chem.* **2009**, *48*, 3308–3322. [CrossRef]
65. Schweinfurth, D.; Sommer, M.G.; Atanasov, M.; Demeshko, S.; Hohlock, S.; Meyer, F.; Neese, F.; Sarkar, B. The ligand field of the azido ligand: Insights into bonding parameters and magnetic anisotropy in a Co(II)-Azido complex. *J. Am. Chem. Soc.* **2015**, *137*, 1993–2005. [CrossRef]
66. Papaefstathiou, G.S.; Perlepes, S.P.; Escuer, A.; Vicente, R.; Font-Bardia, M.; Solans, X. Unique single-atom binding of pseudohalogeno ligands to four metal ions induced by their trapping into high-nuclearity cages. *Angew. Chem. Int. Ed.* **2001**, *40*, 884–886. [CrossRef]
67. Papaefstathiou, G.S.; Escuer, A.; Vicente, R.; Font-Bardia, M.; Solans, X.; Perlepes, S.P. Reactivity in polynuclear transition metal chemistry as a means to obtain high-spin molecules: Substitution of μ_4-OH^- by η^1,μ_4-N_3^- increases nine times the ground state S value of a nonanuclear Nickel(II) cage. *Chem. Commun.* **2001**, 2414–2415. [CrossRef]
68. Boudalis, A.K.; Donnadieu, B.; Nastopoulos, V.; Clemente-Juan, J.M.; Mari, A.; Sanakis, Y.; Tuchagues, J.-P.; Perlepes, S.P. A Nonanuclear Iron(II) single-molecule magnet. *Angew. Chem. Int. Ed.* **2004**, *43*, 2266–2270. [CrossRef]
69. Boudalis, A.K.; Sanakis, Y.; Clemente-Juan, J.M.; Donnadieu, B.; Nastopoulos, V.; Mari, A.; Coppel, Y.; Tuchagues, S.P.; Perlepes, S.P. A family of enneanuclear Iron(II) single-molecule magnets. *Chem. Eur. J.* **2008**, *14*, 2514–2526.

70. Stamatatos, T.C.; Rentschler, E. Organic chelate-free and azido-rich metal clusters and coordination polymers from the use of Me$_3$SiN$_3$: A new synthetic route to complexes with beautiful structures and diverse magnetic properties. *Chem. Commun.* **2019**, *55*, 11–26. [CrossRef]
71. Alexandropoulos, D.I.; Vignesh, K.R.; Stamatatos, T.C.; Dunbar, K.R. Rare "Janus"-faced {Fe$^{II}_7$} single-molecule magnet exhibiting intramolecular ferromagnetic interactions. *Chem. Sci.* **2019**, *10*, 1626–1633. [CrossRef]
72. Alexandropoulos, D.I.; Cunha-Silva, L.; Escuer, A.; Stamatatos, T.C. New classes of ferromagnetic materials with exclusively end-on Azido Bridges: From single-molecule magnets to 2D molecule-based magnets. *Chem. Eur. J.* **2014**, *20*, 13860–13864. [CrossRef]
73. Alexandropoulos, D.I.; Fournet, A.; Cunha-Silva, L.; Christou, G.; Stamatatos, T.C. "Molecular Nanoclusters": A 2-nm-Sized {Mn$_{29}$} cluster with a spherical structure. *Inorg. Chem.* **2016**, *55*, 12118–12121.
74. Coronado, E.; Dunbar, K.R. Preface for the forum of molecular magnetism: The role of inorganic chemistry. *Inorg. Chem.* **2009**, *48*, 3293–3295. [CrossRef] [PubMed]
75. Ludi, A. Prussian blue, an inorganic evergreen. *J. Chem. Educ.* **1981**, *58*, 1013.
76. Waldmann, O. A criterion for the anisotropy barrier in single-molecule magnets. *Inorg. Chem.* **2007**, *46*, 10035–10037. [CrossRef] [PubMed]
77. Sokol, J.J.; Hee, A.G.; Long, J.R. A cyano-bridged single-molecule magnet: Slow magnetic relaxation in a trigonal prismatic MnMo$_6$(CN)$_{18}$ cluster. *J. Am. Chem. Soc.* **2002**, *124*, 7656–7657. [PubMed]
78. Schelter, E.J.; Prosvirin, A.V.; Dunbar, K.R. Molecular cube of Re^{II} and Mn^{II} that exhibits single-molecule magnetism. *J. Am. Chem. Soc.* **2004**, *126*, 15004–15005. [CrossRef] [PubMed]
79. Schelter, E.J.; Karadas, F.; Avendano, C.; Prosvirin, A.V.; Wernsdorfer, W.; Dunbar, K.R. A family of mixed-metal cyanide cubes with alternating octahedral and tetrahedral corners exhibiting a variety of magnetic behaviors including single molecule magnetism. *J. Am. Chem. Soc.* **2007**, *129*, 8139–8149. [PubMed]
80. Freedman, D.E.; Jenkins, D.M.; Lavarone, A.T.; Long, J.R. A redox-switchable single-molecule magnet incorporating [Re(CN)$_7$]$^{3-}$. *J. Am. Chem. Soc.* **2008**, *130*, 2884–2885. [PubMed]
81. Zadrozny, J.M.; Freedman, D.E.; Jenkins, D.M.; Harris, T.D.; Lavarone, A.T.; Mathionière, C.; Clérac, R.; Long, J.R. Slow Magnetic relaxation and charge-transfer in cyano-bridged coordination clusters incorporating [Re(CN)$_7$]$^{3-/4-}$. *Inorg. Chem.* **2010**, *49*, 8886–8896. [CrossRef] [PubMed]
82. Feng, X.; Liu, J.; Harris, T.D.; Hill, S.; Long, J.R. Slow magnetic relaxation induced by a large transverse zero-field splitting in a Mn^{II}Re^{IV}(CN)$_2$ single-chain magnet. *J. Am. Chem. Soc.* **2012**, *134*, 7521–7529. [CrossRef]
83. Pinkowicz, D.; Southerland, H.I.; Avendaño, C.; Prosvirin, A.; Sanders, C.; Wernsdorfer, W.; Pedersen, K.S.; Dreiser, J.; Clérac, R.; Nehrkorn, J.; et al. Cyanide single-molecule magnets exhibiting solvent dependent reversible "on" and "off" exchange bias behavior. *J. Am. Chem. Soc.* **2015**, *137*, 14406–14422.
84. Wang, X.-Y.; Avendaño, C.; Dunbar, K.R. Molecular magnetic materials based on 4d and 5d transition metals. *Chem. Soc. Rev.* **2011**, *40*, 3213–3238.
85. Pinkowicz, D.; Rodgajny, R.; Nowicka, B.; Chorazy, S.; Reczyński, M.; Sieklucka, B. Magnetic clusters based on octacyanidometallates. *Inorg. Chem. Front.* **2015**, *2*, 10–27. [CrossRef]
86. Rebilly, J.-N.; Mallah, T. Synthesis of Single-Molecule Magnets Using Metallocyanates. In *Single-Molecule Magnets and Related Phenomena*; Winpenny, R.E.P., Ed.; Springer: Berlin, Germany, 2006; pp. 103–131.
87. Schelter, E.J.; Bera, J.K.; Basca, J.; Galán-Mascarós, J.R.; Dunbar, K.R. New paramagnetic Re(II) compounds with nitrile and cyanide ligands prepared by homolytic scission of dirhenium complexes. *Inorg. Chem.* **2003**, *42*, 4256–4258.
88. Bennett, M.V.; Long, J.R. New cyanometalate building units: Synthesis and characterization of [Re(CN)$_7$]$^{3-}$ and [Re(CN)$_8$]$^{3-}$. *J. Am. Chem. Soc.* **2003**, *125*, 2394–2395. [CrossRef]
89. Atanasov, M.; Zadrozny, J.M.; Long, J.R.; Neese, F. A theoretical analysis of chemical bonding, vibronic coupling and magnetic anisotropy in linear Iron(II) complexes with single-molecule magnet behavior. *Chem. Sci.* **2013**, *4*, 139–156. [CrossRef]
90. Beagley, B.; Pritchard, R.G.; Eaborn, C.; Washburne, S.S. A gas-phase electron diffraction study of tris(trimethylsilyl)methane. A C–H bond of high p-character. *J. Chem. Soc. Chem. Commun.* **1981**, 710–711. [CrossRef]

91. Longshaw, A.I.; Carland, M.W.; Krenske, E.H.; Coote, M.L.; Sherburn, M.S. Tris(trimethylsilyl)methane is not an effective mediator of radical reactions. *Tetrahedron Lett.* **2007**, *48*, 5585–5588. [CrossRef]
92. Cook, M.A.; Eaborn, C.; Jukes, A.E.; Walton, D.R.M. [Tris(trimethylsilyl)methyl]lithim: An alkyllithium compound of unusual stability. *J. Organomet. Chem.* **1970**, *24*, 529–535. [CrossRef]
93. LaPointe, A.M. Fe[C(SiMe$_3$)$_3$]$_2$: Synthesis and reactivity of a monomeric homoleptic Iron(II) alkyl complex. *Inorg. Chim. Acta* **2003**, *345*, 359–362. [CrossRef]
94. Zadrozny, J.M.; Atanasov, M.; Bryan, A.M.; Lin, C.-Y.; Rekken, B.D.; Power, P.P.; Neese, F.; Long, J.R. Slow magnetization dynamics in a series of two-coordinate Iron(II) complexes. *Chem. Sci.* **2013**, *4*, 125–138.
95. Kramers, H.A. A general theory of paramagnetic rotation in crystals. *Proc. R. Acad. Sci. Amst.* **1930**, *33*, 959–972.
96. Zadrozny, J.M.; Xiao, D.J.; Atanasov, M.; Long, G.J.; Grandjean, F.; Neese, F.; Long, J.R. Magnetic blocking in a linear Iron(I) complex. *Nat. Chem.* **2013**, *5*, 577–581. [CrossRef]
97. Bill, E. Iron lines up. *Nat. Chem.* **2013**, *5*, 556–557. [CrossRef]
98. Zadrozny, J.M.; Xiao, D.X.; Long, J.R.; Atanasov, M.; Neese, F.; Grandjean, F.; Long, G.J. Mössbauer spectroscopy as a probe of magnetization dynamics in the linear Iron(I) and Iron(II) complexes [Fe(C(SiMe$_3$)$_3$)$_2$]$^{1-/0}$. *Inorg. Chem.* **2013**, *52*, 13123–13131. [CrossRef]
99. Rau, I.G.; Baumann, S.; Rusponi, S.; Donati, F.; Stepanow, S.; Gragnaniello, L.; Dreiser, J.; Piamonteze, C.; Nolting, F.; Gangopadhyay, S.; et al. Reaching the magnetic anisotropy limit of a 3d metal atom. *Science* **2014**, *344*, 988–992. [CrossRef]
100. Atanasov, M.; Aravena, D.; Suturina, E.; Bill, E.; Maganas, D.; Neese, F. First principles approach to the electronic structure, magnetic anisotropy and spin relaxation in mononuclear 3d-transition metal single molecule magnets. *Coord. Chem. Rev.* **2015**, *289–290*, 177–214. [CrossRef]
101. Bunting, P.C.; Atanasov, M.; Damgaard-Møller, E.; Perfetti, M.; Crassee, I.; Orlita, M.; Overgaard, J.; van Slageren, J.; Long, J.R. A linear Cobalt(II) complex with maximal orbital angular momentum from a non-Aufbau ground state. *Science* **2018**, *362*, 7319. [CrossRef] [PubMed]
102. Yao, X.-N.; Du, J.-Z.; Zhang, Y.-Q.; Leng, X.-B.; Yang, W.-W.; Jiang, S.-D.; Wang, Z.-X.; Quyang, Z.-W.; Deng, L.; Wang, B.-W.; et al. Two-Coordinate Co(II) Imido complexes as outstanding single-molecule magnets. *J. Am. Chem. Soc.* **2017**, *139*, 373–380. [CrossRef] [PubMed]
103. Vaida, S.; Shulka, P.; Tripathi, S.; Rivière, E.; Mallah, T.; Rajaraman, G.; Shanmugam, M. Substituted versus naked thiourea ligand containing pseudotetrahedral Cobalt(II) complexes: A comparative study on its magnetization relaxation dynamics phenomenon. *Inorg. Chem.* **2018**, *57*, 3371–3386. [CrossRef] [PubMed]
104. Zadrozny, J.M.; Long, J.R. Slow magnetic relaxation at zero field in the tetrahedral complex [Co(SPh)$_4$]$^{2-}$. *J. Am. Chem. Soc.* **2011**, *133*, 20732–20734. [CrossRef] [PubMed]
105. Zadrozny, J.M.; Telser, J.; Long, J.R. Slow magnetic relaxation in the tetrahedral Cobalt(II) complexes [Co(EPh)$_4$]$^{2-}$ (E = O, S, Se). *Polyhedron* **2013**, *64*, 209–217. [CrossRef]
106. Suturina, E.A.; Maganas, D.; Bill, E.; Atanasov, M.; Neese, F. Magneto-Structural correlations in a series of pseudotetrahedral [Co^{II}(XR)$_4$]$^{2-}$ single molecule magnets: An ab initio ligand field study. *Inorg. Chem.* **2015**, *54*, 9948–9961. [CrossRef]
107. Suturina, E.; Nehrkorn, J.; Zadrozny, J.M.; Liu, J.; Atanasov, M.; Weyhermüller, T.; Maganas, D.; Hill, S.; Schnegg, A.; Bill, E.; et al. Magneto-Structural correlations in pseudotetrahedral forms of the [Co(SPh)$_4$]$^{2-}$ complex probed by magnetometry, MCD spectroscopy, advanced EPR techniques, and ab initio electronic structure calculations. *Inorg. Chem.* **2017**, *56*, 3102–3118. [CrossRef]
108. Saber, M.R.; Dunbar, K.R. Ligand effects on the magnetic anisotropy of tetrahedral Cobalt complexes. *Chem. Commun.* **2014**, *50*, 12266–12269. [CrossRef]
109. Vaidya, S.; Upadhyay, A.; Singh, S.K.; Gupta, T.; Tewary, S.; Langley, S.K.; Walsh, J.P.S.; Murray, K.S.; Rajaraman, G.; Shanmugan, M. A synthetic strategy for switching the single ion anisotropy in tetrahedral Co(II) complexes. *Chem. Commun.* **2015**, *51*, 3739–3742. [CrossRef]
110. Mezei, G.; Zaleski, C.M.; Pecoraro, V.L. Structural and functional evolution of metallacrowns. *Chem. Rev.* **2007**, *107*, 4933–5003. [CrossRef] [PubMed]

111. Baskin, J.M.; Bertozzi, C.R. Copper-Free click chemistry: Bioorthogonal reagents for tagging azides. *Aldrichim. Acta* **2010**, *43*, 15–23.
112. Kolb, H.C.; Finn, M.G.; Sharpless, K.B. Click chemistry: Diverse chemical function from a few good reactions. *Angew. Chem. Int. Ed.* **2001**, *40*, 2004–2021. [CrossRef]
113. Happ, P.; Plenk, C.; Rentschler, E. 12-MC-4 metallacrowns as versatile tools for SMM research. *Coord. Chem. Rev.* **2015**, *289–290*, 238–260. [CrossRef]
114. Plenk, C.; Krause, J.; Beck, M.; Rentschler, E. Rational linkage of magnetic molecules using click chemistry. *Chem. Commun.* **2015**, *51*, 6524–6527. [CrossRef]
115. Plenk, C.; Krause, J.; Rentschler, E. A Click-Funtionalized single-molecule magnet based on Cobalt(II) and its analogous Manganese(II) and Zinc(II) compounds. *Eur. J. Inorg. Chem.* **2015**, *2015*, 370–374. [CrossRef]
116. Bolcar, M.A.; Aubin, S.M.J.; Folting, K.; Hendrickson, D.N.; Christou, G. A new Manganese cluster topology capable of yielding high-spin species: Mixed-valence $[Mn_7(OH)_3Cl_3(hmp)_9]^{2+}$ with $S \geq 10$. *Chem. Commun.* **1997**, 1485–1486. [CrossRef]
117. Boscovic, C.; Brechin, E.K.; Streib, W.E.; Folting, K.; Hendrickson, D.N.; Christou, G. A new class of single-molecule magnets: Mixed-valent $[Mn_{12}O_8Cl_4(O_2CPh)_8(hmp)_6]$. *Chem. Commun.* **2001**, 467–468. [CrossRef]
118. You, J.; Yamaguchi, A.; Nakano, M.; Krzystek, J.; Streib, W.E.; Brunel, L.-C.; Ishimoto, H.; Christou, G.; Hendrickson, D.N. Mixed-Valence tetranuclear single-molecule magnets. *Inorg. Chem.* **2001**, *40*, 4604–4616. [CrossRef]
119. Brechin, E.K.; Boskovic, C.; Wernsdorfer, W.; You, J.; Yamaguchi, A.; Sañudo, E.C.; Concolino, T.R.; Rheingold, A.L.; Ishimoto, H.; Hendrickson, D.N.; et al. Quantum tunneling of magnetization in a new $[Mn_{18}]^{2+}$ single-molecule magnet with $S = 13$. *J. Am. Chem. Soc.* **2002**, *124*, 9710–9711. [CrossRef]
120. Sañudo, E.C.; Brechin, E.K.; Boskovic, C.; Wernsdorfer, W.; Yoo, J.; Yamaguchi, A.; Concolino, T.R.; Abboud, K.A.; Rheingold, A.L.; Ishimoto, H.; et al. $[Mn_{18}]^{2+}$ and $[Mn_{21}]^{4+}$ single-molecule magnets. *Polyhedron* **2003**, *22*, 2267–2271. [CrossRef]
121. Yang, E.-C.; Wernsdorfer, W.; Hill, S.; Edwards, R.S.; Nakano, M.; Maccagnano, S.; Zakharov, L.N.; Rheingold, A.L.; Christou, G.; Hendrickson, D.N. Exchange bias in Ni_4 single-molecule magnets. *Polyhedron* **2003**, *22*, 1727–1733. [CrossRef]
122. Boskovic, C.; Brechin, E.K.; Streib, W.E.; Folting, K.; Bollinger, J.C.; Hendrickson, D.N.; Christou, G. Single-Molecule magnets: A new family of Mn_{12} clusters of formula $[Mn_{12}O_8X_4(O_2CPh)_8L_6]$. *J. Am. Chem. Soc.* **2002**, *124*, 3725–3736. [PubMed]
123. Harden, N.C.; Bolcar, M.A.; Wernsdorfer, W.; Abboud, K.A.; Streib, W.E.; Christou, G. Heptanuclear and decanuclear Manganese complexes with the anion of 2-hydroxymethylpyridine. *Inorg. Chem.* **2003**, *42*, 7067–7076. [CrossRef] [PubMed]
124. Sañudo, E.C.; Wernsdorfer, W.; Abboud, K.A.; Christou, G. Synthesis, structure, and magnetic properties of a Mn_{21} single-molecule magnet. *Inorg. Chem.* **2004**, *43*, 4137–4144.
125. Brechin, E.K.; Sañudo, E.C.; Wernsdorfer, W.; Boskovic, C.; Yoo, J.; Hendrickson, D.N.; Yamaguchi, A.; Ishimoto, I.; Concolino, T.E.; Rheingold, A.L.; et al. Single-Molecule magnets: Structure and properties of $[Mn_{18}O_{14}(O_2CMe)_{18}(hep)_4(hepH)_2(H_2O)_2](ClO_4)_2$ with $S = 13$. *Inorg. Chem.* **2005**, *44*, 502–511.
126. Stamatatos, T.C.; Abboud, K.A.; Wernsdorfer, W.; Christou, G. High-Nuclearity, high-symmetry, high-spin molecules: A Mixed-Valence Mn_{10} cage possessing rare T symmetry and an $S = 22$ ground state. *Angew. Chem. Int. Ed.* **2006**, *45*, 4134–4137.
127. Stamatatos, T.C.; Boudalis, A.K.; Pringouri, K.V.; Raptopoulou, C.P.; Terzis, A.; Wolowska, J.; McInnes, E.J.L.; Perlepes, S.P. Mixed-Valence Cobalt(II/III) carboxylate clusters: $Co^{II}_4Co^{III}_2$ and $Co^{II}Co^{III}_2$ complexes from the use of 2-(Hydroxymethyl)pyridine. *Eur. J. Inorg. Chem.* **2007**, 5098–5104. [CrossRef]
128. Stamatatos, T.C.; Poole, K.M.; Abboud, K.A.; Wernsdorfer, W.; O'Brien, T.A.; Christou, G. High-Spin Mn_4 and Mn_{10} molecules: Large spin changes with structure in mixed-valence $Mn^{II}_4Mn^{III}_6$ clusters with azide and alkoxide-based ligands. *Inorg. Chem.* **2008**, *47*, 5006–5021.

129. Efthymiou, C.G.; Papatriantafyllopoulou, C.; Alexopoulou, N.I.; Raptopoulou, C.P.; Boča, R.; Mrozinski, J.; Bakalbassis, E.G.; Perlepes, S.P. A mononuclear complex and a cubane cluster from the initial use of 2-(hydroxymethyl)pyridine in Nickel(II) carboxylate chemistry. *Polyhedron* **2009**, *28*, 3373–3381. [CrossRef]
130. Taguchi, T.; Daniels, M.R.; Abboud, K.A.; Christou, G. Mn_4, Mn_6 and Mn_{11} clusters from the use of bulky diphenyl(pyridine-2-yl)methanol. *Inorg. Chem.* **2009**, *48*, 9235–9245.
131. Taguchi, T.; Wernsdorfer, W.; Abboud, K.A.; Christou, G. Mn_8 and Mn_{16} clusters from the use of 2-(Hydroxymethyl)pyridine, and comparison with the products from bulkier chelates: A new high nuclearity single-molecule magnet. *Inorg. Chem.* **2010**, *49*, 10579–10589. [CrossRef] [PubMed]
132. Papatriantafyllopoulou, C.; Abboud, K.A.; Christou, G. Carboxylate-Free $Mn^{III}_2Ln^{III}_2$ (Ln = Lanthanide) and $Mn^{III}_2Y^{III}_2$ complexes from the use of (2-hydroxymethyl)pyridine: Analysis of spin frustration effects. *Inorg. Chem.* **2011**, *50*, 8959–8966. [CrossRef] [PubMed]
133. Abbasi, P.; Quinn, K.; Alexandropoulos, D.I.; Damjanovich, M.; Wernsdorfer, W.; Escuer, A.; Mayans, J.; Pilkington, M.; Stamatatos, T.C. Transition metal single-molecule magnets: A {Mn_{31}} Nano-sized Cluster with a large energy barrier of ~60 K and magnetic hysteresis at ~5 K. *J. Am. Chem. Soc.* **2017**, *139*, 15644–15647. [CrossRef] [PubMed]
134. March, J. *Advanced Organic Chemistry*, 4th ed.; Wiley: New York, NY, USA, 1992; pp. 882–883.
135. Wilkinson, S.G. Alcohols. In *Comprehensive Organic Chemistry*; Barton, D., Ollis, D., Eds.; Pergamon Press: Oxford, UK, 1979; Volume 1, Chapter 4.1; pp. 579–706.
136. Tasiopoulos, A.J.; Perlepes, S.P. Diol-type ligands as central 'players' in the chemistry of high-spin molecules and single-molecule magnets. *Dalton Trans.* **2008**, 5537–5555. [CrossRef]
137. Papatriantafyllopoulou, C.; Moushi, E.E.; Christou, G.; Tasiopoulos, A.J. Filling the gap between the quantum and classical worlds of nanoscale magnetism: Giant molecular aggregates based on paramagnetic 3d metal ions. *Chem. Soc. Rev.* **2016**, *45*, 1597–1628. [CrossRef]
138. Moushi, E.E.; Lampropoulos, C.; Wernsdorfer, W.; Nastopoulos, V.; Christou, G.; Tasiopoulos, A.J. A large [$Mn_{10}Na$]$_4$ loop of four linked Mn_{10} loops. *Inorg. Chem.* **2007**, *46*, 3795–3797.
139. Moushi, E.E.; Lampropoulos, C.; Wernsdorfer, W.; Nastopoulos, V.; Christou, G.; Tasiopoulos, A.J. Inducing single-molecule magnetism in a family of loop-of-loops aggregates: Heterometallic $Mn_{10}Na_4$ Clusters and the homometallic Mn_{44} analogue. *J. Am. Chem. Soc.* **2010**, *132*, 16146–16155.
140. Charalambous, M.; Moushi, E.E.; Nguyen, T.N.; Papatriantafyllopoulou, C.; Nastopoulos, V.; Christou, G.; Tasiopoulos, A.J. Giant heterometallic [$Mn_{36}Ni_4$]$^{0/2-}$ and [$Mn_{32}Co_8$] "loop-of-loops-and-supertetrahedra" molecular aggregates. *Front. Chem.* **2019**, *7*, 96. [CrossRef]
141. Charalambous, M.; Moushi, E.E.; Papatriantafyllopoulou, C.; Wernsdorfer, W.; Nastopoulos, V.; Christou, G.; Tasiopoulos, A.J. A $Mn_{36}Ni_4$ 'loop-of-loops-and-sypertetrahedra' aggregate possessing a high S_T = 26 ± 1 spin ground state. *Chem. Commun.* **2012**, *48*, 5140–5142. [CrossRef]
142. Milios, C.J.; Piligkos, S.; Brechin, E.K. Ground state spin-switching via targeted structural distortion: Twisted single-molecule magnets from derivatised salicylaldoximes. *Dalton Trans.* **2008**, 1809–1817. [CrossRef] [PubMed]
143. Jones, L.F.; Inglis, R.; Cochrane, M.E.; Mason, K.; Collins, A.; Parsons, S.; Perlepes, S.P.; Brechin, E.K. New structural types and different oxidation levels in the family of Mn_6-oxime single-molecule magnets. *Dalton Trans.* **2008**, 6205–6210. [CrossRef] [PubMed]
144. Milios, C.J.; Inglis, R.; Vinslava, A.; Prescimone, A.; Parsons, S.; Perlepes, S.P.; Christou, G.; Brechin, E.K. Turning up the spin, turning on single-molecule magnetism: From S = 1 to S = 7 in a [Mn_8] cluster via ligand induced structural distortion. *Chem. Commun.* **2007**, 2738–2740. [CrossRef] [PubMed]
145. Inglis, R.; Jones, L.F.; Milios, C.J.; Datta, S.; Collins, A.; Parsons, S.; Wernsdorfer, W.; Hill, S.; Perlepes, S.P.; Piligkos, S.; et al. Attempting to understand (and control) the relationship between structure and magnetism in an extended family of Mn_6 single-molecule magnets. *Dalton Trans.* **2009**, 3403–3412. [CrossRef] [PubMed]
146. Milios, C.J.; Vinslava, A.; Whittaker, A.G.; Parsons, S.; Wernsdorfer, W.; Christou, G.; Perlepes, S.P.; Brechin, E.K. Microwave-Assisted synthesis of a hexanuclear Mn^{III} single-molecule magnet. *Inorg. Chem.* **2006**, *45*, 5272–5274. [CrossRef]

147. Milios, C.J.; Inglis, R.; Bagai, R.; Wernsdorfer, W.; Collins, A.; Moggach, S.; Parsons, S.; Perlepes, S.P.; Christou, G.; Brechin, E.K. Enhancing SMM properties in a family of [Mn$_6$] clusters. *Chem. Commun.* **2007**, 3476–3478. [CrossRef]

148. Milios, C.J.; Raptopoulou, C.P.; Terzis, A.; Lloret, F.; Vicente, R.; Perlepes, S.P.; Escuer, A. hexanuclear Manganese(III) single-molecule magnets. *Angew. Chem. Int. Ed.* **2004**, *43*, 210–212. [CrossRef]

149. Milios, C.J.; Inglis, E.; Vinslava, A.; Bagai, E.; Wernsdorfer, W.; Parsons, S.P.; Perlepes, S.P.; Christou, G.; Brechin, E.K. Toward a magnetostructural correlation for a family of Mn$_6$ SMMs. *J. Am. Chem. Soc.* **2007**, *129*, 12505–12511. [CrossRef]

150. Cordero, B.; Roubeau, O.; Teat, S.J.; Escuer, A. Building of a novel single molecule magnet by assembly of anisotropic {Mn$_3$(μ_3-O)(salox)$_3$}$^+$ triangles. *Dalton Trans.* **2011**, *40*, 7127–7129. [CrossRef]

151. Inglis, R.; Katsenis, A.D.; Collins, A.; White, F.; Milios, C.J.; Papaefstathiou, G.S.; Brechin, E.K. Assembling molecular triangles into discrete and infinite architectures. *Cryst. Eng. Commun.* **2010**, *12*, 2064–2072. [CrossRef]

152. Stoumpos, C.C.; Inglis, R.; Karotsis, G.; Jones, L.F.; Collins, A.; Parsons, S.; Milios, C.J.; Papaefstathiou, G.S.; Brechin, E.K. Supramolecular entanglement from interlocked molecular nanomagnets. *Cryst. Growth Des.* **2009**, *9*, 24–27. [CrossRef]

153. Inglis, R.; Papaefstathiou, G.S.; Wernsdorfer, W.; Brechin, E.K. Ferromagnetic [Mn$_3$] single-molecule magnets and their supramolecular networks. *Aust. J. Chem.* **2009**, *62*, 1108–1118. [CrossRef]

154. Manoli, M.; Inglis, R.; Piligkos, S.; Yanha, L.; Wernsdorfer, W.; Brechin, E.K.; Tasiopoulos, A.J. A hexameric [Mn$^{III}_{18}$Na$_6$] wheel based on [Mn$^{III}_3$O]$^{7+}$ sub-units. *Chem. Commun.* **2016**, *52*, 12829–12832. [CrossRef] [PubMed]

155. Manoli, M.; Inglis, R.; Manos, M.J.; Nastopoulos, V.; Wernsdorfer, W.; Brechin, E.K.; Tasiopoulos, A.J. A [Mn$_{32}$] double-decker wheel. *Angew. Chem. Int. Ed.* **2011**, *50*, 4441–4444. [CrossRef]

156. Milios, C.J.; Stamatatos, T.C.; Perlepes, S.P. The coordination chemistry of pyridyl oximes. *Polyhedron* **2006**, *25*, 134–194. [CrossRef]

157. Abele, E.; Abele, R.; Lukevics, E. Pyridine oximes: Synthesis, reactions, and biological activity. *Chem. Heterocycl. Compd.* **2003**, *39*, 825–865. [CrossRef]

158. Pavlov, A.A.; Savkina, S.A.; Belov, A.S.; Nelyubina, Y.V.; Efimov, N.N.; Voloshin, Y.Z.; Novikov, V.V. Trigonal prismatic tris-pyridineoxime transition metal complexes: A Cobalt(II) compound with high magnetic anisotropy. *Inorg. Chem.* **2017**, *56*, 6943–6951. [CrossRef]

159. Escuer, A.; Vlahopoulou, G.; Maunter, F.A. Use of 6-methylpyridine-2-carbaldehydeoxime in Nickel(II) carboxylate chemistry: Synthetic, structural and magnetic properties of penta and hexanuclear complexes. *Dalton Trans.* **2011**, *40*, 10109–10116. [CrossRef]

160. Esteban, J.; Ruiz, E.; Font-Bardia, M.; Calvet, T.; Escuer, A. Triangular nickel complexes derived from 2-Pyridylcyanoxime: An approach to the mangetic properties of the [Ni$_3$(μ_3-OH){pyC(R)NO}$_3$]$^{2+}$ core. *Chem. Eur. J.* **2012**, *18*, 3637–3648. [CrossRef]

161. Escuer, A.; Vlahopoulou, G.; Perlepes, S.P.; Mautner, F.A. Trinuclear, tetranuclear, and polymeric Cu^{II} complexes from the first use of 2-Pyridylcyanoxime in transition metal chemistry: Synthetic, structural, and magnetic studies. *Inorg. Chem.* **2011**, *50*, 2468–2478. [CrossRef]

162. Papatriantafyllopoulou, C.; Stamatatos, T.C.; Wernsdorfer, W.; Teat, S.J.; Tasiopoulos, A.J.; Escuer, A.; Perlepes, S.P. Combining Azide, carboxylate, and 2-Pyridyloximate ligands in transition-metal chemistry: Ferromagnetic Ni$^{II}_5$ clusters with a bowtie skeleton. *Inorg. Chem.* **2010**, *49*, 10486–10496. [CrossRef]

163. Alexandropoulos, D.I.; Papatriantafyllopoulou, C.; Aromi, G.; Roubeau, O.; Teat, S.J.; Perlepes, S.P.; Christou, G.; Stamatatos, T.C. The Highest-Nuclearity manganese/oximate complex: An unusual Mn$^{II/III}_{15}$ cluster with an S = 6 ground state. *Inorg. Chem.* **2010**, *49*, 3962–3964. [CrossRef]

164. Lampropoulos, C.; Stamatatos, T.C.; Manos, M.J.; Tasiopoulos, A.J.; Abboud, K.A.; Christou, G. New mixed-valence Mn$^{II/III}_6$ complexes bearing oximato and azido ligands: Synthesis, and structural and magnetic characterization. *Eur. J. Inorg. Chem.* **2010**, 2244–2253. [CrossRef]

165. Milios, C.J.; Piligkos, S.; Bell, A.R.; Laye, R.H.; Teat, S.J.; Vicente, R.; McInnes, E.; Escuer, A.; Perlepes, S.P.; Winpenny, R.E.P. A rare mixed-valence state manganese (II/IV) tetranuclear cage formed using phenyl 2-pyridyl ketone oxime and azide as ligands. *Inorg. Chem. Commun.* **2006**, *9*, 638–641. [CrossRef]

166. Milios, C.J.; Stamatatos, T.C.; Kyritsis, P.; Terzis, A.; Raptopoulou, C.P.; Vicente, R.; Escuer, A.; Perlepes, S.P. Phenyl 2-Pyridyl ketone and its oxime in manganese carboxylate chemistry: Synthesis, characterization, X-ray studies and magnetic properties of mononuclear, trinuclear and octanuclear complexes. *Eur. J. Inorg. Chem.* **2004**, 2885–2901. [CrossRef]
167. Adebayo, O.A.; Abboud, K.A.; Christou, G. Mn_3 single-molecule magnets and Mn_6/Mn_9 clusters from the use of Methyl 2-Pyridyl ketone oxime in manganese phosphinate and phosphonate chemistry. *Inorg. Chem.* **2017**, *56*, 11352–11364. [CrossRef] [PubMed]
168. Stamatatos, T.C.; Foguet-Albiol, D.; Lee, S.-C.; Stoumpos, C.C.; Raptopoulou, C.P.; Terzis, A.; Wernsdorfer, W.; Hill, S.O.; Perlepes, S.P.; Christou, G. "Switching On" the properties of single-molecule magnetism in triangular Manganese(III) complexes. *J. Am. Chem. Soc.* **2007**, *129*, 9484–9499. [CrossRef] [PubMed]
169. Cano, J.; Cauchy, T.; Ruiz, E.; Milios, C.J.; Stoumpos, C.C.; Stamatatos, T.C.; Perlepes, S.P.; Christou, G.; Brechin, E.K. On the origin of ferromagnetism in oximato-based $[Mn_3O]^{7+}$ triangles. *Dalton Trans.* **2008**, 234–240. [CrossRef]
170. Mowson, A.M.; Nguyen, T.N.; Abboud, K.A.; Christou, G. Dimeric and tetrameric supramolecular aggregates of single-molecule magnets via carboxylate substitution. *Inorg. Chem.* **2013**, *52*, 12320–12322. [CrossRef] [PubMed]
171. Nguyen, T.N.; Abboud, K.A.; Christou, G. MOF-like supramolecular network of Mn_3 single-molecule magnets formed by extensive π–π stacking. *Polyhedron* **2016**, *103*, 150–156. [CrossRef]
172. Nguyen, T.N.; Wernsdorfer, W.; Abboud, K.A.; Christou, G. A supramolecular aggregate of four exchange-biased single-molecule magnets. *J. Am. Chem. Soc.* **2011**, *133*, 20688–20691. [CrossRef] [PubMed]
173. Nguyen, T.N.; Wernsdorfer, W.; Shiddiq, M.; Abboud, K.A.; Hill, S.; Christou, G. Supramolecular aggregates of single-molecule magnets: Exchange-biased quantum tunneling of magnetization in a rectangular $[Mn_3]_4$ tetramer. *Chem. Sci.* **2016**, *7*, 1156–1173. [PubMed]
174. Nguyen, T.N.; Shiddiq, M.; Ghosh, T.; Abboud, K.A.; Hill, S.; Christou, G. Covalently linked dimer of Mn_3 Single-Molecule magnets and retention of its structure and quantum properties in solution. *J. Am. Chem. Soc.* **2015**, *137*, 7160–7168.
175. Vincent, J.B.; Chang, H.-R.; Folting, K.; Huffman, J.C.; Christou, G.; Hendrickson, D.N. Preparation and physical properties of trinuclear oxo-centered manganese complexes of the general formulation $[Mn_3O(O_2CR)_6L_3]^{0,+}$ (R = Me or Ph; L = a neutral donor group) and the crystal structures of $[Mn_3O(O_2CMe)_6(pyr)_3](pyr)$ and $[Mn_3O(O_2CPh)_6(pyr)_2(H_2O)]\cdot 0.5MeCN$. *J. Am. Chem. Soc.* **1987**, *109*, 5703–5711.
176. Efthymiou, C.; Winterlich, M.; Papatriantafyllopoulou, C. Breakthrough in radical-bridged single-molecule magnets. In *Single-Molecule Magnets: Molecular Architectures and Building Blocks for Spintronics*; Holynska, M., Ed.; Wiley-VCH: Weinheim, Germany, 2019; pp. 315–351.

 © 2020 by the authors. Licensee MDPI, Basel, Switzerland. This article is an open access article distributed under the terms and conditions of the Creative Commons Attribution (CC BY) license (http://creativecommons.org/licenses/by/4.0/).

Review

Iron Oxide Nanoparticles: An Alternative for Positive Contrast in Magnetic Resonance Imaging

Irene Fernández-Barahona [1,2,†], María Muñoz-Hernando [1,3,†], Jesus Ruiz-Cabello [2,4,5,6], Fernando Herranz [1,4] and Juan Pellico [4,7,*]

1. NanoMedMol Group, Instituto de Química Medica (IQM), Consejo Superior de Investigaciones Científicas (CSIC), 28006 Madrid, Spain; ifernandezbarahona@ucm.es (I.F.-B.); maria.munoz@cnic.es (M.M.-H.); fherranz@iqm.csic.es (F.H.)
2. Facultad de Farmacia, Universidad Complutense de Madrid, 28040 Madrid, Spain; jruizcabello@cicbiomagune.es
3. Centro Nacional de Investigaciones Cardiovasculares Carlos III (CNIC), 28029 Madrid, Spain
4. Ciber de Enfermedades Respiratorias (CIBERES), 28029 Madrid, Spain
5. Center for Cooperative Research in Biomaterials (CIC biomaGUNE), Basque Research and Technology Alliance (BRTA), 20014 Donostia San Sebastián, Spain
6. IKERBASQUE, Basque Foundation for Science, 48013 Bilbao, Spain
7. School of Biomedical Engineering & Imaging Sciences, King's College London, St. Thomas' Hospital, London SE1 7EH, UK
* Correspondence: juan.pellico@kcl.ac.uk; Tel.: +44-(0)-7712870441
† These authors contributed equally to this work.

Received: 31 March 2020; Accepted: 8 April 2020; Published: 10 April 2020

Abstract: Iron oxide nanoparticles have been extensively utilised as negative (T_2) contrast agents in magnetic resonance imaging. In the past few years, researchers have also exploited their application as positive (T_1) contrast agents to overcome the limitation of traditional Gd^{3+} contrast agents. To provide T_1 contrast, these particles must present certain physicochemical properties with control over the size, morphology and surface of the particles. In this review, we summarise the reported T_1 iron oxide nanoparticles and critically revise their properties, synthetic protocols and application, not only in MRI but also in multimodal imaging. In addition, we briefly summarise the most important nanoparticulate Gd and Mn agents to evaluate whether T_1 iron oxide nanoparticles can reach Gd/Mn contrast capabilities.

Keywords: iron oxide nanoparticles; magnetic resonance imaging; positive contrast agents

1. Introduction

Iron oxide nanoparticles (IONPs) are one of the most used nanomaterials in biomedicine. Among the reasons justifying this interest, their biocompatibility and magnetic properties are probably the most important. These properties have boosted their use in hyperthermia cancer treatment and, as imaging probes, in magnetic resonance imaging (MRI). When IONPs are prepared using "traditional" synthetic methods they show superparamagnetic properties. In other words, these nanoparticles show a very strong magnetic response when placed under the influence of a magnetic field, turning to zero when the magnetic field is off. Because of this, when placed inside MRI equipment, IONPs act as "small magnets", suppressing the signal and, therefore, appearing as a dark spot, the so-called negative contrast. Due to the strong magnetic response, the concentration needed for an in vivo application is often low. However, based on this, IONPs have been, for a long time, the never fulfilled eternal promise to change the current clinical scenario in MRI. Currently, Gd-based compounds are the standard probes when an MRI scan is performed. It is well-known that, under certain circumstances, Gd compounds show important toxicity. This is particularly important for patients suffering from kidney problems. Besides the toxicity problems, Gd probes normally have a small molecular weight and are, after injection, rapidly extravasated, excluding them from many applications that require long circulating times. If Gd-based probes present these problems, why have IONPs not displaced them from clinical practice? Basically, because the signal provided by Gd compounds is much more useful for in vivo diagnosis than that provided by IONPs for many diseases. This is due to the dark, negative signal that traditional IONPs generate. Frequently, in many diseases, hypointense (dark) areas appear naturally in an MR image. If the image probe generates a dark signal over a dark background, diagnosis gets complicated. For this reason, in recent years, researchers have searched for an alternative that can join the good physicochemical properties of IONPs with the outstanding imaging properties of Gd compounds. This has led to numerous synthetic developments producing extremely small iron oxide nanoparticles that, being more paramagnetic than superpramagnetic, are capable of generating bright, positive contrast in MRI. Here, we will critically review these developments, highlighting achievements and considering what is left to accomplish to reach a point at which the use of IONPs in clinics is as frequent as the use of gadolinium compounds.

2. T_1-BASED MR

2.1. Spin Density and Relaxation Times

Magnetic resonance imaging (MRI) uses differences in spin density or relaxation properties (T_1, longitudinal or spin lattice relaxation time, T_2, transversal or spin-spin relaxation time, T_2^*, apparent transversal relaxation time) to generate signal-to-noise ratio and contrast between different soft tissues in an image. Apart from these relaxation times that are interrogating different features of molecular dynamics and physical mechanisms, each soft tissue (liver, brain, heart, lung, etc.) will have a different chemical composition and different spin densities to encode an image using MRI. Water or fat are the usual predominant content of tissues; therefore, spin densities of tissues are often given by the concentration of water weighted by the relaxation properties and the acquisition parameters and type of sequence used. T_1 is normally much longer than T_2 and does not exhibit proportionality among them. Table 1 shows typical relaxation time values for different tissues at normal magnetic fields for clinical applications.

Table 1. Typical relaxation times values of some tissues at clinical magnetic fields.

	Tissue	1.5 T		3 T	
		T_1 (ms)	T_2 (ms)	T_1 (ms)	T_2 (ms)
Brain	Grey matter	1150	100	1600	70
	White matter	800	80	1100	60
	CSF	4500	2200		
Skeletal muscle		1000	35	1400	30
Fat		250	60		
Blood		1400	290	1900	275
Liver		580	55	810	56
Cardiac muscle		1030	42	1400	47

2.2. T_1-Weighted or Positive Contrast Using Gradient and Spin Echo Sequences

For T_1-contrast, imaging is performed by emphasising the differences in longitudinal components of magnetisation. This information can be coded with typical spin-echo or gradient-echo sequences. In the spin-echo sequences, two radiofrequency (RF) pulses (one at 90° followed by another at 180° or refocusing pulse) are used to acquire the signal. T_1-weighted imaging in a spin-echo sequence is determined by the repetition time (i.e., the temporal distance between two consecutive 90-degree pulses, also called T_R) whilst echo times (named T_E hereinafter) are kept at the lowest possible value. The spin-echo sequences have the value of being immune to off-resonance artefacts caused by B_0

inhomogeneities and to magnetic susceptibility shifts due to heterogenous tissues (such as multiple air-tissue interfaces in lung tissue or in some brain or abdominal regions) or to the presence of magnetic impurities (such as the presence of iron oxide nanoparticles).

In contrast to this alternative, a gradient-echo [1] image is formed with a single pulse (flip angle (θ) generally inferior to 90°), combined with the application of two lobes, also called dephasing and rephasing gradients, and the absence of a 180° refocusing pulse, per each RF excitation. Differently to the spin-echo methods, local susceptibilities are not refocused in this approach, so image quality is normally inferior and signal reduction is anticipated at gradient-echo sequences.

For our purpose here, to help to understand the contents of this review, for T_1-contrast or positive contrast with nanoparticles, keeping a short T_E in the pulse sequence should not be a problem for most MRI systems. Longer T_E enables a T_2^* weighting contrast, which for certain nanoparticles can still be recommendable. However, for positive contrast or T_1-contrast, T_E should be kept as short as possible. The selection of the best θ to maximise the T_1 contrast will then be critical for final optimisation. This flip angle finally enables T_1 weighting and must be investigated to allow fast imaging (short T_R) with this possibility. To do so, we need to take into consideration the expected T_1 and T_2 values of the different tissues for the magnetic field used in the study (see Table 1). As we can see from this table, T_1 values generally will increase with B_0, whereas T_2 remain constant, so the properties as contrast agent are normally very different at low (<1.5 T) and high fields (>3 T).

3. Nanoparticles for Positive Contrast MRI

Despite the high resolution of anatomical features presented by MRI, sensitivity is one of the weak points. MRI relies on the differences in tissue proton density and therefore differences in tissues' relaxation times to generate contrast. These differences allow discrimination between bone, air, and soft tissues in vivo. Nevertheless, discrimination of certain tissues and diseased areas gets complicated when these differences do not generate enough contrast. For this reason, contrast agents are regularly used to facilitate the diagnosis and characterisation of pathologies at cellular and molecular levels [2]. T_1 or positive contrast agents shorten the longitudinal relaxation time in the areas or tissues where accumulation occurs, making brighter images. The majority of clinically available T_1 contrast agents are paramagnetic metallo-chelates, composed of Gd, whose potential toxicity and short circulating times have driven the quest for improved positive contrast agents that overcome these limitations [3].

Nanostructured materials exhibit unique properties by virtue of their size. At this scale, quantum effects dominate material behaviour, conferring the size-dependent magnetic, electrical and optical behavior of nanomaterials. At the nanoscale, materials present an enhanced surface-to-volume ratio, extremely useful for bioconjugation purposes and targeted imaging. There is a broad variety of nanoparticulate T_1 contrast agents mainly based on the incorporation of paramagnetic Gd or Mn (Table 2). Some of the most remarkable are summarised below.

3.1. Paramagnetic Gd_2O_3 Nanoparticles

Gadolinium-based contrast agents are the gold standard T_1 contrast agents for MRI; however, their toxicity and reduced number of applications boosted the quest for alternatives. Most of Gd-based T_1 contrast agents are based on organic molecules chelating Gd^{3+} ions to prevent toxicity. However, inorganic nanoparticles made up of Gd are increasingly common. Gadolinium oxide [4,5], gadolinium fluoride [6,7] and gadolinium phosphate [8] NPs have been synthesised and tried as T_1 contrast agents. Most recent examples are focused on targeted T_1 Gd_2O_3 NPs [9], dual T_1–T_2 MRI probes [10], hybrid MRI/fluorescent probes [11,12] and theranostic (theraphy + diagnostic) probes [13] for tumour imaging. A Gd_2O_3-NP hybrid CT/MRI probe functionalised with bisphosphonate was used by Mastrogiacomo et al. to visualise calcium phosphate bone cement [14]. Dai et al. carried out a comparison between their PEGylated-Gd_2O_3 nanoparticles and the commercially available Magnevist, observing that their NPs presented a long half-life in blood and efficient MRI contrast, lower hepatic and renal toxicity and greater accumulation at the tumour site [15].

3.2. Paramagnetic MnO Nanoparticles

Manganese ions have emerged as a potential alternative to gadolinium as T_1 enhancer; however, their toxicity, affecting the central nervous and the cardiovascular systems, has determined its clinical application [16,17]. Manganese oxide nanoparticles have emerged as the most suitable alternative to overcome these toxicity issues [18]. Size and shape control allow fine tuning of the relaxivity values of these nanoparticles. PEGylated MnO NPs have been demonstrated by several groups to be useful and non-toxic T_1 contrast agents [19,20]. Their use for tumour detection is fairly popular, both in a non-specific and specific manner. Wang et al. achieved MnO NP accumulation in gliomas by elongating the circulation time of their MnO NPs, functionalising their surface with cysteine [21]. Chen et al. also managed to visualise mouse gliomas using MnO NPs, however using a glioma-specific moiety: folic acid [22]. Gallo et al. achieved M21 tumour visualisation in mice using RGD (arginine-glycine-aspartate peptide)-functionalised MnO nanoparticles [23]. Renal carcinoma T_1 MR imaging was accomplished by Li et al., targeting MnO NPs using AS1411 aptamer [24]. Manganese oxide NPs have also been used as theranostic platforms for drug delivery and photothermal therapy at the tumour site [25–27]. MnO NPs have served as a platform to synthesise hybrid molecular imaging probes for both PET (Positron Emission Tomography)/MRI for tumour vasculature imaging using ^{64}Cu [28] and fluorescence/MRI with Cy7.5 for lymph node mapping [29].

3.3. Organic Nanostructured Materials

Organic nanostructured materials have a long history as contrast agents for MRI. However, their use has been limited, as the relaxivity values of most of them, such as albumin- and dextran-based MR probes, are often insufficiently high.

Dendrimers are polymeric molecules with monomers branching out radially from a central core, forming a tree-like architecture [30]. As their synthesis is stepwise, resultant structures present narrow polydispersity, and terminal groups on the surface of the dendrimer can be included in a controlled manner [31]. The most commonly used monomers are polyamidoamine (PAMAM), polypropylimine (PPI), poly(ether imine) (PETIM) and poly-L-lysine (PLL) [32]. Due to plentiful anchoring sites for paramagnetic ions in their structure, they are appropriate nanoplatforms to integrate paramagnetic ions in their structure. Gadolinium (Gd^{3+}) is one of the most repeatedly used ions in dendrimers for T_1 MRI. The macromolecular size of these ions increases the rotational correlation times of integrated Gd^{3+}, resulting in relaxivities larger than most of the clinically approved Gd-based contrast agents with low molecular weight. The use of these macromolecular probes has been demonstrated in lymphatic imaging [33], tumour detection [34], liver fibrosis staging [35], and colon cancer and brain theranostics [36]. Mn^{2+}-based compounds have been also proposed as non-toxic alternatives to Gd chelates to image hepatocellular carcinoma and atherosclerosis [37,38]. Fan et al. have recently shown that Cu^{2+} can be integrated in dendrimeric structures to form a platform for tumour/metastasis imaging and chemotherapy [39].

3.4. Silica Based Nanoparticles

One of the main applications for SiO_2 nanoparticles is drug delivery that, combined with the possibility of carrying paramagnetic ions in their pores or surface, yields excellent theranostic agents. Kim et al. used Mn-doped silica NPs to detect hepatocellular carcinoma [40]. Li et al. made use of a theranostic agent based on mesoporous manganese silica NPs loaded with doxorubicin to image and treat a breast cancer xenograft murine model [41]. Gd^{3+}-containing silica NPs have also been used in vivo and/or in vitro as a T_1 contrast agent [42–44]. Recently, Carniato et al. summarised the most remarkable examples of Gd-based mesoporous silica nanoparticles for MRI. In this work, the influence over the relaxometric properties of important factors such as the porosity, the localisation of the paramagnetic chelate and the surface properties of the mesoporous silica nanoparticles are discussed in detail [45]. For instance, Davis et al. reported mesoporous silica nanoparticles doped with Gd in the inner or outer part of the pore and in the surface of the particle showing great differences in the relaxivity values depending on the Gd localisation [46].

Table 2. Composition, hydrodynamic diameter (D_H, nm) and r_1 (mM^{-1} s^{-1}) at magnetic field B_0 (T) of some reported T_1 based nanoparticles.

Nanomaterial	Composition	D_H Size (nm)	r_1 (mM^{-1} s^{-1})	B_0 (T)	Ref.
Paramagnetic inorganic NPs-Gadolinium	Core-shell Gd$_2$O$_3$@polisiloxane	3.3 ± 0.8	8.8	7	[4]
	D-glucuronic acid-coated Gd$_2$O$_3$	1	9.9	1.5	[5]
	Citrate-coated GdF$_3$, AEP-coated GdF$_3$/LaF$_3$	129.3	8.8 ± 0.2	3	[6]
	PAA$_{25}$-stabilized GdF$_3$/CeF$_3$ NPAs	70	40 ± 2	1.5	[7]
	PGP/dextran-K01	23.2 ± 7.8	13.9	0.5	[8]
	ES-GON-PAA	<2	70.2 ± 1.8	1.5	[9]
	Gd$_2$O$_3$@PCD-FA	131 ± 4.6	3.95	3	[10]
	Gd$_2$O$_3$-Fl-PEG-BBN	52.3	4.23	3	[13]
	Bisphosphonate-functionalised Gd$_2$O$_3$	70	15.41	3	[14]
	PEG-Gd$_2$O$_3$	36.35 ± 1.9	29	3	[15]
Paramagnetic inorganic NPs-Manganese	MnO@PDn	24.8 ± 0.2	4.4	1.41	[19]
	mPEG-SA-dopamine-MnO	120	16.14	3	[20]
	L-cysteine-functionalised PEG-coated Mn$_3$O$_4$	213.3 ± 2.4	3.66	0.5	[21]
	FA-TETT-MnO	122	4.83	7	[22]
	MnO@AUA@PEG$_{5000}$@RGD	56.7 ± 13.2	1.44	9.4	[23]
	PEG-MnO	15.08 ± 2.7	12.94	3	[24]
	Mn-LDH	48	9.48	-	[25]
	MnCO$_3$@polydopamine	173	8.3	7	[27]
	NOTA-Mn$_3$O$_4$@PEG-TRC105	32.6 ± 4.5	0.54	4.7	[28]
	Mn$_3$O$_4$@PEG-Cy$_{7.5}$	10 ± 2.3	0.53	7	[29]
Dendrimers	PAMAM G5-BnDOTA-Gd	6.5	12.98	3	[33]
	Folic acid-G5-DOTA-Gd	-	26 ± 0.06	2	[34]
	Den-cRGD-DOTA-Gd	13.2	7.1 ± 0.3	4.7	[35]
	Gd^{3+}-G$_2$-Gd-Aspargine	90		1.5	[36]
	(Au)$_{100}$G5.NH$_2$-Fl-DOTA(Mn)-HA	245.3	5.42	0.5	[37]
	PAMAM G8-DTPA-Mn	13.3 ± 1.2	3.5 ± 0.1	1.5	[38]
	G5.NHAc-Pyr/Cu(II)	153.2 ± 4.6	0.7024	0.5	[39]
Liposomes	DPPC/DPPG Gd-Liposomes	72 ± 6	1.13	0.5	[47]
	MCO-I-68-Gd/DNA liposomes	150			[48]
	Mab-Gd-SLs	129.9 ± 40.9	8.06	1.5	[49]
	RGD- and ATWLPPR- functionalised Gd-liposomes	89.9	~6	3	[50]
	RGD-CPGd-L	128	4.24	11.7	[51]
	THI0567-targeted liposomal-Gd	150–250	2 × 10^5/particle	1	[52]
Silica NPs	Mn-SiO$_2$	25 ± 2	6.7	3	[40]
	Doxorubicin-loaded SiO$_2$@MnSiO$_3$	150	4.34	3	[41]
	Silyated Gd complex-coated [Ru(bpy)$_3$]Cl$_2$	37	19.7	3	[53]
	Gd-Si-DTTA	75	28.8	3	[44]
	Gd-DOTA-MSNs	66.3 ± 6.6	33.57 ± 1.29	7	[46]
	Gd-DTPA-334	20 ± 2	18.7	0.5	[54]
	SRPs	8.3	11.9	1.5	[55]
Carbon nanotubes (CNTs)	Gd ultrashort single-walled CNTs	-	90	1.5	[56]
	Gd-MWNT	-	6.61	7	[57]
	PAA-GNTs	-	150	1.5	[58]
	MWNT/GdL	-	50.3	0.5	[59]
Metal–organic frameworks (MOFs)	Eu-, Gd-, Tb- doped MOFs	100 × 35	35.8	3	[60]
	Core-shell PB@MIL-100(Fe)	100	1.3	3	[61]
	C(RGDfK)-MnMOFs	50–100 × 750	4.0	9.4	[62]
	PCN-222(Mn)	241	35.3	1	[63]

3.5. Liposomes

Liposomes are spherical structures formed by one or several concentric lipid bilayers with an aqueous phase inside [64]. Due to their amphiphilic composition, they can integrate hydrophobic and/or hydrophilic molecules. This, added to their outstanding biocompatibility, has boosted their use as nanocarriers in drug delivery and molecular imaging. There are numerous examples of liposome-based MRI probes carrying paramagnetic agents, mostly composed of Gd^{3+} and Mn^{2+} in the aqueous lumen [47,65–68]. Alternatively, some other liposomal contrast agents used in MRI carry the paramagnetic molecule in their lipid bilayer [48,69–72]. Recent applications of these probes include imaging of tumours [49–51,73], atherosclerotic plaque [52] and blood brain barrier permeability [74].

4. Iron Oxide Nanoparticles for MRI

Iron oxide nanoparticles have been mostly used as negative (T_2) contrast agent. Their superparamagnetic behaviour, driven by the magnetic mono-domains at nanometric scale in the appropriate iron oxide phase, along with the great saturation magnetisation values, provide excellent T_2 shortening in MRI images [75]. The limitations of T_2-driven diagnosis has stimulated the development of iron oxide-based T_1 agents [76]. IONPs act as T_1 agent when certain physicochemical properties are fulfilled. These features, along with the synthetic method to achieve them, are summarised below.

4.1. Physicochemical Properties

IONP components, the core and coating, play an important role in the contrast behaviour. The composition, size and shape of both components must be controlled since they determine the iron oxide phase, crystallinity, magnetic properties and the hydrodynamic size. All of these properties are key to develop an IONP-based T_1 contrast agent.

In terms of size, we have to consider both the size of the core and the hydrodynamic size (core + coating). As a rule of thumb, in IONPs, the smaller the core the better the positive contrast. When the core size is decreased (<5 nm) the magnetic single domains decrease, leading to a spin canting effect, which provides five oriented and unpaired Fe^{3+}/Fe^{2+} electrons, making the particle more paramagnetic than superparamagnetic (Figure 1). Smaller cores usually imply less crystallinity, with a subsequent decrease in the saturation magnetisation values (similar to paramagnetic materials). Therefore, maghemite (γ-Fe_2O_3) is often preferred for T_1 contrast rather than magnetite (Fe_3O_4) where the crystallinity is usually higher.

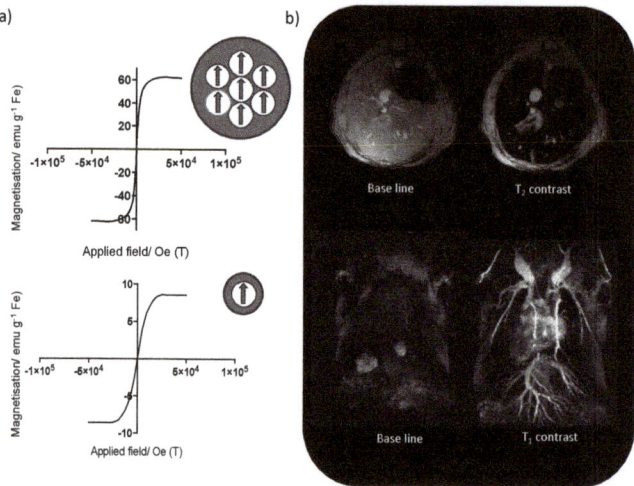

Figure 1. (a) Change in the magnetic behaviour of iron oxide nanoparticles with the decrease of the core size, from superparamagnetic (top) to paramagnetic (bottom), (b) Mouse liver T_2-weighted MRI using iron oxide nanoparticles with bigger core size (top), T_1-weighted MR angiography using iron oxide nanoparticles with smaller core sizes (bottom).

Core size and composition are essential but so is an appropriate coating. The stability of the nanoparticles is crucial to avoid aggregation, which triggers multiple particles to act as a single magnetic domain and, hence, increases the T_2 effect. In this regard, different coatings such as small molecules, macromolecules or proteins have served as stabilisers [77–80]. In addition, the size of the coating has been demonstrated to be a key factor in the T_1 effect [81]. Large size coatings (i.e., large hydrodynamic size, >30–40 nm) restrict water access to the nanoparticle core increasing the outer-sphere contribution to the relaxation mechanism. Coatings providing ultrasmall hydrodynamic sizes (<7 nm) usually confer poor stability, with aggregation increasing the T_2 effect. This is, however, an interesting challenge due to the renal clearance of ultrasmall IONPs, which increase their translational potential. Coatings rendering medium hydrodynamic sizes (10–20 nm) provide good colloidal stability with a thin coating that increases the water exchange rate, boosting the T_1 effect.

Although these are general considerations, there is no general rule established to describe the best-case scenario for each T_1-IONP. An optimal T_1-weighted sequence, taking into account the physicochemical and relaxometric properties of the nanoparticles and selecting the experimental parameters to highlight them appropriately for each magnetic field, is mandatory. A great number of works have shown a wide variety of formulations, sizes and coatings with the magnetic field, pulse sequences, and acquisition parameters not always optimised for T_1-weighted imaging. Table 3 summarises some of the reported T_1-IONPs, depicting these differences and the influence over the relaxometric properties.

Table 3. Hydrodynamic diameter (D_H, nm), core size (nm) and r_1, r_2 (mM^{-1} s^{-1}) at magnetic field B_0 (T) of reported T_1-IONPs.

Sample	D_H (nm)	Core Size (nm)	r_1 (mM^{-1} s^{-1})	r_2 (mM^{-1} s^{-1})	B_0 (T)	(ref)
Cubic IONP	18	11	3.4	36.8	3	[82]
MDBC-USPIO	24	3.4	4.8	22.56	1.5	[83]
Pegylated SPIONs	10.1	5.4	19.7	39.5	1.5	[84]
Fe$_3$O$_4$@SiO$_2$	30 - 40	4	1.2	7.8	3	[85]
SPION	20 ± 7	5–10	13.31	40.90	1.4	[86]
ESIONs	-	3	4.78	29.25	3	[87]
IONAs	17	9	5.1	21.3	3	[88]
Cat-MDBC/USNP	20	3.4 ± 1.8	6.8	37.1	1.4	[89]
UMIONs	7.5	3.3 ± 0.5	8.3	35.1	4.7	[90]
GSH-IO NPs	4.19 ± 0.31	3.72 ± 0.12	3.63	8.28	4.7	[91]
Fe$_3$O$_4$-PEG-RGD	212.5	2.7 ± 0.2	1.4	-	0.5	[92]
UTIO-nanowhiskers	-	2 × 20	6.13	11.15	1.4	[93]
C-ESION120	7.9	4.2	11.9	22.9	1.5	[81]
ES-MION3	-	3.6	8.8	22.7	1.5	[94]
Ultrasmall Fe$_3$O$_4$	-	1.9	1.41	2.87	7	[95]
Fe$_2$O$_3$-water	8 ± 2	4.9 ± 0.6	17.6	35.8	1.5	[96]
Fe$_2$O$_3$-Citrate	18 ± 4	5 ± 1	14.5	66.9	1.5	[96]
Fe$_3$O$_4$-PMAA-PTTM	-	4.34 ± 1.54	24.2	67.2	0.5	[77]
Fe$_3$O$_4$-PEG1100	10–15	4	7.3	17.5	1.4	[97]
PEG750-VSION	19.8	3.5 ± 0.6	1.74	40.6	9.4	[98]
PEG2000-VSION	22.2	3.5 ± 0.6	1.12	31.1	9.4	[98]
Ultrasmall Fe$_3$O$_4$	5.8	1.7	8.20	16.67	1.4	[79]
Ultrasmall Fe$_3$O$_4$	5.8	2.2	6.15	28.62	1.4	[79]
Metal-Doped IONPs						
Cu4-NP	16.1	3.5	15.7	32.8	1.5	[99]
EuIO-14 nanocubes	14.0 ± 1.9	14.0 ± 1.9	36.79 ± 1.16	97.52 ± 2.16	0.5	[100]
ZnFe$_2$O$_4$	-	4	7.93	14.64	1.5	[101]
NiFe$_2$O$_4$	-	5	6.85	12.92	1.5	[101]
Zn$_{0.3}$Fe$_{2.7}$O$_4$@SiO$_2$	-	18	615	1657	0.13×10^{-3}	[102]

4.2. Synthesis

Co-precipitation and thermal decomposition are the most frequently used methods to produce iron oxide nanoparticles. In addition, the extensive work on iron oxide nanoparticles over the last decade is bringing new procedures every year for the synthesis of nanoparticles of different sizes, shapes and composition and, hence, magnetic behaviour. In this section, the most remarkable methods are briefly described pointing out those which render nanoparticles with positive contrast capabilities.

4.2.1. Co-Precipitation

The co-precipitation protocol is the most used throughout the literature. It is based on the precipitation of iron oxides throughout a mixture of aqueous solutions of ferrous and ferric salts at a 2:1 ratio under basic conditions. The success of the co-precipitation method lies in its simplicity, flexibility and the hydrophilicity of the nanoparticles. However, it also shows some associated drawbacks such as the lack of control over the uniformity of the nanoparticles [103]. Modifications in the co-precipitation protocol can render ultrasmall iron oxide nanoparticles (USPIO) for T_1-MRI. A high increase in the temperature of the reaction, combined with different polymers, has been used to develop nanoparticles between 3 and 8 nm with associated longitudinal relaxivities of up to 9 mM^{-1} s^{-1} at 4.7 T and 31 mM^{-1} s^{-1} at 1 T [90,104].

4.2.2. Thermal Decomposition

This well-known method, based on the decomposition of organic precursors at very high temperatures, has been widely utilised because of the high uniformity, crystallinity and control over the size of the nanoparticles. These nanoparticles are hydrophobic and therefore only stable in hydrophobic solvents, which implies an extra reaction step to stabilise the nanoparticles in physiological media. Despite this inconvenience, the control over the hydrodynamic size allows the obtention of USPIO with potential in positive contrast [87]. Wei et al. described the thermal decomposition of Fe(oleate)$_3$ in the presence of 1-tetradecene, 1-hexadecene and 1-octadecene, rendering nanoparticles from 2.5 to 7.0 nm of a maghemite core oxidised with trimethylamine N-oxide. The nanoparticles, stabilised in water using a ligand exchange reaction with a zwitterionic dopamine sulfonate (ZDS), showed a modest r_1 = 5.2 mM^{-1} s^{-1} at 1.5 T and 1.5 mM^{-1} s^{-1} at 7 T [78]. Another remarkable example describes the synthesis of USPIO by decomposition of Fe(acac)$_3$ at 300 °C in the presence of oleic acid, oleylamine, hexadecanediol and oleyl alcohol as solvent. Ligand exchange with different bisphosphonate-based ligands provided nanoparticles with a magnetic core of 3.6 nm and a maximum r_1 = 11 mM^{-1} s^{-1} at 1.5 T [98].

4.2.3. Polyol Synthesis

Polyol synthesis has been gaining attention in the recent years for the synthesis of IONPs. This method allows for an easy scaling-up of the reaction in a single-step reaction producing hydrophilic nanoparticles, although aggregation often happens. The method essentially consists of the reduction of the organometallic precursor in the presence of different polyols such as trimethylene glycol, propylene glycol or ethylene glycol. The role of the polyol has been reported not only as stabiliser but also as reducing agent [105].

Concerning IONPs with T_1 capabilities, the reduction of Fe(acac)$_3$ in diethylene glycol at 200 °C, under an inert atmosphere provided nanoparticles with 8 nm of hydrodynamic size and 3 nm of core size. These small particles exhibited r_1 = 12 mM^{-1} s^{-1} at 1.41 T with a low r_2/r_1 = 2.4 [106]. Another reported example depicts a one-pot reaction of 5.4 nm IONPs by reduction of Fe(acac)$_3$ in the presence of triethylene glycol and HOOC-PEG-COOH at 260 °C. Under these conditions, the synthesised particles showed a r_1 = 19.7 mM^{-1} s^{-1} at 1.5 T with a r_2/r_1 ratio of 2.0 [84].

4.2.4. Microwave Assisted Synthesis

Due to the simplicity, fast kinetics and reproducibility of the reactions, microwave-assisted synthesis of IONPs has grown lately. The use of microwaves ensures a very fast and homogeneous heating in the sample, which translates in a narrow size distribution of the nanoparticles [107]. Regarding the synthesis of IONPs for T_1 contrast, our group has been deeply involved in the use of microwaves. Essentially, extremally small iron oxide cores (~2.5 nm) can be achieved with $FeCl_3$ as iron source. The reaction is conducted at 100 °C for 10 min in the presence of hydrazine hydrate (reducing agent) and a surfactant, usually dextran or sodium citrate. In a first approach, a r_1 = 5.97 mM^{-1} s^{-1} at 1.5 T was obtained using (FITC)-dextran or dextran 6KDa as surfactant [108,109]. After these first approaches, better longitudinal relaxivities were observed using sodium citrate as surfactant. In this case, an increase in the temperature of the reaction turns over the contrast capabilities from T_2 to T_1 with an r_1 = 11.9 mM^{-1} s^{-1} at 1.5 T when the reaction is performed at 120 °C [81]. Very recently, Fernandez-Barahona et al. increased this value up to 15.7 mM^{-1} s^{-1}, doping the iron oxide core with 4% mol of Cu [99].

5. In Vivo Applications

The use of iron oxide nanoparticles as contrast agents for MRI has usually been associated with T_2-MRI. This is one of the main reasons why the use of iron oxide-based contrast agents for MRI is far from standard in clinical practice. The dark (T_2) signal makes them difficult to distinguish from naturally occurring hypointense areas in many diseases—caused by calcium deposits, bleeding, other metals or any signal void present in the area of interest. Given this, the need for nontoxic positive contrast agents has led to intense research in the development of iron oxide nanoparticles that produce high T_1-weighted MRI signals, becoming therefore, a very active topic.

5.1. Iron-Based and Iron Oxide Nanoparticles

In recent years, several research groups have been developing new iron-based probes, mostly for T_1-MRI but also for dual contrast MRI. Regarding T_1-MRI, some researchers have been able to show, in vivo, a positive contrast effect. In 2013 Ju et al. [110] elaborated non-toxic synthetic melanin-like nanoparticles complexed with paramagnetic Fe^{3+} ions and stabilised with PEG (PEGylated Fe^{3+}-MelNPs), which were inspired by the MRI signal-enhancing capability of natural melanin. These nanoparticles showed relaxivity values of r_1 = 17 mM^{-1} s^{-1} and r_2 = 18 mM^{-1} s^{-1} at 3 T, which are higher than those of existing T_1-MRI contrast agents based on gadolinium (Gd) or manganese (Mn). In vivo, PEGylated Fe^{3+}-MelNPs showed a positive signal enhancement in the spleen and liver of healthy mice within 0.5 and 1.5 h, respectively, after intravenous injection. During the same year, Peng et al. [111] synthesised antiferromagnetic α-iron oxide-hydroxide (α-FeOOH) nanocolloids, with diameters of 2–3 nm, which were placed inside the mesopores of worm-like silica nanoparticles. These nanocomposites exhibited a low r_2/r_1 ratio of 1.9, making them suitable as T_1-weighted contrast agents. The in vivo experiments carried out showed a positive enhancement in the brain, bladder and kidneys of healthy mice. Moreover, in 2015, Iqbal et al. [85] produced biocompatible silica-coated superparamagnetic IONPs with diameters of ~4 nm, which showed in vivo T_1 contrast enhancement in the heart, liver, kidney and bladder of healthy mice. In addition, Macher et al. [93] developed an innovative T_1 MRI contrast agent, ultrathin iron oxide nanowhiskers. These nanostructures, with dimensions of 2 × 20 nm, possessed a high surface-to-volume ratio, leading to a strong paramagnetic signal, a property suitable for T_1 contrast. In vivo experiments showed a positive contrast enhancement in rat models after intraperitoneal (IP) and subcutaneous injection of the nanowhiskers. In 2019, Tao et al. [77] synthesised (poly(acrylic acid)-poly(methacrylic acid) iron oxide nanoparticles (PMAA-PTTM-IONPs) 4.34 nm in diameter with a low r_2/r_1 ratio of 2.78, adequate for T_1-weighted contrast agents. In vivo experiments in mouse models showed biocompatibility and T_1 contrast enhancement in liver and kidney.

Even though the previously described probes were shown to provide contrast enhancement during in vivo T_1-MRI, no specific in vivo application was described for them; thus, only their biodistribution was tested. Nevertheless, different research groups have developed probes for particular applications such as magnetic resonance angiography (MRA) and tumour imaging.

MRA has been shown to be a very helpful technique in clinical imaging. With its use, several diseases could be detected including myocardial infarction, renal failure, atherosclerotic plaque, thrombosis and tumour angiogenesis. Regarding research carried out on MRA, in 2011 Kim et al. [87] synthesised ultrasmall iron oxide nanoparticles (ESION, size < 4 nm) capped using poly(ethylene glycol)-derivatised phosphine oxide (PO-PEG) ligands, which enabled clear observation of various blood vessels, with sizes down to 0.2 mm, during in vivo T_1-MRI in rat models. They were able to maintain the bright signal of blood vessels for 1 h on dynamic time-resolved MR angiography, showing that this type of probe can be used for T_1 enhanced blood pool MRI (Figure 2).

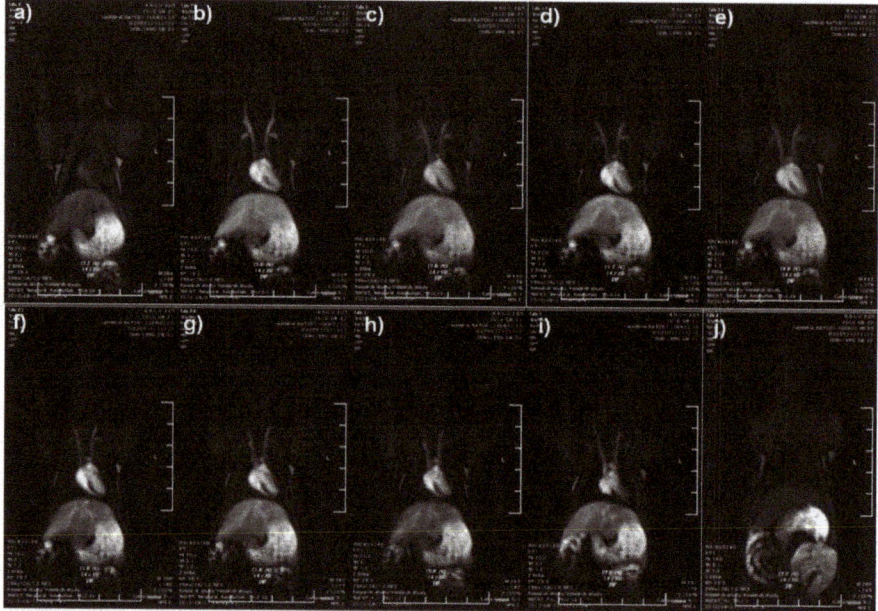

Figure 2. ESION-enhanced in vivo MR images with dynamic time-resolved MR sequence acquired at (**a**) 0 s (**b**) 30 s, (**c**) 1 min, (**d**) 2 min, (**e**) 3min, (**f**) 5 min, (**g**) 10 min, (**h**) 30 min, (**i**) 60 min, and (**j**) 1 day after the injection. Reproduced with permission from [87], published by ACS, 2011.

Furthermore, in 2014 Chan et al. [83] produced multidentate block-copolymer-stabilised ultrasmall superparamagnetic iron oxide nanoparticles (MDBC-USPIOs), with diameters <5 nm and a r_2/r_1 ratio of 4.74 to test them as a promising T_1-positive contrast agent for in vivo MRI. Results from in vivo MRI showed a strong blood signal enhancement after their intravenous injection in mouse models. During the same year, Liu et al. [91] developed glutathione-coated iron oxide nanoparticles (GSH-IONPs) with sizes of around 3.75 nm and a r_2/r_1 ratio of 2.28 as a novel T_1-MRI contrast agent. The in vivo results showed a strong vascular enhancement at the carotid artery and superior sagittal sinus of healthy mice models, making it a promising contrast agent for thrombus detection.

In 2015, Bhavesh et al. [108] proposed an extremely fast microwave synthesis of fluorescein-labelled dextran-coated extremely small IONPs for their use as a contrast agent for T_1-weighted MRI. This method yielded very small NPs with hydrodynamic diameters of 21.5 nm and with a r_2/r_1 ratio of 4.7 suitable for T_1-MRI. In vivo MRA of healthy mice showed a clear depiction of the main vascular architecture; furthermore, high quality visualisation of small vessels was maintained even 90 min post NP injection, highlighting the advantage of these NPs to be used as a contrast agent for blood pool imaging applications. In 2017, Pellico et al. [81] described the changes produced on the relaxometric properties of IONPs when the thickness of their organic coating is modified. For that purpose, this group synthesised all IONPs using a microwave-driven synthesis method; however, in order to change the coating layer thickness, different syntheses were carried out heating at different temperatures. Results showed that IONPs with a thinner coating yielded an excellent positive T_1-MRI contrast, more specifically NPs synthesised at 120 °C. These NPs, with a hydrodynamic size of 7.9 nm and a r_2/r_1 ratio of 1.9 showed great contrast enhancement in both body and brain MRA of healthy mice, providing the possibility of better visualising their vasculature. In 2018, Vangijzegem et al. [98] synthesised very small PEGylated iron oxide nanoparticles (VSIONPs) with core sizes of 3.5 nm that showed positive contrast in in vivo MRA, enabling the heart chamber and the vena cava of healthy mice to be observed.

Regarding research on T_1-weighted MR tumour imaging, in 2014 Wu et al. [112] produced mesoporous silica NPs with drug-labelled USPIONs confined within the mesoporous matrix (Fe-MSNs) as a pH-responsive theranostic platform. The tumour accumulation, driven by the enhanced permeability and retention effect (EPR), was confirmed by the T_1 enhancement in the affected site. Moreover, the unique metal–ligand coordination bonding between Fe species and the anticancer drug molecules provided the carrier with a pH-responsive drug release feature, triggering a controlled drug release under the acidic microenvironment in the tumour area. Moreover, Shen et al. [94] in 2017 synthesised extremely small 3.6 nm magnetic iron oxide nanoparticles (ES-MIONs) functionalised with dimeric RGD peptide (RGD$_2$) and PEG methyl ether (mPEG), which were afterwards loaded with the anticancer drug doxorubicin hydrochloride (DOX), as a tumour targeting theranostic platform. In this probe, the ES-MIONs serve as the contrast agent for T_1-MRI, while RGD$_2$ is used for tumour targeting and DOX for chemotherapy. In vivo experiments on tumour-bearing mice showed an enhanced T_1-MRI signal when using this platform together with partial regression of tumours due to active targeting and chemotherapy. Furthermore, Li et al. [88] in 2019 developed dynamically reversible iron oxide nanoparticle assemblies (IONAs) consisting of ES-IONs cross-linked by small molecular aldehyde derivative ligands. The linkage of these ligands is cleaved in the presence of acidic environments, such as tumours, releasing ES-IONs that are capable of producing contrast in T_1-weighted MRI (Figure 3a). To demonstrate the specificity of the tandem linkage-cleavage, pH-insensitive cross-linked iron oxide nanoparticles (Ins-IONAs) and micelle-like pH sensitive polymer-assisted iron oxide nanoparticle assemblies (PIONAs) of similar sizes were used as controls (Figure 3a). In vivo experiments on tumour-bearing mice clearly indicated that the IONAs could amplify the T_1-MRI signal, whilst Ins-IONAs and PIONAs showed significantly lower intensities (Figure 3b).

In addition to the undertaken research to find new probes that serve as contrast agents for T_1-weighted MRI, there have been some research groups that have focused on developing probes for dual contrast MRI, that is, probes that are capable of producing both negative (T_2) and positive (T_1) contrast. In MRI, T_1-weighted images typically provide better spatial resolution, while T_2-weighted images can provide enhanced detection of lesions. For this reason, dual contrast probes could potentially provide better imaging information leading to higher diagnostic accuracy. Regarding research in this field, Jung et al. [86] in 2014 synthesised 5 nm SPIONs that, as shown using in vivo MRA experiments in rat models, were capable of producing both positive and negative contrast in MRA by using different acquisition pulses. On the one hand, ultrashort echo (UTE) sequence, which differently to the sequences explained in Section 2 use a radial sampling of the k-space to minimise T_E, positively enhanced vascular signals in MR angiography, providing highly resolved vessel structures. In addition, typical gradient sequences such as fast low angle shot (FLASH) acquisition yielded strong negative vessel contrast, resulting in a higher number of discernible vessel branches. Moreover, in 2018 Alipour et al. [82] developed 11 nm silica coated cubic SPIONs as a dual-mode contrast agent for T_1 and T_2 MRI. In vivo investigations on a 3 T MRI scanner demonstrated both positive and negative contrast enhancement 70 min post intravenous injection in healthy rat models.

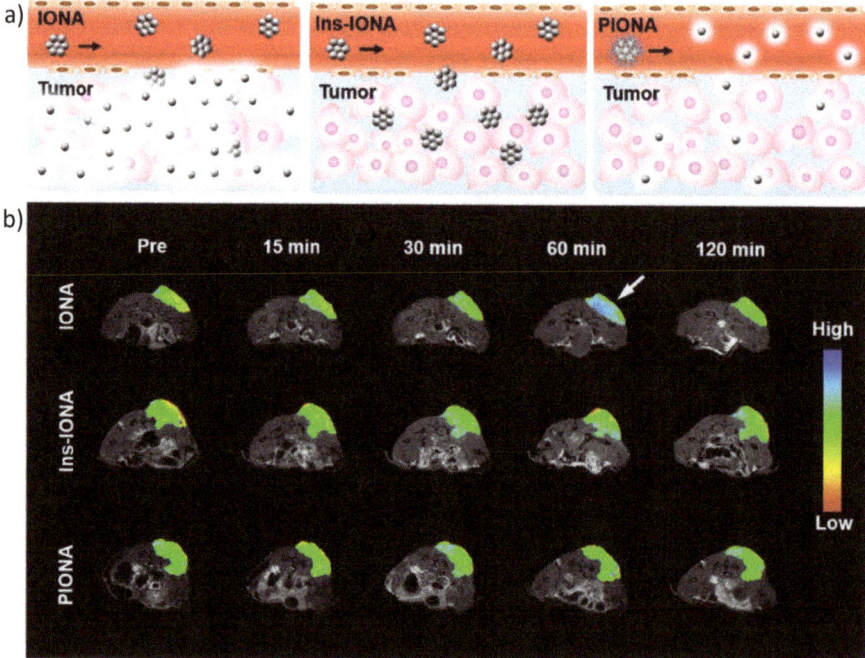

Figure 3. (a) Schematic illustration of the structural changes of iron oxide nanoparticle assemblies (IONAs), Ins–IONAs and polymer-assisted IONAs (PIONAs) in blood flow. The IONAs and Ins–IONAs are stable in blood circulation. After accumulation in tumour, IONAs further disassemble into dispersed ESIONs for non-linear amplification of MR imaging, while Ins–IONAs remain in the quenched T_1 MR state. PIONAs gradually dissociate during the blood circulation even before their accumulation in tumour. (b) T_1-weighted MR images of tumour-bearing mice before and after i.v. injection of IONAs, Ins–IONAs and PIONAs. Modified and reproduced with permission from [88], published by ACS, 2019.

5.2. Doped Iron Oxide Nanoparticles

Throughout the past years, some metals have been integrated into IONPs in order to get better T_1 enhancement. Among different metals, focus remains on gadolinium (Gd), which are not included since our goal is focusing on alternatives, and manganese (Mn); however, there are some studies that use other metals such as copper (Cu) and europium (Eu).

Manganese-based contrast agents, similarly to Gd contrast agents, can significantly enhance the positive contrast in T_1-weighted MRI. Furthermore, they can serve as contrast agents for visualising sub-anatomic structures and activities of brain and heart by taking advantage of Mn^{2+} as a biological calcium analogue. Nevertheless, their high toxicity limits their clinical applications, and for that reason, it is highly desirable to develop alternative non-toxic manganese contrast agents for clinical MRI. Several research groups have been recently developing manganese-containing iron oxide nanoparticles as alternative positive MRI contrast agents. In 2013, Li et al. [113] synthesised nontoxic ultrasmall manganese ferrite ($MnFe_2O_4$) nanoparticles of 2-3 nm that were shown to enhance the T_1 contrast in the liver, kidneys and brain of healthy mice during in vivo MRI. Similarly, in 2014 Huang et al. [114] produced Mn-doped IONPs of around 5 nm in diameter capable of functioning as a contrast agent for T_1-weighted MRI having a low r_2/r_1 ratio of 2.6. Post injection, in vivo T_1-MRI showed a brighter signal in the liver regions of healthy mice. Furthermore, in 2015 Zhang et al. [115] synthesised bovine serum albumin-coated manganese-doped IONPs (MnIO-BSA), 5 nm in size and with a low r_2/r_1 ratio of 2.18, as a theranostic platform for tumour targeting. The probe, once accumulated in the tumour, was able to provide contrast in T_1-weighted MRI and be heated using an external NIR light source for photothermal therapy (PTT) purposes. The in vivo MRI experiments using a tumour-bearing mouse model exhibited significant signal enhancement (about two times) at the tumour site. Furthermore, it was demonstrated that hyperthermia caused by the photothermal effect of the MnIO-BSA nanoparticles under NIR laser irradiation resulted in significant death of the cancer cells.

Only a few studies use copper or europium as doping metals. In 2019, our group synthesised extremely small citrate-coated iron oxide nanoparticles doped with Cu as a T_1-weighted contrast agent [99]. After studying several Cu doping amounts, IONPs doped with 4% mol of Cu (Cu4-NPs) were chosen as the best probe. Cu4-NPs presented a hydrodynamic diameter of ~15 nm and a core size of ~3.5 nm. Moreover, they showed a high r_1 value of 15.7 mM^{-1} s^{-1} at 1.5 T and a low r_2/r_1 ratio of 2.1. Interestingly, the reason for this increase in the r_1 values was due to the distribution of Cu atoms within the iron oxide structure. In vivo MRA showed that Cu4-NPs provided high-quality images with fine details of the vasculature up to 30 min post injection (Figure 4). In order to explore a different probe application, RGD molecules were conjugated to the surfaces of Cu4-NPs for active tumour targeting. In vivo T_1-MRI experiments on tumour-bearing mice showed an increased positive signal on the tumour after intravenous injection of the RGD-Cu4-NP, further confirming their use as a positive contrast agent. Europium was used by Yang et al. [100] in 2015; this group produced citrate-coated europium-doped iron oxide nanocubes, 14 nm in size, with high r_1 values of 36.8 mM^{-1} s^{-1} at 0.5 T and a low r_2/r_1 ratio of 2.65 for T_1-weighted MRI. In vivo results showed an increase in signal brightness in the heart of healthy mice.

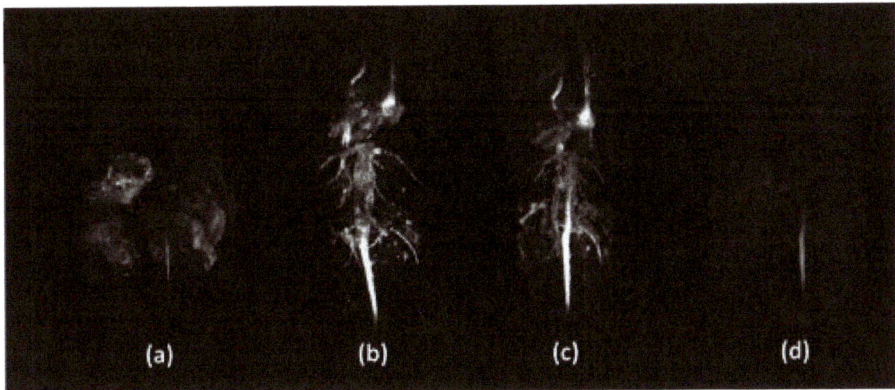

Figure 4. MRI (T_1-weighted imaging) body angiography in healthy mice, before (**a**) and after the intravenous injection, 0.04 mmol Fe kg^{-1}, of Cu4-NPs at (**b**) 15 min (**c**) 30 min (**d**) 45 min. Reproduced with permission from [99].

5.3. Multimodal T_1-Iron Oxide Nanoparticles

The combination of MRI with complementary imaging techniques such as PET or SPECT (Single Photon Emission Computed Tomography), allows protocols to be developed that exploit their synergy. Although a great number of multimodal IONPs have been described for T_2-weighted MRI, only a few examples have been reported for IONPs acting as positive contrast agents [116].

These examples are focused on the combination of biocompatible iron oxide nanoparticles with radionuclides, such as Galium-68 (^{68}Ga) or Technetium-99m (^{99m}Tc) for PET and SPECT, respectively.

^{68}Ga has been successfully used with IONPs in several multimodal experiments. In 2016, Pellico et al. developed a T_1-weighted MRI/PET platform consisting of extremely small ^{68}Ga-doped IONPs synthesised using a microwave-driven protocol, which were diversely functionalised for their use in several applications. For example, ^{68}Ga core-doped IONPs functionalised with RGD were synthesised for tumour angiogenesis targeting. This probe presented a hydrodynamic size of 20.6 nm with core sizes of 2.5 nm. Furthermore, it showed a r_1 value of 5.7 mM^{-1} s^{-1} at 1.5 T and a low r_2/r_1 ratio of 3.9, making it suitable for T_1-weighted MRI. In vivo MRI experiments in tumour-bearing mouse models showed a brighter signal in the tumour 24 h post NP injection.[114] In addition, in 2018 the group synthesised a biorthogonal nano-radiotracer for in vivo pretargeted molecular imaging of atherosclerosis. This probe was based on the in vivo tetrazine ligation of an imaging and a targeting moiety, which were functionalised with a tetrazine (Tz) and a trans-cyclooctene (TCO), respectively. The imaging part consisted of ^{68}Ga-doped IONPs that provided simultaneous PET and T_1-MRI signals, while the atherosclerosis targeting part consisted of an oxidised LDL (Low-Density Lipoprotein)-targeting IgM antibody (E-06). ^{68}Ga-IONP-Tz presented a hydrodynamic diameter of 15.5 nm with a core size of 2.8 nm. Moreover, they showed a high r_1 value of 7.1 mM^{-1} s^{-1} at 1.5 T and a low r_2/r_1 ratio of 2.5, making them useful as contrast agents for T_1-weighted MRI. Ex vivo T_1-MRI of the aorta of an atherosclerotic mouse model showed a brighter signal on the atherosclerotic plaque [117].

Regarding research carried out on ^{99m}Tc-doped IONPs, in 2013 Sandiford et al. [118] produced PEGylated bisphosphonate-coated USPIOs to use as a multimodal platform in T_1-MRI and SPECT. This probe showed a high r_1 value of 9.5 mM^{-1} s^{-1} at 3 T and a low r_2/r_1 ratio of 2.9, enabling their use as positive contrast agents. In vivo MRI experiments in healthy BALB/C mice showed a strong T_1 effect post NP injection resulting in a substantial increase in the signal from blood, making vessels, the heart compartments and other highly vascularised organs such as the spleen visible (Figure 5a). This biodistribution was also observed using SPECT/CT with the nanoparticles circulating 40 min after i.v. injection (Figure 5b).

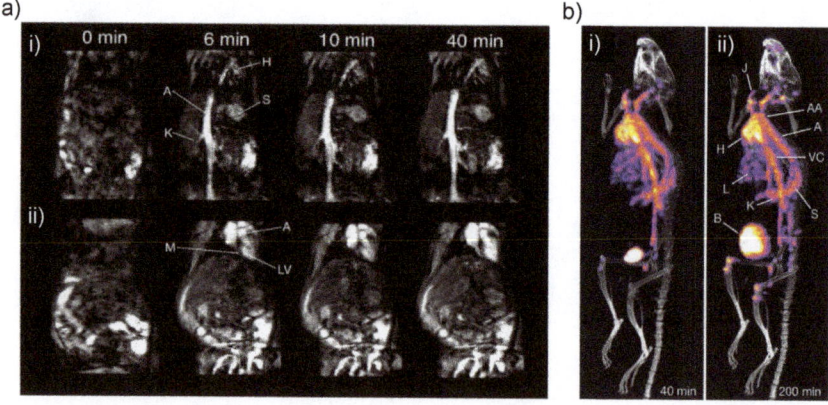

Figure 5. (a) In vivo MRI studies with PEG(5)-BP-USPIO: T_1-weighted images showing the increase in signal from blood in the vessels and the heart (ii) at different time points (t = 0 min, pre-injection). (b) Maximum intensity projection SPECT-CT images after i.v. injection of radiolabelled (99mTc) PEG(5)-BP-USPIO at the first (i, 40 min) and last (ii, 200 min) time points. Labels: H = heart, S = spleen, K = kidney, A = aorta, M = myocardium, LV = left ventricle, J = jugular vein, AA = aortic arch, VC = vena cava, L = liver, B = bladder. Modified and reproduced with permission from [118].

Besides the use of T_1-IONPs for dual MRI(T_1)/PET or SPECT imaging, G. Wang et al. reported an example for MRI(T_1)/CT imaging. In this case, ultrasmall magnetite nanoparticles were combined with Au nanocages rendering F-AuNC@Fe$_3$O$_4$. The ultrasmall Fe$_3$O$_4$ particles (2.2 nm) showed a r_1 = 6.3 mM^{-1} s^{-1}, providing a T_1-enhancement in MRI, whilst a strong X-ray attenuation was observed due to the AuNC [119].

6. Conclusions and Future Perspectives

In the last decade, significant effort has been made in the design of iron oxide formulations for T_1 contrast due to the associated toxicity in the application of the traditional Gd^{3+} contrast agents. We have summarised the most important features to optimise IONPs as a T_1 contrast agent, i.e., ultrasmall core size with moderate crystallinity (usually maghemite (γ-Fe_2O_3)) and high colloidal stability with hydrodynamic sizes ranging from 5 to 20 nm. With the focus on the physicochemical characteristics of T_1-IONPs such as iron phase, core size, type of coating or the surface charge, there is still room for improvement using the described synthetic procedures or novel strategies. Although the optimisation of these properties brings associated higher T_1 performance, the longitudinal relaxivities are still far from those obtained in some Gd^{3+}-based nanoparticles such as gadofullerens, gadolinium oxide particles, Gd-nanocages or Gd^{3+} conjugated to mesoporous silica nanoparticles [120]. It is noteworthy that the application of Gd^{3+}-based agents has been thoroughly studied for many years, whilst the potential of IONPs as T_1 agents is an emerging field of research. Therefore, there is still a lack of consensus to determine the best scenario for T_1-IONPs. To date, the variability of the particles showing different physicochemical properties and MRI responses is high. However, a broad number of applications have been reported including not only MRI but also multimodal imaging for molecular imaging applications or metal-doped particles for therapeutic purposes. In addition, other issues such as biocompatibility, pharmacokinetics or delivery pathways must be studied in advance to guarantee their clinical translation. In conclusion, there is great potential in the development of T_1-IONPs with several concerns to be evaluated in detail. Further studies must bring a clear conclusion about the best properties and then adapt the formulations to clinical requirements.

Funding: F.H. was funded through the Spanish Ministry for Economy and Competitiveness (MEyC), grant number [SAF2016-79593-P] and the Ministry of Science, grant number [RED2018-102469-T]. J.R-C. was funded through the Spanish Ministry for Economy and Competitiveness (MEyC), grant number [SAF2017-84494-C2-R], and the Gobierno Vasco, Dpto. Industria, Innovación, Comercio y Turismo under the ELKARTEK Program (Grant No. KK-2019/bmG19). CIC biomaGUNE is supported by the Maria de Maeztu Units of Excellence Program from the Spanish State Research Agency—Grant No. MDM-2017-0720.

Conflicts of Interest: The authors declare no conflict of interest.

References

1. Markl, M.; Leupold, J. Gradient echo imaging. *J. Magn. Reson. Imaging* **2012**, *35*, 1274–1289. [CrossRef] [PubMed]
2. Lohrke, J.; Frenzel, T.; Endrikat, J.; Alves, F.C.; Grist, T.M.; Law, M.; Lee, J.M.; Leiner, T.; Li, K.-C.; Nikolaou, K.; et al. 25 Years of Contrast-Enhanced MRI: Developments, Current Challenges and Future Perspectives. *Adv. Ther.* **2016**, *33*, 1–28. [CrossRef] [PubMed]
3. Na, H.B.; Hyeon, T. Nanostructured T_1 MRI contrast agents. *J. Mater. Chem.* **2009**, *19*, 6267–6273. [CrossRef]
4. Bridot, J.-L.; Faure, A.-C.; Laurent, S.; Riviere, C.; Billotey, C.; Hiba, B.; Janier, M.; Josserand, V.; Coll, J.-L.; Elst, L.V.; et al. Hybrid Gadolinium Oxide Nanoparticles: Multimodal Contrast Agents for in Vivo Imaging. *J. Am. Chem. Soc.* **2007**, *129*, 5076–5084. [CrossRef] [PubMed]
5. Park, J.Y.; Baek, M.J.; Choi, E.S.; Woo, S.; Kim, J.H.; Kim, T.J.; Jung, J.C.; Chae, K.S.; Chang, Y.; Lee, G.H. Paramagnetic Ultrasmall Gadolinium Oxide Nanoparticles as Advanced T_1 MRI Contrast Agent: Account for Large Longitudinal Relaxivity, Optimal Particle Diameter, and In Vivo T_1 MR Images. *ACS Nano* **2009**, *3*, 3663–3669. [CrossRef]
6. Evanics, F.; Diamente, P.R.; van Veggel, F.C.J.M.; Stanisz, G.J.; Prosser, R.S. Water-Soluble GdF_3 and GdF_3/LaF_3 NanoparticlesPhysical Characterization and NMR Relaxation Properties. *Chem. Mater.* **2006**, *18*, 2499–2505. [CrossRef]
7. Cheung, E.N.M.; Alvares, R.D.; Oakden, W.; Chaudhary, R.; Hill, M.L.; Pichaandi, J.; Mo, G.C.H.; Yip, C.; Macdonald, P.M.; Stanisz, G.J.; et al. Polymer-Stabilized Lanthanide Fluoride Nanoparticle Aggregates as Contrast Agents for Magnetic Resonance Imaging and Computed Tomography. *Chem. Mater.* **2010**, *22*, 4728–4739. [CrossRef]

8. Hifumi, H.; Yamaoka, S.; Tanimoto, A.; Citterio, D.; Suzuki, K. Gadolinium-Based Hybrid Nanoparticles as a Positive MR Contrast Agent. *J. Am. Chem. Soc.* **2006**, *128*, 15090–15091. [CrossRef]
9. Shen, Z.; Fan, W.; Yang, Z.; Liu, Y.; Bregadze, V.I.; Mandal, S.K.; Yung, B.C.; Lin, L.; Liu, T.; Tang, W.; et al. Exceedingly Small Gadolinium Oxide Nanoparticles with Remarkable Relaxivities for Magnetic Resonance Imaging of Tumors. *Small* **2019**, *15*, 1903422. [CrossRef]
10. Mortezazadeh, T.; Gholibegloo, E.; Alam, N.R.; Dehghani, S.; Haghgoo, S.; Ghanaati, H.; Khoobi, M. Gadolinium(III) oxide nanoparticles coated with folic acid-functionalized poly(β-cyclodextrin-*co*-pentetic acid) as a biocompatible targeted nano-contrast agent for cancer diagnostic: In vitro and in vivo studies. *Magn. Reson. Mater. Phys. Biol. Med.* **2019**, *32*, 487–500. [CrossRef]
11. Kumar, S.; Meena, V.K.; Hazari, P.P.; Sharma, S.K.; Sharma, R.K. Rose Bengal attached and dextran coated gadolinium oxide nanoparticles for potential diagnostic imaging applications. *Eur. J. Pharm. Sci.* **2018**, *117*, 362–370. [CrossRef] [PubMed]
12. Qiu, Q.; Wen, Y.; Dong, H.; Shen, A.; Zheng, X.; Li, Y.; Feng, F. A highly sensitive living probe derived from nanoparticle-remodeled neutrophils for precision tumor imaging diagnosis. *Biomater. Sci.* **2019**, *7*, 5211–5220. [CrossRef] [PubMed]
13. Cui, D.; Lu, X.; Yan, C.; Liu, X.; Hou, M.; Xia, Q.; Xu, Y.; Liu, R. Gastrin-releasing peptide receptor-targeted gadolinium oxide-based multifunctional nanoparticles for dual magnetic resonance/fluorescent molecular imaging of prostate cancer. *Int. J. Nanomed.* **2017**, *12*, 6787–6797. [CrossRef] [PubMed]
14. Mastrogiacomo, S.; Kownacka, A.E.; Dou, W.; Burke, B.P.; de Rosales, R.T.; Heerschap, A.; Jansen, J.A.; Archibald, S.J.; Walboomers, X.F. Bisphosphonate Functionalized Gadolinium Oxide Nanoparticles Allow Long-Term MRI/CT Multimodal Imaging of Calcium Phosphate Bone Cement. *Adv. Healthc. Mater.* **2018**, *7*, 1800202. [CrossRef] [PubMed]
15. Dai, Y.; Wu, C.; Wang, S.; Li, Q.; Zhang, M.; Li, J.; Xu, K. Comparative study on in vivo behavior of PEGylated gadolinium oxide nanoparticles and Magnevist as MRI contrast agent. *Nanomed. Nanotechnol. Biol. Med.* **2018**, *14*, 547–555. [CrossRef]
16. Jiang, Y.; Zheng, W. Cardiovascular toxicities upon manganese exposure. *Cardiovasc. Toxicol.* **2005**, *5*, 345–354. [CrossRef]
17. Sidoryk-Wegrzynowicz, M.; Aschner, M. Manganese toxicity in the central nervous system: The glutamine/glutamate-γ-aminobutyric acid cycle. *J. Intern. Med.* **2013**, *273*, 466–477. [CrossRef]
18. Wolf, G.; Baum, L. Cardiovascular toxicity and tissue proton T_1 response to manganese injection in the dog and rabbit. *Am. J. Roentgenol.* **1983**, *141*, 193–197. [CrossRef]
19. Chevallier, P.; Walter, A.; Garofalo, A.; Veksler, I.; Lagueux, J.; Begin-Colin, S.; Felder-Fleschc, D.; Fortin, M.-A. Tailored biological retention and efficient clearance of pegylated ultra-small MnO nanoparticles as positive MRI contrast agents for molecular imaging. *J. Mater. Chem. B* **2014**, *2*, 1779–1790. [CrossRef]
20. Huang, H.; Yue, T.; Xu, Y.; Xu, K.; Xu, H.; Liu, S.; Yu, J.; Huang, J. PEGylation of MnO nanoparticles via catechol-Mn chelation to improving T_1-weighted magnetic resonance imaging application. *J. Appl. Polym. Sci.* **2015**, *132*, 2–9. [CrossRef]
21. Wang, P.; Yang, J.; Zhou, B.; Hu, Y.; Xing, L.; Xu, F.; Shen, M.; Zhang, G.; Shi, X. Antifouling manganese oxide nanoparticles: Synthesis, characterization, and applications for enhanced MR imaging of tumors. *ACS Appl. Mater. Interfaces* **2017**, *9*, 47–53. [CrossRef] [PubMed]
22. Chen, N.; Shao, C.; Qu, Y.; Li, S.; Gu, W.; Zheng, T.; Ye, L.; Yu, C. Folic Acid-Conjugated MnO Nanoparticles as a T_1 Contrast Agent for Magnetic Resonance Imaging of Tiny Brain Gliomas. *ACS Appl. Mater. Interfaces* **2014**, *6*, 19850–19857. [CrossRef] [PubMed]
23. Gallo, J.; Alam, I.S.; Lavdas, I.; Wylezinska-Arridge, M.; Aboagye, E.O.; Long, N.J. RGD-targeted MnO nanoparticles as T_1 contrast agents for cancer imaging-the effect of PEG length in vivo. *J. Mater. Chem. B* **2014**, *2*, 868–876. [CrossRef]
24. Li, J.; Wu, C.; Hou, P.; Zhang, M.; Xu, K. One-pot preparation of hydrophilic manganese oxide nanoparticles as T_1 nano-contrast agent for molecular magnetic resonance imaging of renal carcinoma in vitro and in vivo. *Biosens. Bioelectron.* **2018**, *102*, 1–8. [CrossRef] [PubMed]
25. Li, B.; Gu, Z.; Kurniawan, N.; Chen, W.; Xu, Z.P. Manganese-Based Layered Double Hydroxide Nanoparticles as a T_1-MRI Contrast Agent with Ultrasensitive pH Response and High Relaxivity. *Adv. Mater.* **2017**, *29*, 1700373. [CrossRef] [PubMed]

26. McDonagh, B.H.; Singh, G.; Hak, S.; Bandyopadhyay, S.; Augestad, I.L.; Peddis, D.; Sandvig, I.; Sandvig, A.; Glomm, W.R. L-DOPA-Coated Manganese Oxide Nanoparticles as Dual MRI Contrast Agents and Drug-Delivery Vehicles. *Small* **2016**, *12*, 301–306. [CrossRef] [PubMed]
27. Cheng, Y.; Zhang, S.; Kang, N.; Huang, J.; Lv, X.; Wen, K.; Ye, S.; Chen, Z.; Zhou, X.; Ren, L. Polydopamine-Coated Manganese Carbonate Nanoparticles for Amplified Magnetic Resonance Imaging-Guided Photothermal Therapy. *ACS Appl. Mater. Interfaces* **2017**, *9*, 19296–19306. [CrossRef]
28. Zhan, Y.; Shi, S.; Ehlerding, E.B.; Graves, S.A.; Goel, S.; Engle, J.W.; Liang, J.; Tian, J.; Cai, W. Radiolabeled, Antibody-Conjugated Manganese Oxide Nanoparticles for Tumor Vasculature Targeted Positron Emission Tomography and Magnetic Resonance Imaging. *ACS Appl. Mater. Interfaces* **2017**, *9*, 38304–38312. [CrossRef]
29. Zhan, Y.; Zhan, W.; Li, H.; Xu, X.; Cao, X.; Zhu, S.; Liang, J.; Chen, X. In vivo dual-modality fluorescence and magnetic resonance imaging-guided lymph node mapping with good biocompatibility manganese oxide nanoparticles. *Molecules* **2017**, *22*, 2208. [CrossRef]
30. McMahon, M.T.; Bulte, J.W.M. Two decades of dendrimers as versatile MRI agents: A tale with and without metals. *Wiley Interdiscip. Rev. Nanomed. Nanobiotechnol.* **2018**, *10*, e1496. [CrossRef]
31. Dias, A.P.; da Silva Santos, S.; da Silva, J.V.; Parise Filho, R.; Ferreira, E.I.; El Seoud, O.; Giarolla, J. Dendrimers in the context of nanomedicine. *Int. J. Pharm.* **2020**, *573*, 118814. [CrossRef] [PubMed]
32. Hu, J.; Hu, K.; Cheng, Y. Tailoring the dendrimer core for efficient gene delivery. *Acta Biomater.* **2016**, *35*, 1–11. [CrossRef] [PubMed]
33. Opina, A.C.; Wong, K.J.; Griffiths, G.L.; Turkbey, B.I.; Bernardo, M.; Nakajima, T.; Kobayashi, H.; Choyke, P.L.; Vasalatiy, O. Preparation and long-term biodistribution studies of a PAMAM dendrimer G5–Gd-BnDOTA conjugate for lymphatic imaging. *Nanomedicine* **2015**, *10*, 1423–1437. [CrossRef] [PubMed]
34. Swanson, S.D.; Kukowska-Latallo, J.F.; Patri, A.K.; Chen, C.; Ge, S.; Cao, Z.; Kotlyar, A.; East, A.T.; Baker, J.R. Targeted gadolinium-loaded dendrimer nanoparticles for tumor-specific magnetic renosance contrast enhancement. *Int. J. Nanomedicine* **2008**, *3*, 201–210.
35. Li, F.; Yan, H.; Wang, J.; Li, C.; Wu, J.; Wu, S.; Rao, S.; Gao, X.; Jin, Q. Non-invasively differentiating extent of liver fibrosis by visualizing hepatic integrin αvβ3 expression with an MRI modality in mice. *Biomaterials* **2016**, *102*, 162–174. [CrossRef]
36. Alamdari, N.H.; Alaei-Beirami, M.; Shandiz, S.A.S.; Hejazinia, H.; Rasouli, R.; Saffari, M.; Ebrahimi, S.E.S.; Assadi, A.; Ardestani, M.S. Gd^{3+}-Asparagine-Anionic Linear Globular Dendrimer Second-Generation G2 Complexes: Novel Nanobiohybrid Theranostics. *Contrast Media Mol. Imaging* **2017**, *2017*, 3625729.
37. Wang, R.; Luo, Y.; Yang, S.; Lin, J.; Gao, D.; Zhao, Y.; Liu, J.; Shi, X.; Wang, X. Hyaluronic acid-modified manganese-chelated dendrimer-entrapped gold nanoparticles for the targeted CT/MR dual-mode imaging of hepatocellular carcinoma. *Sci. Rep.* **2016**, *6*, 33844. [CrossRef]
38. Nguyen, T.H.; Bryant, H.; Shapsa, A.; Street, H.; Mani, V.; Fayad, Z.A.; Frank, J.A.; Tsimikas, S.; Briley-Saebo, K.C. Manganese G8 dendrimers targeted to oxidation-specific epitopes: In vivo MR imaging of atherosclerosis. *J. Magn. Reson. Imaging* **2015**, *41*, 797–805. [CrossRef]
39. Fan, Y.; Zhang, J.; Shi, M.; Li, D.; Lu, C.; Cao, X.; Peng, C.; Mignani, S.; Majoral, J.-P.; Shi, X. Poly (amidoamine) Dendrimer-Coordinated Copper(II) Complexes as a Theranostic Nanoplatform for the Radiotherapy-Enhanced Magnetic Resonance Imaging and Chemotherapy of Tumors and Tumor Metastasis. *Nano Lett.* **2019**, *19*, 1216–1226. [CrossRef]
40. Kim, S.M.; Im, G.H.; Lee, D.G.; Lee, J.H.; Lee, W.J.; Lee, I.S. Mn^{2+}-doped silica nanoparticles for hepatocyte-targeted detection of liver cancer in T_1-weighted MRI. *Biomaterials* **2013**, *34*, 8941–8948. [CrossRef]
41. Li, X.; Zhao, W.; Liu, X.; Chen, K.; Zhu, S.; Shi, P.; Chen, Y.; Shi, J. Mesoporous manganese silicate coated silica nanoparticles as multi-stimuli-responsive T_1-MRI contrast agents and drug delivery carriers. *Acta Biomater.* **2016**, *30*, 378–387. [CrossRef] [PubMed]
42. Pellico, J.; Ellis, C.M.; Miller, J.; Davis, J.J. Water gated contrast switching with polymer–silica hybrid nanoparticles. *Chem. Commun.* **2019**, *55*, 8540–8543. [CrossRef] [PubMed]
43. Wartenberg, N.; Fries, P.; Raccurt, O.; Guillermo, A.; Imbert, D.; Mazzanti, M. A Gadolinium Complex Confined in Silica Nanoparticles as a Highly Efficient T_1/T_2 MRI Contrast Agent. *Chem. Eur. J.* **2013**, *19*, 6980–6983. [CrossRef] [PubMed]
44. Taylor, K.M.L.; Kim, J.S.; Rieter, W.J.; An, H.; Lin, W.; Lin, W. Mesoporous Silica Nanospheres as Highly Efficient MRI Contrast Agents. *J. Am. Chem. Soc.* **2008**, *130*, 2154–2155. [CrossRef]

45. Carniato, F.; Tei, L.; Botta, M. Gd-Based Mesoporous Silica Nanoparticles as MRI Probes. *Eur. J. Inorg. Chem.* **2018**, *2018*, 4936–4954. [CrossRef]
46. Davis, J.J.; Huang, W.-Y.; Davies, G.-L. Location-tuned relaxivity in Gd-doped mesoporous silica nanoparticles. *J. Mater. Chem.* **2012**, *22*, 22848–22850. [CrossRef]
47. Fossheim, S.L.; Fahlvik, A.K.; Klaveness, J.; Muller, R.N. Paramagnetic liposomes as MRI contrast agents: Influence of liposomal physicochemical properties on the in vitro relaxivity. *Magn. Reson. Imaging* **1999**, *17*, 83–89. [CrossRef]
48. Leclercq, F.; Cohen-Ohana, M.; Mignet, N.; Sbarbati, A.; Herscovici, J.; Scherman, D.; Byk, G. Design, Synthesis, and Evaluation of Gadolinium Cationic Lipids As Tools for Biodistribution Studies of Gene Delivery Complexes. *Bioconjug. Chem.* **2003**, *14*, 112–119. [CrossRef]
49. Qiu, L.H.; Zhang, J.W.; Li, S.P.; Xie, C.; Yao, Z.W.; Feng, X.Y. Molecular imaging of angiogenesis to delineate the tumor margins in glioma rat model with endoglin-targeted paramagnetic liposomes using 3T MRI. *J. Magn. Reson. Imaging* **2015**, *41*, 1056–1064. [CrossRef]
50. Song, Y.; Li, W.; Meng, S.; Zhou, W.; Su, B.; Tang, L.; Zhao, Y.; Wu, X.; Yin, D.; Fan, M.; et al. Dual integrin $\alpha v\beta 3$ and NRP-1-Targeting Paramagnetic Liposome for Tumor Early Detection in Magnetic Resonance Imaging. *Nanoscale Res. Lett.* **2018**, *13*, 380. [CrossRef]
51. Ren, L.; Chen, S.; Li, H.; Zhang, Z.; Zhong, J.; Liu, M.; Zhou, X. MRI-guided liposomes for targeted tandem chemotherapy and therapeutic response prediction. *Acta Biomater.* **2016**, *35*, 260–268. [CrossRef]
52. Woodside, D.G.; Tanifum, E.A.; Ghaghada, K.B.; Biediger, R.J.; Caivano, A.R.; Starosolski, Z.A.; Khounlo, S.; Bhayana, S.; Abbasi, S.; Craft, J.W., Jr.; et al. Magnetic Resonance Imaging of Atherosclerotic Plaque at Clinically Relevant Field Strengths (1T) by Targeting the Integrin $\alpha 4\beta 1$. *Sci. Rep.* **2018**, *8*, 3733. [CrossRef] [PubMed]
53. Rieter, W.J.; Kim, J.S.; Taylor, K.M.; An, H.; Lin, W.; Tarrant, T.; Lin, W. Hybrid Silica Nanoparticles for Multimodal Imaging. *Angew. Chem. Int. Ed.* **2007**, *46*, 3680–3682. [CrossRef] [PubMed]
54. Lechevallier, S.; Mauricot, R.; Gros-Dagnac, H.; Chevreux, S.; Lemercier, G.; Phonesouk, E.; Golzio, M.; Verelst, M. Silica-Based Nanoparticles as Bifunctional and Bimodal Imaging Contrast Agents. *Chempluschem* **2017**, *82*, 770–777. [CrossRef] [PubMed]
55. Mignot, A.; Truillet, C.; Lux, F.; Sancey, L.; Louis, C.; Denat, F.; Boschetti, F.; Bocher, L.; Gloter, A.; Stéphan, O.; et al. A Top-Down Synthesis Route to Ultrasmall Multifunctional Gd-Based Silica Nanoparticles for Theranostic Applications. *Chem. Eur. J.* **2013**, *19*, 6122–6136. [CrossRef]
56. Gizzatov, A.; Hernández-Rivera, M.; Keshishian, V.; Mackeyev, Y.; Law, J.J.; Guven, A.; Sethi, R.; Qu, F.F.; Muthupillai, R.; da Graça Cabreira-Hansen, M.; et al. Surfactant-free Gd^{3+}-ion-containing carbon nanotube MRI contrast agents for stem cell labeling. *Nanoscale* **2015**, *7*, 12085–12091. [CrossRef]
57. Servant, A.; Jacobs, I.; Bussy, C.; Fabbro, C.; Da Ros, T.; Pach, E.; Ballesteros, B.; Prato, M.; Nicolay, K.; Kostarelos, K. Gadolinium-functionalised multi-walled carbon nanotubes as a T_1 contrast agent for MRI cell labelling and tracking. *Carbon* **2016**, *97*, 126–133. [CrossRef]
58. Moghaddam, S.E.; Hernández-Rivera, M.; Zaibaq, N.G.; Ajala, A.; da Graça Cabreira-Hansen, M.; Mowlazadeh-Haghighi, S.; Willerson, J.T.; Perin, E.C.; Muthupillai, R.; Wilson, L.J. A New High-Performance Gadonanotube-Polymer Hybrid Material for Stem Cell Labeling and Tracking by MRI. *Contrast Media Mol. Imaging* **2018**, *2018*, 2853736. [CrossRef]
59. Richard, C.; Doan, B.T.; Beloeil, J.C.; Bessodes, M.; Tóth, É.; Scherman, D. Noncovalent Functionalization of Carbon Nanotubes with Amphiphilic Gd^{3+} Chelates: Toward Powerful T_1 and T_2 MRI Contrast Agents. *Nano Lett.* **2008**, *8*, 232–236. [CrossRef]
60. Rieter, W.J.; Taylor, K.M.L.; An, H.; Lin, W.; Lin, W. Nanoscale Metal–Organic Frameworks as Potential Multimodal Contrast Enhancing Agents. *J. Am. Chem. Soc.* **2006**, *128*, 9024–9025. [CrossRef]
61. Wang, D.; Zhou, J.; Chen, R.; Shi, R.; Zhao, G.; Xia, G.; Li, R.; Liu, Z.; Tian, J.; Wang, H.; et al. Controllable synthesis of dual-MOFs nanostructures for pH-responsive artemisinin delivery, magnetic resonance and optical dual-model imaging-guided chemo/photothermal combinational cancer therapy. *Biomaterials* **2016**, *100*, 27–40. [CrossRef] [PubMed]
62. Taylor, K.M.L.; Rieter, W.J.; Lin, W. Manganese-Based Nanoscale Metal–Organic Frameworks for Magnetic Resonance Imaging. *J. Am. Chem. Soc.* **2008**, *130*, 14358–14359. [CrossRef] [PubMed]

63. He, M.; Chen, Y.; Tao, C.; Tian, Q.; An, L.; Lin, J.; Tian, Q.; Yang, H.; Yang, S. Mn–Porphyrin-Based Metal–Organic Framework with High Longitudinal Relaxivity for Magnetic Resonance Imaging Guidance and Oxygen Self-Supplementing Photodynamic Therapy. *ACS Appl. Mater. Interfaces* **2019**, *11*, 41946–41956. [CrossRef] [PubMed]
64. Mulder, W.J.M.; Strijkers, G.J.; van Tilborg, G.A.F.; Griffioen, A.W.; Nicolay, K. Lipid-based nanoparticles for contrast-enhanced MRI and molecular imaging. *NMR Biomed.* **2006**, *19*, 142–164. [CrossRef]
65. Magin, R.L.; Wright, S.M.; Niesman, M.R.; Chan, H.C.; Swartz, H.M. Liposome delivery of NMR contrast agents for improved tissue imaging. *Magn. Reson. Med.* **1986**, *3*, 440–447. [CrossRef]
66. Koenig, S.H.; Brown, R.D.; Kurland, R.; Ohkit, S. Relaxivity and binding of Mn^{2+} ions in solutions of phosphatidylserine vesicles. *Magn. Reson. Med.* **1988**, *7*, 133–142. [CrossRef]
67. Devoisselle, J.M.; Vion-Dury, J.; Galons, J.P.; Confort-Gouny, S.; Coustaut, D.; Canioni, P.; Cozzone, P.J. Entrapment of Gadolinium-DTPA in Liposomes: Characterization of Vesicles by P-31 NMR Spectroscopy. *Investig. Radiol.* **1988**, *23*, 719–724. [CrossRef]
68. Tilcock, C.; Unger, E.; Cullis, P.; MacDougall, P. Liposomal Gd-DTPA: Preparation and characterization of relaxivity. *Radiology* **1989**, *171*, 77–80. [CrossRef]
69. Kabalka, G.W.; Davis, M.A.; Moss, T.H.; Buonocore, E.; Hubner, K.; Holmberg, E.; Maruyama, K.; Huang, L. Gadolinium-labeled liposomes containing various amphiphilic Gd-DTPA derivatives: Targeted MRI contrast enhancement agents for the liver. *Magn. Reson. Med.* **1991**, *19*, 406–415. [CrossRef]
70. Kabalka, G.W.; Buonocore, E.; Hubner, K.; Davis, M.; Huang, L. Gadolinium-labeled liposomes containing paramagnetic amphipathic agents: Targeted MRI contrast agents for the liver. *Magn. Reson. Med.* **1988**, *8*, 89–95. [CrossRef]
71. Trubetskoy, V.S.; Cannillo, J.A.; Milshtein, A.; Wolf, G.L.; Torchilin, V.P. Controlled delivery of Gd-containing liposomes to lymph nodes: Surface modification may enhance MRI contrast properties. *Magn. Reson. Imaging* **1995**, *13*, 31–37. [CrossRef]
72. Bertini, I.; Bianchini, F.; Calorini, L.; Colagrande, S.; Fragai, M.; Franchi, A.; Gallo, O.; Gavazzi, C.; Luchinat, C. Persistent contrast enhancement by sterically stabilized paramagnetic liposomes in murine melanoma. *Magn. Reson. Med.* **2004**, *52*, 669–672. [CrossRef] [PubMed]
73. Sipkins, D.A.; Cheresh, D.A.; Kazemi, M.R.; Nevin, L.M.; Bednarski, M.D.; Li, K.C. Detection of tumor angiogenesis in vivo by αvβ3-targeted magnetic resonance imaging. *Nat. Med.* **1998**, *4*, 623–626. [CrossRef] [PubMed]
74. Aryal, M.; Papademetriou, I.; Zhang, Y.Z.; Power, C.; McDannold, N.; Porter, T. MRI Monitoring and Quantification of Ultrasound-Mediated Delivery of Liposomes Dually Labeled with Gadolinium and Fluorophore through the Blood-Brain Barrier. *Ultrasound Med. Biol.* **2019**, *45*, 1733–1742. [CrossRef] [PubMed]
75. Qin, J.; Laurent, S.; Jo, Y.S.; Roch, A.; Mikhaylova, M.; Bhujwalla, Z.M.; Muller, R.N.; Muhammed, M.A. High-Performance Magnetic Resonance Imaging T2 Contrast Agent. *Adv. Mater.* **2007**, *19*, 1874–1878. [CrossRef]
76. Bao, Y.; Sherwood, J.A.; Sun, Z. Magnetic iron oxide nanoparticles as T_1 contrast agents for magnetic resonance imaging. *J. Mater. Chem. C* **2018**, *6*, 1280–1290. [CrossRef]
77. Tao, C.; Zheng, Q.; An, L.; He, M.; Lin, J.; Tian, Q.; Yang, S. T_1-Weight Magnetic Resonance Imaging Performances of Iron Oxide Nanoparticles Modified with a Natural Protein Macromolecule and an Artificial Macromolecule. *Nanomaterials* **2019**, *9*, 170. [CrossRef]
78. Wei, H.; Bruns, O.T.; Kaul, M.G.; Hansen, E.C.; Barch, M.; Wiśniowska, A.; Chen, O.; Chen, Y.; Li, N.; Okada, S.; et al. Exceedingly small iron oxide nanoparticles as positive MRI contrast agents. *Proc. Natl. Acad. Sci. USA* **2017**, *114*, 2325–2330. [CrossRef]
79. Wang, G.; Zhang, X.; Skallberg, A.; Liu, Y.; Hu, Z.; Mei, X.; Uvdal, K. One-step synthesis of water-dispersible ultra-small Fe_3O_4 nanoparticles as contrast agents for T_1 and T_2 magnetic resonance imaging. *Nanoscale* **2014**, *6*, 2953. [CrossRef]
80. Illés, E.; Szekeres, M.; Tóth, I.Y.; Farkas, K.; Földesi, I.; Szabó, Á.; Iván, B.; Tombácz, E. PEGylation of Superparamagnetic Iron Oxide Nanoparticles with Self-Organizing Polyacrylate-PEG Brushes for Contrast Enhancement in MRI Diagnosis. *Nanomaterials* **2018**, *8*, 776. [CrossRef]

81. Pellico, J.; Ruiz-Cabello, J.; Fernández-Barahona, I.; Gutiérrez, L.; Lechuga-Vieco, A.V.; Enríquez, J.A.; Morales, M.P.; Herranz, F. One-Step Fast Synthesis of Nanoparticles for MRI: Coating Chemistry as the Key Variable Determining Positive or Negative Contrast. *Langmuir* **2017**, *33*, 10239–10247. [CrossRef] [PubMed]
82. Alipour, A.; Soran-Erdem, Z.; Utkur, M.; Sharma, V.K.; Algin, O.; Saritas, E.U.; Demir, H.V. A new class of cubic SPIONs as a dual-mode T_1 and T_2 contrast agent for MRI. *Magn. Reson. Imaging* **2018**, *49*, 16–24. [CrossRef]
83. Chan, N.; Laprise-Pelletier, M.; Chevallier, P.; Bianchi, A.; Fortin, M.A.; Oh, J.K. Multidentate block-copolymer-stabilized ultrasmall superparamagnetic iron oxide nanoparticles with enhanced colloidal stability for magnetic resonance imaging. *Biomacromolecules* **2014**, *15*, 2146–2156. [CrossRef] [PubMed]
84. Hu, F.; Jia, Q.; Li, Y.; Gao, M. Facile synthesis of ultrasmall PEGylated iron oxide nanoparticles for dual-contrast T_1- and T_2-weighted magnetic resonance imaging. *Nanotechnology* **2011**, *22*, 245604. [CrossRef] [PubMed]
85. Iqbal, M.Z.; Ma, X.; Chen, T.; Ren, W.; Xiang, L.; Wu, A. Silica-coated super-paramagnetic iron oxide nanoparticles (SPIONPs): A new type contrast agent of T_1 magnetic resonance imaging (MRI). *J. Mater. Chem. B* **2015**, *3*, 5172–5181. [CrossRef]
86. Jung, H.; Park, B.; Lee, C.; Cho, J.; Suh, J.; Park, J.; Kim, Y.; Kim, J.; Cho, G.; Cho, H. Dual MRI T_1 and $T_2^{(*)}$ contrast with size-controlled iron oxide nanoparticles. *Nanomed. Nanotechnol. Biol. Med.* **2014**, *10*, 1679–1689. [CrossRef] [PubMed]
87. Kim, B.H.; Lee, N.; Kim, H.; An, K.; Park, Y.I.; Choi, Y.; Shin, K.; Lee, Y.; Kwon, S.G.; Na, H.B.; et al. Large-scale synthesis of uniform and extremely small-sized iron oxide nanoparticles for high-resolution T_1 magnetic resonance imaging contrast agents. *J. Am. Chem. Soc.* **2011**, *133*, 12624–12631. [CrossRef]
88. Li, F.; Liang, Z.; Liu, J.; Sun, J.; Hu, X.; Zhao, M.; Liu, J.; Bai, R.; Kim, D.; Sun, X.; et al. Dynamically Reversible Iron Oxide Nanoparticle Assemblies for Targeted Amplification of T_1-Weighted Magnetic Resonance Imaging of Tumors. *Nano Lett.* **2019**, *19*, 4213–4220. [CrossRef]
89. Li, P.; Chevallier, P.; Ramrup, P.; Biswas, D.; Vuckovich, D.; Fortin, M.A.; Oh, J.K. Mussel-Inspired Multidentate Block Copolymer to Stabilize Ultrasmall Superparamagnetic Fe_3O_4 for Magnetic Resonance Imaging Contrast Enhancement and Excellent Colloidal Stability. *Chem. Mater.* **2015**, *27*, 7100–7109. [CrossRef]
90. Li, Z.; Yi, P.W.; Sun, Q.; Lei, H.; Li Zhao, H.; Zhu, Z.H.; Smith, S.C.; Lan, M.B.; Lu, G.Q. Ultrasmall Water-Soluble and Biocompatible Magnetic Iron Oxide Nanoparticles as Positive and Negative Dual Contrast Agents. *Adv. Funct. Mater.* **2012**, *22*, 2387–2393. [CrossRef]
91. Liu, C.L.; Peng, Y.K.; Chou, S.W.; Tseng, W.H.; Tseng, Y.J.; Chen, H.C.; Hsiao, J.K.; Chou, P.T. One-Step, Room-Temperature Synthesis of Glutathione-Capped Iron-Oxide Nanoparticles and their Application in In Vivo T_1-Weighted Magnetic Resonance Imaging. *Small* **2014**, *10*, 3962–3969. [CrossRef] [PubMed]
92. Luo, Y.; Yang, J.; Yan, Y.; Li, J.; Shen, M.; Zhang, G.; Mignanic, S.; Shi, X. RGD-functionalized ultrasmall iron oxide nanoparticles for targeted T_1-weighted MR imaging of gliomas. *Nanoscale* **2015**, *7*, 14538–14546. [CrossRef] [PubMed]
93. Macher, T.; Totenhagen, J.; Sherwood, J.; Qin, Y.; Gurler, D.; Bolding, M.S.; Bao, Y. Ultrathin iron oxide nanowhiskers as positive contrast agents for magnetic resonance imaging. *Adv. Funct. Mater.* **2015**, *25*, 490–494. [CrossRef]
94. Shen, Z.; Chen, T.; Ma, X.; Ren, W.; Zhou, Z.; Zhu, G.; Zhang, A.; Liu, Y.; Song, J.; Li, Z.; et al. Multifunctional Theranostic Nanoparticles Based on Exceedingly Small Magnetic Iron Oxide Nanoparticles for T_1-Weighted Magnetic Resonance Imaging and Chemotherapy. *ACS Nano* **2017**, *11*, 10992–11004. [CrossRef] [PubMed]
95. Shen, L.H.; Bao, J.F.; Wang, D.; Wang, Y.X.; Chen, Z.W.; Ren, L.; Zhou, X.; Ke, X.B.; Chen, M.; Yang, A.Q. One-step synthesis of monodisperse, water-soluble ultra-small Fe_3O_4 nanoparticles for potential bio-application. *Nanoscale* **2013**, *5*, 2133–2141. [CrossRef]
96. Taboada, E.; Rodríguez, E.; Roig, A.; Oró, J.; Roch, A.; Muller, R.N. Relaxometric and Magnetic Characterization of Ultrasmall Iron Oxide Nanoparticles with High Magnetization. Evaluation as Potential T_1 Magnetic Resonance Imaging Contrast Agents for Molecular Imaging. *Langmuir* **2007**, *23*, 4583–4588. [CrossRef]
97. Tromsdorf, U.I.; Bruns, O.T.; Salmen, S.C.; Beisiegel, U.; Weller, H. A Highly Effective, Nontoxic T_1 MR Contrast Agent Based on Ultrasmall PEGylated Iron Oxide Nanoparticles. *Nano Lett.* **2009**, *9*, 4434–4440. [CrossRef]

98. Vangijzegem, T.; Stanicki, D.; Boutry, S.; Paternoster, Q.; Vander Elst, L.; Muller, R.N.; Laurent, S. VSION as high field MRI T_1 contrast agent: Evidence of their potential as positive contrast agent for magnetic resonance angiography. *Nanotechnology* **2018**, *29*, 265103. [CrossRef]
99. Fernández-Barahona, I.; Gutiérrez, L.; Veintemillas-Verdaguer, S.; Pellico, J.; Morales MD, P.; Catala, M.; del Pozo, M.A.; Ruiz-Cabello, J.; Herranz, F. Cu-Doped Extremely Small Iron Oxide Nanoparticles with Large Longitudinal Relaxivity: One-Pot Synthesis and in Vivo Targeted Molecular Imaging. *ACS Omega* **2019**, *4*, 2719–2727. [CrossRef]
100. Yang, L.; Zhou, Z.; Liu, H.; Wu, C.; Zhang, H.; Huang, G.; Ai, H.; Gao, J. Europium-engineered iron oxide nanocubes with high T_1 and T_2 contrast abilities for MRI in living subjects. *Nanoscale* **2015**, *7*, 6843–6850. [CrossRef]
101. Zeng, L.; Ren, W.; Zheng, J.; Cui, P.; Wu, A. Ultrasmall water-soluble metal-iron oxide nanoparticles as T_1-weighted contrast agents for magnetic resonance imaging. *Phys. Chem. Chem. Phys.* **2012**, *14*, 2631–2636. [CrossRef] [PubMed]
102. Yin, X.; Russek, S.E.; Zabow, G.; Sun, F.; Mohapatra, J.; Keenan, K.E.; Boss, M.A.; Zeng, H.; Liu, P.J.; Viert, A.; et al. Large T_1 contrast enhancement using superparamagnetic nanoparticles in ultra-low field MRI. *Sci. Rep.* **2018**, *8*, 11863. [CrossRef] [PubMed]
103. Ali, A.; Hira Zafar, M.Z.; ul Haq, I.; Phull, A.R.; Ali, J.S.; Hussain, A. Synthesis, characterization, applications, and challenges of iron oxide nanoparticles. *Nanotechnol. Sci. Appl.* **2016**, *9*, 49–67. [CrossRef] [PubMed]
104. Tao, C.; Chen, Y.; Wang, D.; Cai, Y.; Zheng, Q.; An, L.; Lin, J.; Tian, Q.; Yang, S. Macromolecules with Different Charges, Lengths, and Coordination Groups for the Coprecipitation Synthesis of Magnetic Iron Oxide Nanoparticles as T_1 MRI Contrast Agents. *Nanomaterials* **2019**, *9*, 699. [CrossRef] [PubMed]
105. Joseyphus, R.J.; Kodama, D.; Matsumoto, T.; Sato, Y.; Jeyadevan, B.; Tohji, K. Role of polyol in the synthesis of Fe particles. *J. Magn. Magn. Mater.* **2007**, *310*, 2393–2395. [CrossRef]
106. Hu, F.; MacRenaris, K.W.; Waters, E.A.; Liang, T.; Schultz-Sikma, E.A.; Eckermann, A.L.; Meade, T.J. Ultrasmall, Water-Soluble Magnetite Nanoparticles with High Relaxivity for Magnetic Resonance Imaging. *J. Phys. Chem. C* **2009**, *113*, 20855–20860. [CrossRef]
107. Fernández-Barahona, I.; Muñoz-Hernando, M.; Herranz, F. Microwave-Driven Synthesis of Iron-Oxide Nanoparticles for Molecular Imaging. *Molecules* **2019**, *24*, 1224. [CrossRef]
108. Bhavesh, R.; Lechuga-Vieco, A.V.; Ruiz-Cabello, J.; Herranz, F. T_1-MRI Fluorescent Iron Oxide Nanoparticles by Microwave Assisted Synthesis. *Nanomaterials* **2015**, *5*, 1880–1890. [CrossRef]
109. Pellico, J.; Ruiz-Cabello, J.; Saiz-Alía, M.; del Rosario, G.; Caja, S.; Montoya, M.; Fernández de Manuel, L.; Morales, M.P.; Gutiérre, L.; Galiana, B.; et al. Fast synthesis and bioconjugation of ^{68}Ga core-doped extremely small iron oxide nanoparticles for PET/MR imaging. *Contrast Media Mol. Imaging* **2016**, *11*, 203–210. [CrossRef]
110. Ju, K.Y.; Lee, J.W.; Im, G.H.; Lee, S.; Pyo, J.; Park, S.B.; Lee, J.H.; Lee, J.K. Bio-inspired, melanin-like nanoparticles as a highly efficient contrast agent for T_1-weighted magnetic resonance imaging. *Biomacromolecules* **2013**, *14*, 3491–3497. [CrossRef]
111. Peng, Y.K.; Liu, C.L.; Chen, H.C.; Chou, S.W.; Tseng, W.H.; Tseng, Y.J.; Kang, C.C.; Hsiao, J.K.; Chou, P.T. Antiferromagnetic iron nanocolloids: A new generation in vivo T_1 mri contrast agent. *J. Am. Chem. Soc.* **2013**, *135*, 18621–18628. [CrossRef] [PubMed]
112. Wu, M.; Meng, Q.; Chen, Y.; Xu, P.; Zhang, S.; Li, Y.; Zhang, L.; Wang, M.; Yao, H.; Shi, J. Ultrasmall confined Iron oxide nanoparticle MSNs as a pH-responsive theranostic platform. *Adv. Funct. Mater.* **2014**, *24*, 4273–4283. [CrossRef]
113. Li, Z.; Wang, S.X.; Sun, Q.; Zhao, H.L.; Lei, H.; Lan, M.B. Ultrasmall Manganese Ferrite Nanoparticles as Positive Contrast Agent for Magnetic Resonance Imaging. *Adv. Healthc. Mater.* **2013**, *2*, 958–964. [CrossRef] [PubMed]
114. Huang, G.; Li, H.; Chen, J.; Zhao, Z.; Yang, L.; Chi, X.; Chen, Z.; Wang, X.; Gao, J. Tunable T_1 and T_2 contrast abilities of manganese-engineered iron oxide nanoparticles through size control. *Nanoscale* **2014**, *6*, 10404–10412. [CrossRef] [PubMed]
115. Zhang, M.; Cao, Y.; Wang, L.; Ma, Y.; Tu, X.; Zhang, Z. Manganese doped iron oxide theranostic nanoparticles for combined T_1 magnetic resonance imaging and photothermal therapy. *ACS Appl. Mater. Interfaces* **2015**, *7*, 4650–4658. [CrossRef] [PubMed]
116. Lee, N.; Yoo, D.; Ling, D.; Cho, M.H.; Hyeon, T.; Cheon, J. Iron Oxide Based Nanoparticles for Multimodal Imaging and Magnetoresponsive Therapy. *Chem. Rev.* **2015**, *115*, 10637–10689. [CrossRef] [PubMed]

117. Pellico, J.; Fernández-Barahona, I.; Benito, M.; Gaitán-Simón, Á.; Gutiérrez, L.; Ruiz-Cabello, J.; Herranz, F. Unambiguous detection of atherosclerosis using bioorthogonal nanomaterials. *Nanomed. Nanotechnol. Biol. Med.* **2019**, *17*, 26–35. [CrossRef]
118. Sandiford, L.; Phinikaridou, A.; Protti, A.; Meszaros, L.K.; Cui, X.; Yan, Y.; Frodsham, G.; Williamson, P.A.; Gaddum, N.; Botnar, R.M.; et al. Bisphosphonate-anchored pegylation and radiolabeling of superparamagnetic iron oxide: Long-circulating nanoparticles for in vivo multimodal (T_1 MRI-SPECT) imaging. *ACS Nano* **2013**, *7*, 500–512. [CrossRef]
119. Wang, G.; Gao, W.; Zhang, X.; Mei, X. Au Nanocage Functionalized with Ultra-small Fe_3O_4 Nanoparticles for Targeting T_1–T_2 Dual MRI and CT Imaging of Tumor. *Sci. Rep.* **2016**, *6*, 28258. [CrossRef]
120. Pellico, J.; Ellis, C.M.; Davis, J.J. Nanoparticle-Based Paramagnetic Contrast Agents for Magnetic Resonance Imaging. *Contrast Media Mol. Imaging* **2019**, *2019*, 1845637. [CrossRef]

© 2020 by the authors. Licensee MDPI, Basel, Switzerland. This article is an open access article distributed under the terms and conditions of the Creative Commons Attribution (CC BY) license (http://creativecommons.org/licenses/by/4.0/).

Communication

Magnetic Composite Submicron Carriers with Structure-Dependent MRI Contrast

Anastasiia A. Kozlova [1],*, Sergey V. German [2,3], Vsevolod S. Atkin [4], Victor V. Zyev [5], Maxwell A. Astle [6], Daniil N. Bratashov [1], Yulia I. Svenskaya [4] and Dmitry A. Gorin [3]

1. Biomedical Photoacoustics Lab, Saratov State University, 410012 Saratov, Russia; bratashovdn@info.sgu.ru
2. Laboratory of Objects and Spectroscopy of Nanoobjects of Molecular Spectroscopy Department, Institute of Spectroscopy of the Russian Academy of Sciences (ISAN), 108840 Troitsk, Russia; gsv0709@mail.ru
3. Skolkovo Institute of Science and Technology, Skolkovo Innovation Center, 121205 Moscow, Russia; gorinda@mail.ru
4. Research and Educational Institute of Nanostructures and Biosystems, Saratov State University, 410012 Saratov, Russia; atkin.vsevolod@gmail.com (V.S.A.); yulia_svenskaya@mail.ru (Y.I.S.)
5. Scientific Research Institute of Fundamental and Clinical Uronephrology, Saratov State Medical University, 410012 Saratov, Russia; zuev.viktor.sgmu@gmail.com
6. School of Chemistry, University of Nottingham, University Park, Nottingham NG7 2RD, UK; maxwell.astle@nottingham.ac.uk
* Correspondence: anastasia.kozlova245@yandex.ru; Tel.: +7-906-308-4282

Received: 11 December 2019; Accepted: 27 January 2020; Published: 30 January 2020

Abstract: Magnetic contrast agents are widely used in magnetic resonance imaging in order to significantly change the signals from the regions of interest in comparison with the surrounding tissue. Despite a high variety of single-mode T_1 or T_2 contrast agents, there is a need for dual-mode contrast from the one agent. Here, we report on the synthesis of magnetic submicron carriers, containing Fe_3O_4 nanoparticles in their structure. We show the ability to control magnetic resonance contrast by changing not only the number of magnetite nanoparticles in one carrier or the concentration of magnetite in the suspension but also the structure of the core–shell itself. The obtained data open up the prospects for dual-mode T_1/T_2 magnetic contrast formation, as well as provides the basis for future investigations in this direction.

Keywords: magnetically-guided drug delivery systems; magnetite nanoparticles; magnetic resonance imaging (MRI); polymeric core–shells; magnetic submicron core–shells; drug delivery

1. Introduction

The image quality and resolution in magnetic resonance imaging (MRI) depend on the working mode, which choice is determined by the type of administrated contrast agent that reduces the longitudinal (T_1) or transverse (T_2) relaxation times. T_1 contrast agents are more preferable for clinical applications as they provide a stronger (brighter) signal of the regions of interest in comparison to the surrounding tissue and, therefore, provide easy detection and further analysis. Gadolinium (Gd) complexes have been widely used as T_1 MRI contrast agents for decades. However, in recent years there has been an ongoing debate about whether the benefits of their application generally outweigh the associated risks, as Gd-based complexes, especially in the case of linear (open-chained) ligands, can accumulate in brain tissue [1,2] as well as in the bone [3,4] and central nervous system [5] even in healthy patients. For patients with renal failure, these agents (based on linear Gd-complexes) constitute a high risk, because the moderate amount of them remains within the body for more than 24 h [6] and could not be effectively removed from the organism [7]. However, a recent study [8] reports about gadolinium-based contrast agents' ability to cross the blood cell membrane and their retention in erythrocytes and leukocytes. Although the toxicity of gadolinium depends on its stability in a particular complex, the risk of neural tissue retention exists for all gadolinium-based agents [4].

Graphene oxide, doped with Gd, was used to provide T_1 contrast and diminish toxicity [9]. However, in case of safe and effective choice, iron oxide nanoparticles are still preferable to use as a contrast agent. They degrade in the organism and could be cleared out as soon as 12 h after administration [10]. Moreover, ultrasmall (<4 nm) iron oxide nanoparticles provide high T_1 contrast, r_1 relaxivity reaches 20 mM^{-1} s^{-1} with the use of various ligands [7,10]. At this dimension, particles have reduced magnetic volume along with increased surface area and surface defects providing paramagnetic properties [11]. Iron oxide nanoparticle incorporation into stable material such as SiO_2 provides T_1 contrast enhancement compared to the colloid state, nearly equal to commercially available agents [12]. Nevertheless, the T_1 contrast increase leads to a reduction in T_2 contrast agent efficiency.

Both T_1 and T_2 MRI contrast agents are used to visualize drug delivery carriers and track them during the time. Despite T_1-weighted images are preferred for clinical practice, T_2 mode allows for investigation in inflamed areas of fluid-rich tissues. Therefore, recent MRI studies have been aimed at the development of agents and drug delivery systems providing dual-mode T_1/T_2 contrast. The most common way to create dual contrast agents is to combine materials with opposite contrast properties [13,14], such as iron oxide nanoparticles and gadolinium [15]. Another way is the use of iron oxide particles with the appropriate size and magnetization: Nanoparticles of more than 10 nm in diameter are typically T_2 contrast agents, where the size less than 4 nm provides high T_1 contrast [16]. However, nanoparticles of 3–10 nm in diameter were shown to possess dual MRI contrast, depending on the size [16], shape [17], and surface functionalization [7]. In vivo dual T_1/T_2 MRI contrast after contrast agent administration was shown in [18] for both liver and kidney of the study mouse.

In this study, we developed the drug delivery systems with controllable structure-dependent MRI contrast, based on dual-mode contrast Fe_3O_4 nanoparticles. These carriers are able not only to change the contrast after their degradation [19] but also to have various T_1 and T_2 contrast properties depending on the amount and localization of iron oxide nanoparticles.

2. Results

2.1. Magnetite Nanoparticle Synthesis

Magnetite nanoparticle colloid was synthesized by the chemical precipitation method in an inert atmosphere. From our point of view, the size of magnetite around 10 nm is optimal to have contrast for both T_1 and T_2, because the following increasing of nanoparticle size leads to significant reducing of r_1 but decreasing of nanoparticle size induces the significant decreasing of r_2 [20]. The concentration of magnetite colloid was 0.72 mg/mL (from the colorimetric titration method). The transmission electron microscopy (TEM) image of Fe_3O_4 nanoparticles is presented in Figure 1a. The average diameter, measured by dynamic light scattering (DLS, Figure 1b), was 9.3 ± 2.7 nm.

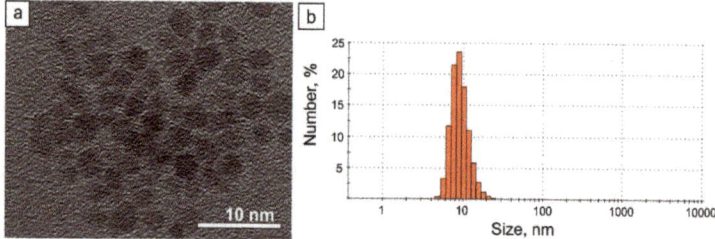

Figure 1. Characterization of synthesized magnetite nanoparticles: (**a**) TEM image and (**b**) size distribution, measured by dynamic light scattering.

2.2. Magnetic Polymeric Submicron Core–Shells

The three different types of magnetic submicron core–shells were formed to study the dependence of the magnetic nanoparticles amount and core–shell structure on their MRI properties. The average size of the submicron core–shells was 576 ± 102 nm. Each type had a shell formed by poly(allylamine hydrochloride) (PAH) and poly(sodium 4-styrenesulfonate) (PSS) by layer-by-layer (LbL) technique [21]. The first type (so-called S, from the shell, Figure 2a,b) contained only one layer of magnetite nanoparticles in the shell. In the previous study [19] we had shown that the variation of the package density of magnetite nanoparticles in the layer of the microcapsule shell changes the resulting MRI contrast. With the less distance between nanoparticles in the layer, the contrast (the ratio of the sample signal to the water signal) and r_1 and r_2 relaxivities decreased. This can be caused by the increasing role or magnetic interactions between nanoparticles when the distance becomes comparable to their size because of close packing [22]. In the current study, the package density of magnetite that provides the highest MRI contrast was chosen according to previous results [19].

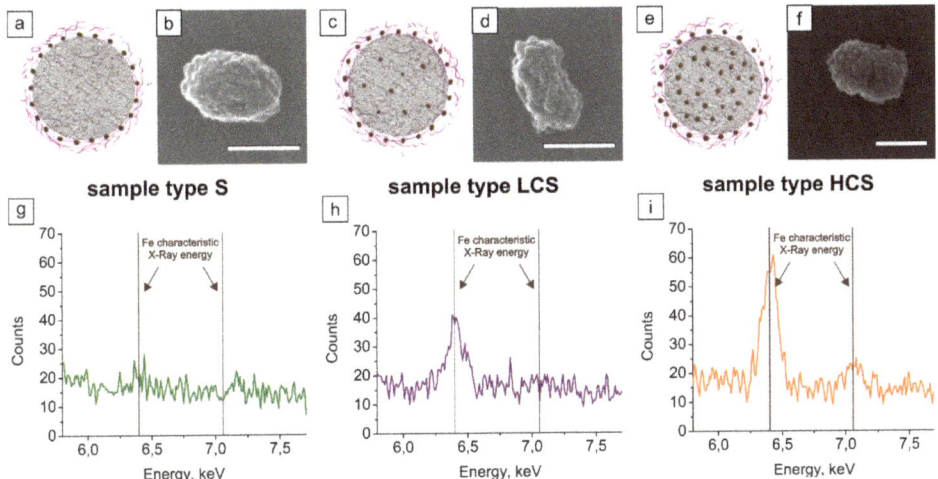

Figure 2. Scheme of (**a**) S, (**c**) LCS, and (**e**) HCS submicron core–shells; SEM images of (**b**) S, (**d**) LCS, and (**f**) HCS and EDX analysis of (**g**) S, (**h**) LCS, and (**i**) HCS samples. Scale bar: 500 nm.

The second type of submicron core–shells (LCS, from low-core plus shell, Figure 2c,d) contained one layer of magnetite in the shell along with Fe_3O_4 nanoparticles in the inner volume of the core–shell, loaded by the freezing-induced loading (FIL) method [23]. Finally, the third type (HCS, high-core+shell, Figure 2e,f) was the submicron core–shells analogous to LCS, but with four FIL loadings of magnetite. SEM images of obtained core-shells are presented in Figure S1 (Supplementary Materials).

The presence of iron (III) oxide nanoparticles was confirmed by energy-dispersive X-Ray spectroscopy (EDX), showing the two peaks at Kα = 6.3996 keV and Kβ = 7.058 keV Fe characteristic energies for the HCS sample with the highest Fe content (Figure 2i). With the decrease in magnetite concentration in the sample, only one peak remains, as demonstrated for the LCS sample in Figure 2h. However, some sensitivity threshold exists, below which the magnetite content is too low to be determined. This happens for the S sample with iron (III) oxide nanoparticles only in the shell, so we cannot see any specific Fe peaks at the EDX spectrum (Figure 2g). The transmission electron microscopy (TEM) images of individual submicron core-shells are presented in Figure 3.

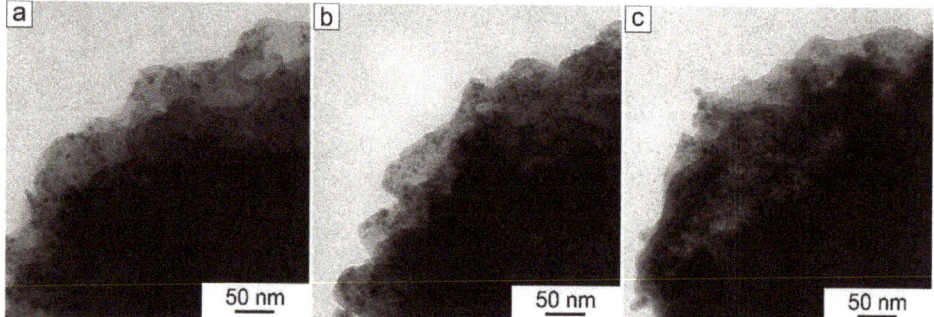

Figure 3. TEM images of (**a**) S, (**b**) LCS, and (**c**) HCS submicron core–shell.

The amount of magnetite loading for each core–shell suspension was measured by colorimetric titration. First, the core–shells were disrupted by the addition of sulfuric acid. Then, the qualitative reaction of Fe ions with ammonium rhodanide was carried out, and the resulting absorption of the solution was measured by a Synergy H1 multi-plate reader. To calculate the magnetite amount, the level of absorption was compared with that for the standard Fe solution. All of the measurements were made at least three times, and the resulting magnetite concentrations are shown in Table 1. The mass amount of Fe_3O_4 in one carrier (from 1.78 ± 0.01 pg for the S sample to 7.95 ± 0.13 pg for the HCS sample) was calculated with respect to the average core–shell number in each sample suspension. The dynamics of loading within the sample strongly correlates with the EDX data.

Table 1. Amount of magnetite nanoparticles loading into submicron core–shells data.

Type of Sample	Core–Shell's Structure	Amount of Fe_3O_4 Loaded, mg	The Mass Amount of Magnetite per Carrier, pg
S	PAH/PSS/PSH/MNPs/PAH/PSS	0.445 ± 0.003	1.78 ± 0.01
LCS	MNPs^{low}@PAH/PSS/PSH/MNPs/PAH/PSS	0.605 ± 0.003	2.42 ± 0.01
HCS	MNPs^{high}@PAH/PSS/PSH/MNPs/PAH/PSS	1.987 ± 0.033	7.95 ± 0.13

In order to measure the movement rate of the obtained core–shells in an external magnetic field, we used photosedimentometry. The setup of the experiment is shown in Figure 4a. The core–shell suspension was irradiated by 660 nm laser, and the transparency was measured during the time (see graphs in Figure 4b). Dependent on the magnetite content, the speed of the core–shells movement to the magnet changed. So, we calculated the average movement rate according to Equation (1):

$$v_i = \frac{S}{t_i}, \tag{1}$$

where S is the distance between the cuvette wall and the laser, t_i is a time of movement, v_i is movement rate, and I the number of a sample. We obtained the movement rates as follows: v_S = 54.5 µm/s, v_{LCS} = 10.2 µm/s, and v_{HCS} = 3.2 µm/s. Therefore, the HCS core–shells with the highest amount of magnetite possess a higher magnetic moment compared with other samples, so it can be controlled better and easier navigated as a drug delivery carrier.

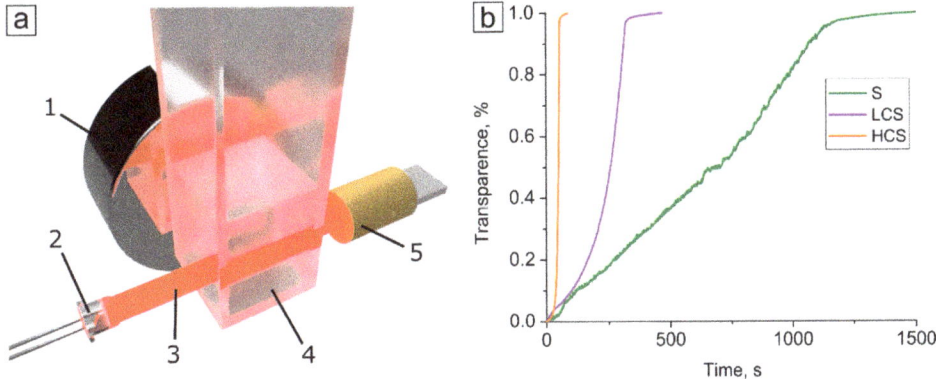

Figure 4. Photosedimentometry of submicron core–shells in water during the time: (**a**) The experimental setup, where 1—permanent magnet, 2—phototransistor, 3—laser beam, 4—cuvette with core–shell suspension, 5—semiconductor laser; (**b**) time dependence of the core–shell suspension transparency in an external magnetic field.

2.3. MRI Study

The obtained core–shell suspensions were investigated by MRI. We obtained T_1 and T_2-weighted images in three different modes: T_1 spin-echo (SE), T_1 gradient echo (FFE), and T_2 turbo spin-echo (TSE). The visualization parameters remained unchanged for all the samples within one visualization mode.

2.3.1. Magnetite Nanoparticle Colloid

The iron oxide nanoparticle colloid possessed a dual T_1/T_2 MRI contrast (Figure 5). This happened because the average size of magnetite nanoparticles was in the range from 4 to 10 nm that provides both T_1 and T_2 contrast [16]. T_1-weighted images demonstrated a significant increase in the signal in comparison to water (surrounding media in the MRI images), stronger for the gradient than for spin-echo mode. The T_2-weighted image had the signal reduction for concentrations less than 0.04 mg/mL, indicating the nanoparticles as a T_2 contrast agent as well.

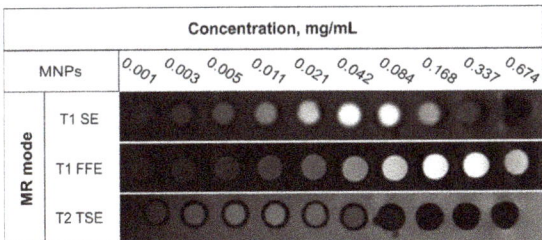

Figure 5. MRI images of magnetite nanoparticle colloid.

2.3.2. Magnetic Submicron Core–Shells

The core–shell suspensions were diluted to various carrier concentrations in the sample prior to MRI visualization since the concentrations of objects were the same for all samples. Thus, we obtained the magnetic resonance images showing the dependence of the sample contrast on its magnetite content (Figure 6). The sample S with the lowest iron(III) oxide content was shown to provide better MRI contrast in the T_1 FFE mode in comparison with LCS and HCS samples. As it can be observed from Figure 3a, the package density of magnetite nanoparticles in the shell of the carrier corresponds to the previously mentioned assumption about the role of magnetic interactions in the resulting MRI contrast.

However, the maximum of T_2 contrast can be observed after the addition of magnetite to the core of the carriers for the LCS sample, although the HCS core–shell suspension demonstrates MRI contrast only in T_1 SE mode.

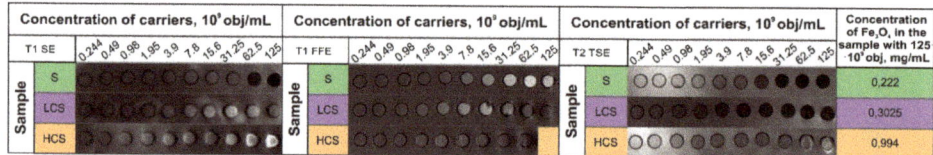

Figure 6. MRI images of magnetic submicron core–shells.

3. Discussion

For both T_1 and T_2 modes we obtained the dependencies of signal intensity, normalized to water, on the magnetite concentration by analyzing the previously shown images (Figures 5 and 6). As for the magnetite colloid, it had a high positive signal intensity (SI) change in both T_1 SE and FFE modes. Although the maximum contrast was around 550% for SE sequence (Figure 7a) and up to 1700% for FFE (Figure 7b), the concentration range, in which the contrast can be observed, is almost four times wider for the gradient mode. However, nearly full decay of a signal was obtained for the T_2 TSE mode that allows using the magnetite colloid as an effective negative contrast agent at concentrations more than 0.04 mg/mL. The forms of contrast curves are typical for the Fe_3O_4 colloid [19].

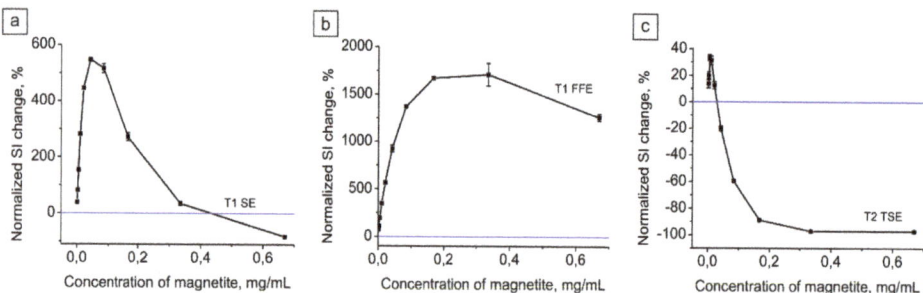

Figure 7. MRI contrast dependence on magnetite colloid concentration: (**a**) T_1 spin-echo and (**b**) T_1 gradient echo; (**c**) T_2 turbo spin-echo. The blue line on each graph–water contrast level.

The obtained MRI images for the core–shell suspensions were also analyzed. Figure 8 shows the dependencies of normalized to water SI change on the concentration of carriers. In T_2 mode (Figure 8c), the full decay of a signal occurred only for the S sample until 0.05 mg/mL of Fe_3O_4 concentration, which is similar to the magnetite colloid behavior (Figure 7c). When the amount of nanoparticles increases (LCS and HCS samples), the T_2 contrast tends to zero. This can be explained by the intensive magnetic interactions between nanoparticles while the distance between them decreases because of their higher concentration and, thus, the density of packaging [19].

For the gradient sequence in T_1 mode, the tendency was generally the same. As it is shown in Figure 8b, the SI change still falls down with more magnetite content in the core–shell. However, the width of the contrast area has the same behavior—the highest for the S sample and the lowest for the HCS one. In the T_1 spin-echo mode the tendency was surprisingly opposite. The S sample showed the lowest SI change no more than 30%, but it significantly increased with the loading of magnetite into the core. Moreover, the HCS sample provided 209% SI change on average even at high magnetite concentration that provides its usage as a positive contrast agent at a wide concentration range along with an ability of control and navigation by an external magnetic field.

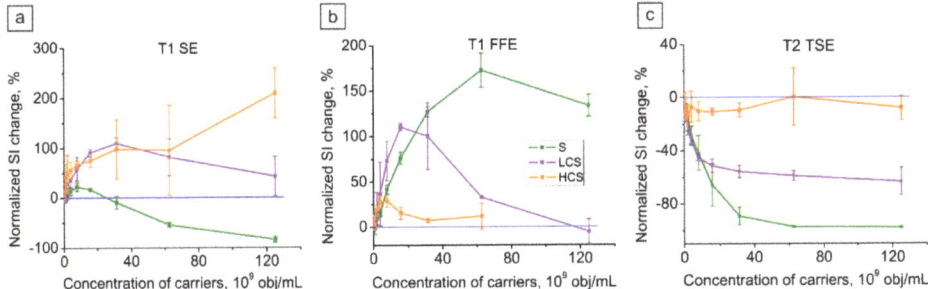

Figure 8. MRI contrast dependence on Fe concentration in submicron core–shells suspension for (**a**) T_1 spin-echo, (**b**) T_1 gradient-echo, and (**c**) T_2 turbo spin-echo: green curve—S sample; purple curve—LCS, and orange curve—HCS submicron core–shells. The blue line on each graph–water contrast level.

To understand whether the characteristic form of magnetite colloid contrast curve remains unchanged after its loading into the core–shells, we calculated the curve form change for each sample according to the following Equation (2):

$$C_i = \frac{N_i}{N_{max}}, \qquad (2)$$

where C_i is the curve form change and N_i is the normalized SI change for the current sample concentration; N_{max} is the maximum normalized SI change for the sample.

The obtained curves are presented in Figure 9. As can be seen from Figure 9a, for the T_1 spin-echo mode the more the amount of Fe_3O_4 presents in the sample, the more the concentration at which the contrast can be observed. The T_1 gradient mode (Figure 9b) demonstrates the opposite behavior, and here with the addition of magnetite to the sample the concentration at which the maximum contrast occurs decreases along with the decrease in contrast properties (Figure 8b). However, in T_2 mode S and LCS samples, unlike the magnetite colloid, possess only negative contrast even for low concentrations, which makes them effective negative contrast agents. However, there is some threshold concentration that exists, after which the sample loses the characteristic for magnetite contrast curve form. For example, this occurred for HCS suspension in T_1 SE and T_2 TSE modes.

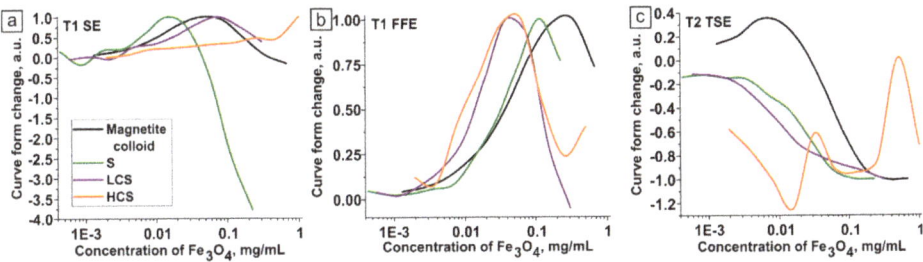

Figure 9. Curve form changes for magnetite colloid and core–shell suspensions in (**a**) T1 SE, (**b**) T1 FFE, and (**c**) T2 TSE modes.

4. Materials and Methods

4.1. Materials

Iron (III) chloride hexahydrate (99.8%, Sigma-Aldrich, St. Louis, MO, United States), iron (II) chloride tetrahydrate (99.8%, Sigma-Aldrich), sodium hydroxide (99.8%, Fluka, Buchs, Switzerland), citric acid (99.8%, Sigma-Aldrich), sodium carbonate, calcium chloride, poly(allylamine hydrochloride)

(PAH, average MW ~70 kDa), poly(sodium 4-styrenesulfonate) (PSS, average MW 70 kDa), ethylene diamine tetraacetic acid disodium salt (Fluka), sodium chloride, glycerin (Reachem, Moscow, Russia) were used without additional purification.

4.2. Magnetite Nanoparticle Synthesis

Magnetic nanoparticles were obtained by chemical precipitation from di- and trivalent salts of iron in the presence of the base [24,25]. Initially, 1.3 g of $FeCl_3 \cdot 6H_2O$ and 0.478 g of $FeCl_2 \cdot 4H_2O$ were dissolved in water under room temperature with mixing. Then, 170 mL of 0.1 M NaOH was added to the reaction cell. For further colloids stabilization, 25 mL of citric acid (32 mg/mL) solution was prepared. A nitrogen was bubbled across the closed cells with mixtures of iron salts, sodium hydroxide, and citric acid for 10 min to remove oxygen from the solutions. Further, the iron salts solutions were injected into the sodium hydroxide solution after the heating until 40 °C with active mixing, after that the solution was left under active mixing and nitrogen pressure for 40 s resulting in black sediment formation of magnetite nanoparticles. Additionally, 25 mL of citric acid was added to the suspension under constant mixing and nitrogen pressure. Dialysis of magnetic hydrosol was conducted during three days in a 3 L vial under slow mixing. Mixing of reagents and washing steps were carried out under nitrogen.

4.3. Fabrication of Magnetic Polymeric Submicron Core–Shells

The submicron vaterite particles were synthesized in glycerol from $CaCl_2$ and Na_2CO_3 salt solutions according to the method described in [23]. First, 400 µL of each salt solution were added to 4 g of glycerol at a constant stirring, and the resulted solution was left mixing for 1 h for $CaCO_3$ submicroparticles formation. Then, the suspension was centrifuged, and the submicroparticles pellet was washed three times with deionized water. After that, the FIL stage was performed for the LCS and HCS samples: Prepared magnetite colloid of 2 mL were added to the sample, and it was freezed while constantly stirring [23]. Freezing/thawing cycles were repeated 1 and 4 times for the LCS and HCS samples consequently. Then, the polymeric shells were formed at all three samples by LbL technique [21] from PAH and PSS polyelectrolyte solutions in concentration of 1 mg/mL. One layer of Fe_3O_4 nanoparticles in concentration of 0.56 mg/mL was adsorbed into the shell. For each shell layer formation, 1 mL of absorbing solution was added to the sample. As a result, the following shells were formed on the template surface: PAH/PSS/PAH/MNPs/PSS/PAH/PSS. The core–shells made in the current paper remained unchanged for at least six months and retained their zeta-potential.

4.4. Colorimetric Titration Measurements

The concentrations of magnetite in all the samples were measured by colorimetric titration, based on the qualitative reation of Fe^{3+} ions with ammonium thiocyanate. First, the magnetite colloid/core–shell suspension was dissolved in 1M H_2SO_4 solution. Then, 20 µL of ammonium thiocyanate were added to the sample under analysis, and the absorption level at the 473 nm wavelength was measured. Analogous to it, we measured the absorption of the standard Fe^{3+} solution in 1 M H_2SO_4 with the known concentration. The final Fe^{3+} ion concentration in the aliquot was calculated according to Equation (3):

$$C_{Fe} = \frac{C_{St} \cdot n_{Fe}}{n_{St}}, \qquad (3)$$

where C_{Fe} is the concentration of Fe^{3+} ions in the aliquot, C_{St} is the concentration of Fe^{3+} ions in the standard solution, n_{Fe} and n_{St} are absorption levels of the aliquot and the standard solution consequently. The final magnetite amount was calculated from the received data.

4.5. TEM, DLS, SEM, and EDX Characterization

The measurements of the size distribution of the nanoparticles were performed using a Zetasizer Nano ZS instrument (Malvern Instruments Ltd, Malvern, UK).

Transmission electron microscopy (TEM) of magnetite nanoparticles was performed using a FEGTEM microscope (JEOL, Akishima, Tokyo, Japan) operating at 200 kV. TEM samples were prepared via a drop-casting technique, where samples were dispersed and deposited onto copper grid mounted "lacey" carbon films (Agar). All images were processed using the Gatan Digital Micrograph software.

TEM of submicron core–shells' structure was performed using a Libra 120 Carl Zeiss microscope (Carl Zeiss SMT AG Company, Oberkochen, Germany) operating at 120 kV. The measurement was made on the basis of the Center of collective usage of scientific equipment in physico-chemical biology and nanobiotechnology "Simbioz", Institute of Biochemistry and Physiology of Plants and Microorganisms Russian Academy of Sciences (IBPPM RAS).

Submicron core–shell morphology and chemical composition were investigated using a scanning electron microscope (SEM) MIRA II LMU (Tescan Orsay Holding, Brno, Czech Republic), equipped with INCA Energy energy-dispersive spectroscopy system. The SEM imaging was performed at the operating voltage of 30 kV under standard secondary and back-scattered electron modes. For the study of chemical composition, the energy-dispersive X-ray spectroscopy (EDX) was used.

4.6. Dynamics of Submicron Core–Shell Sedimentation

Submicron core–shell sedimentation rates in an external magnetic field were measured by the photosedimentometry. The suspension of core–shells in water was placed in the transparent cuvette (Figure 4a, number 4), through which the laser beam (Figure 4a, number 3) was going. We used a 660 nm semiconductor laser for the sample excitation. The permanent magnet (Figure 4a, number 1) was applied to the cuvette's wall, so the magnetic field was directed perpendicularly to the laser path. After that the laser goes through the analyzing suspension and the incident beam reaches the phototransistor (Figure 4a, number 2). Using the described system, the time dependencies of the core–shell suspension transparency in an external magnetic field were obtained. The core–shell movement rates were then calculated.

4.7. MRI Study

In vitro MRI study was carried out using a Philips Achieva 1.5 T high field MRI scanner (Philips Healthcare, Best, The Netherlands) equipped with a phased array coil. For obtaining T_1- and T_2-weighted images spin-echo sequences (SE) and gradient-echo sequences (FFE) were applied. The following parameters were used for conducting measurements: The repetition time (TR) was 450 ms and the echo time (TE) was 15 ms for the T_1-weighted pulse SE sequence; the TR was 61 ms and the TE was 2.484 ms for the T_1-weighted pulse FFE sequence; the TR was 3000 ms and the TE was 47.7 ms for the T_2-weighted pulse sequence. In order to increase MR signal in the T_1-weighted image the TR should be decreased, so tissues do not manage to relax after an external influence, and the TE should be decreased too, because the sample highly changes the T_2 relaxation and it can influence on the resulting MR signal in a T_1-weighted image. At the same time, the T_2-weighted image requires to have both long TR and TE [26].

5. Conclusions

In the current study, we demonstrated the fabrication of magnetic composite submicron carriers with an ability to control the magnetic resonance contrast properties by changing their structure. We found that the highest T_2 and T_1 gradient contrast can be observed from the core–shell suspension with magnetite nanoparticles contained only in the shell. The contrast properties in these two modes become worse with the addition of magnetite nanoparticles by freezing-induced loading and, thus, increasing their packaging density and interactions between them. Nevertheless, the HCS core–shell suspension with 4 FIL loading of magnetite surprisingly possessed the 209% normalized SI change even for the high magnetite concentration in T_1 SE mode. Thus, one can combine a high T_1 contrast, not characteristic for the magnetic capsules, and their control by a magnetic field. However, if choosing the appropriate structure, it can be possible to observe both T_1 and T_2 contrasts from the carrier for remote drug delivery and controlled release, all with the verification by MRI visualization, to which our further studies will be devoted.

Supplementary Materials: The following are available online at http://www.mdpi.com/2304-6740/8/2/11/s1, Figure S1: SEM images of (**a**) S, (**b**) LCS, and (**c**) HCS submicron core–shells.

Author Contributions: Conceptualization and investigation, A.A.K., S.V.G., Y.I.S., D.A.G., and D.N.B.; MRI visualization, V.V.Z.; TEM measurements, M.A.A.; SEM measurements and EDX analysis, V.S.A.; Writing—original draft preparation, A.A.K.; Writing—review and editing, Y.I.S., S.V.G., D.A.G., V.S.A., and D.N.B.; Supervision, D.A.G. and D.N.B.; Funding acquisition, D.A.G., D.N.B. and Y.I.S. All authors have read and agreed to the published version of the manuscript.

Funding: This research was supported by the Russian Science Foundation (grant number 18-19-00354), Russian Ministry of Education and Science (project No 16.8144.2017/9.10), by the Russian Foundation for Basic Research (RFBR grant 18-29-08046).

Acknowledgments: The experimental part of this work related to the preparation of remote-controlled magnetic core–shells was supported by the Russian Science Foundation (grant number 18-19-00354). The transmission electron microscopy and magnetic resonance imaging were supported by the Russian Foundation for Basic Research (RFBR grant 18-29-08046). The scanning electron microscopy measurements and photosedimentonetry were supported by the Russian Ministry of Education and Science (project No 16.8144.2017/9.10). The authors would like to thank the Center of collective usage of scientific equipment in physico-chemical biology and nanobiotechnology "Simbioz", Institute of Biochemistry and Physiology of Plants and Microorganisms Russian Academy of Sciences (IBPPM RAS) for their kind help in the transmission electron microscopy measurements.

Conflicts of Interest: The authors declare no conflict of interest.

References

1. Roberts, D.R.; Holden, K.R. Progressive increase of T1 signal intensity in the dentate nucleus and globus pallidus on unenhanced T1-weighted MR images in the pediatric brain exposed to multiple doses of gadolinium contrast. *Brain Dev.* **2016**, *38*, 331–336. [CrossRef] [PubMed]
2. Pullicino, R.; Radon, M.; Biswas, S.; Bhojak, M.; Das, K. A Review of the Current Evidence on Gadolinium Deposition in the Brain. *Clin. Neuroradiol.* **2018**, *28*, 159–169. [CrossRef] [PubMed]
3. White, G.W.; Gibby, W.A.; Tweedle, M.F. Comparison of Gd(DTPA-BMA) (Omniscan) Versus Gd(HP-DO3A) (ProHance) Relative to Gadolinium Retention in Human Bone Tissue by Inductively Coupled Plasma Mass Spectroscopy. *Invest. Radiol.* **2006**, *41*, 272–278. [CrossRef] [PubMed]
4. Layne, K.A.; Dargan, P.I.; Archer, J.R.H.; Wood, D.M. Gadolinium deposition and the potential for toxicological sequelae – A literature review of issues surrounding gadolinium-based contrast agents. *Br. J. Clin. Pharmacol.* **2018**, *84*, 2522–2534. [CrossRef]
5. Vergauwen, E.; Vanbinst, A.-M.; Brussaard, C.; Janssens, P.; De Clerck, D.; Van Lint, M.; Houtman, A.C.; Michel, O.; Keymolen, K.; Lefevere, B.; et al. Central nervous system gadolinium accumulation in patients undergoing periodical contrast MRI screening for hereditary tumor syndromes. *Hered. Cancer Clin. Pract.* **2018**, *16*, 2. [CrossRef]
6. Oksendal, A.N.; Hals, P.-A. Biodistribution and toxicity of MR imaging contrast media. *J. Magn. Reson. Imaging* **1993**, *3*, 157–165. [CrossRef]
7. Bao, Y.; Sherwood, J.A.; Sun, Z. Magnetic iron oxide nanoparticles as T 1 contrast agents for magnetic resonance imaging. *J. Mater. Chem. C* **2018**, *6*, 1280–1290. [CrossRef]
8. Di Gregorio, E.; Furlan, C.; Atlante, S.; Stefania, R.; Gianolio, E.; Aime, S. Gadolinium Retention in Erythrocytes and Leukocytes From Human and Murine Blood Upon Treatment With Gadolinium-Based Contrast Agents for Magnetic Resonance Imaging. *Invest. Radiol.* **2020**, *55*, 30–37. [CrossRef]
9. Zhang, M.; Liu, X.; Huang, J.; Wang, L.; Shen, H.; Luo, Y.; Li, Z.; Zhang, H.; Deng, Z.; Zhang, Z. Ultrasmall graphene oxide based T1 MRI contrast agent for in vitro and in vivo labeling of human mesenchymal stem cells. *Nanomedicine Nanotechnology, Biol. Med.* **2018**, *14*, 2475–2483. [CrossRef] [PubMed]

10. Stark, D.D.; Weissleder, R.; Elizondo, G.; Hahn, P.F.; Saini, S.; Todd, L.E.; Wittenberg, J.; Ferrucci, J.T. Superparamagnetic iron oxide: Clinical application as a contrast agent for MR imaging of the liver. *Radiology* **1988**, *168*, 297–301. [CrossRef] [PubMed]
11. Kim, B.H.; Lee, N.; Kim, H.; An, K.; Park, Y.I.; Choi, Y.; Shin, K.; Lee, Y.; Kwon, S.G.; Na, H.B.; et al. Large-Scale Synthesis of Uniform and Extremely Small-Sized Iron Oxide Nanoparticles for High-Resolution T 1 Magnetic Resonance Imaging Contrast Agents. *J. Am. Chem. Soc.* **2011**, *133*, 12624–12631. [CrossRef] [PubMed]
12. Starsich, F.H.L.; Eberhardt, C.; Keevend, K.; Boss, A.; Hirt, A.M.; Herrmann, I.K.; Pratsinis, S.E. Reduced Magnetic Coupling in Ultrasmall Iron Oxide T 1 MRI Contrast Agents. *ACS Appl. Bio Mater.* **2018**, *1*, 783–791. [CrossRef]
13. Shin, T.-H.; Choi, J.; Yun, S.; Kim, I.-S.; Song, H.-T.; Kim, Y.; Park, K.I.; Cheon, J. T 1 and T 2 Dual-Mode MRI Contrast Agent for Enhancing Accuracy by Engineered Nanomaterials. *ACS Nano* **2014**, *8*, 3393–3401. [CrossRef] [PubMed]
14. Peng, Y.-K.; Lui, C.N.P.; Chen, Y.-W.; Chou, S.-W.; Raine, E.; Chou, P.-T.; Yung, K.K.L.; Tsang, S.C.E. Engineering of Single Magnetic Particle Carrier for Living Brain Cell Imaging: A Tunable T 1 -/T 2 -/Dual-Modal Contrast Agent for Magnetic Resonance Imaging Application. *Chem. Mater.* **2017**, *29*, 4411–4417. [CrossRef]
15. Szpak, A.; Fiejdasz, S.; Prendota, W.; Strączek, T.; Kapusta, C.; Szmyd, J.; Nowakowska, M.; Zapotoczny, S. T1–T2 Dual-modal MRI contrast agents based on superparamagnetic iron oxide nanoparticles with surface attached gadolinium complexes. *J. Nanoparticle Res.* **2014**, *16*, 2678. [CrossRef]
16. Shen, Z.; Wu, A.; Chen, X. Iron Oxide Nanoparticle Based Contrast Agents for Magnetic Resonance Imaging. *Mol. Pharm.* **2017**, *14*, 1352–1364. [CrossRef]
17. Sharma, V.K.; Alipour, A.; Soran-Erdem, Z.; Aykut, Z.G.; Demir, H.V. Highly monodisperse low-magnetization magnetite nanocubes as simultaneous T 1 – T 2 MRI contrast agents. *Nanoscale* **2015**, *7*, 10519–10526. [CrossRef]
18. Li, Z.; Yi, P.W.; Sun, Q.; Lei, H.; Li Zhao, H.; Zhu, Z.H.; Smith, S.C.; Lan, M.B.; Lu, G.Q.M. Ultrasmall Water-Soluble and Biocompatible Magnetic Iron Oxide Nanoparticles as Positive and Negative Dual Contrast Agents. *Adv. Funct. Mater.* **2012**, *22*, 2387–2393. [CrossRef]
19. German, S.V.; Bratashov, D.N.; Navolokin, N.A.; Kozlova, A.A.; Lomova, M.V.; Novoselova, M.V.; Burilova, E.A.; Zyev, V.V.; Khlebtsov, B.N.; Bucharskaya, A.B.; et al. In vitro and in vivo MRI visualization of nanocomposite biodegradable microcapsules with tunable contrast. *Phys. Chem. Chem. Phys.* **2016**, *18*, 32238–32246. [CrossRef]
20. Zeng, J.; Jing, L.; Hou, Y.; Jiao, M.; Qiao, R.; Jia, Q.; Liu, C.; Fang, F.; Lei, H.; Gao, M. Anchoring Group Effects of Surface Ligands on Magnetic Properties of Fe 3 O 4 Nanoparticles: Towards High Performance MRI Contrast Agents. *Adv. Mater.* **2014**, *26*, 2694–2698. [CrossRef]
21. Sukhorukov, G.B.; Donath, E.; Davis, S.; Lichtenfeld, H.; Caruso, F.; Popov, V.I.; Möhwald, H. Stepwise polyelectrolyte assembly on particle surfaces: A novel approach to colloid design. *Polym. Adv. Technol.* **1998**, *9*, 759–767. [CrossRef]
22. Romodina, M.N.; Khokhlova, M.D.; Lyubin, E.V.; Fedyanin, A.A. Direct measurements of magnetic interaction-induced cross-correlations of two microparticles in Brownian motion. *Sci. Rep.* **2015**, *5*, 10491. [CrossRef] [PubMed]
23. German, S.V.; Novoselova, M.V.; Bratashov, D.N.; Demina, P.A.; Atkin, V.S.; Voronin, D.V.; Khlebtsov, B.N.; Parakhonskiy, B.V.; Sukhorukov, G.B.; Gorin, D.A. High-efficiency freezing-induced loading of inorganic nanoparticles and proteins into micron- and submicron-sized porous particles. *Sci. Rep.* **2018**, *8*, 17763. [CrossRef]
24. Massart, R. Preparation of aqueous magnetic liquids in alkaline and acidic media. *IEEE Trans. Magn.* **1981**, *17*, 1247–1248. [CrossRef]
25. German, S.V.; Inozemtseva, O.A.; Markin, A.V.; Metvalli, K.; Khomutov, G.B.; Gorin, D.A. Synthesis of magnetite hydrosols in inert atmosphere. *Colloid J.* **2013**, *75*, 483–486. [CrossRef]
26. Westbrook, C.; Kaut, C. *MRI in Practice*, 4th ed.; Wiley-Blackwell: Oxford, UK, 1998; Volume 37, ISBN1 0632042052. ISBN2 9780632042050.

© 2020 by the authors. Licensee MDPI, Basel, Switzerland. This article is an open access article distributed under the terms and conditions of the Creative Commons Attribution (CC BY) license (http://creativecommons.org/licenses/by/4.0/).

Article

^{1}H NMR Relaxometric Analysis of Paramagnetic Gd$_2$O$_3$:Yb Nanoparticles Functionalized with Citrate Groups

Fabio Carniato * and Giorgio Gatti

Department of Sciences and Technological Innovation, University of Eastern Piedmont, V. le T. Michel 11, 15121 Alessandria, Italy; giorgio.gatti@uniupo.it
* Correspondence: fabio.carniato@uniupo.it; Tel.: +39-0131360217

Received: 31 January 2019; Accepted: 27 February 2019; Published: 4 March 2019

Abstract: Gd$_2$O$_3$ nanoparticles doped with different amount of Yb^{3+} ions and coated with citrate molecules were prepared by a cheap and fast co-precipitation procedure and proposed as potential "positive" contrast agents in magnetic resonance imaging. The citrate was used to improve the aqueous suspension, limiting particles precipitation. The relaxometric properties of the samples were studied in aqueous solution as a function of the magnetic field strength in order to evaluate the interaction of the paramagnetic ions exposed on the surface with the water molecules in proximity. The nanoparticles showed high relaxivity values at a high magnetic field with respect to the clinically used Gd^{3+}-chelates and comparable to those of similar nanosytems. Special attention was also addressed to the investigation of the chemical stability of the nanoparticles in biological fluid (reconstructed human serum) and in the presence of a chelating agent.

Keywords: paramagnetic properties; gadolinium oxide; relaxation agents; nanoparticles

1. Introduction

The magnetic resonance imaging is currently one of the best diagnostic solutions adopted in clinic to identify different kinds of pathologies, due to the intrinsic high spatial resolution associated to the use of low energy radiation. More than 40% of the clinical practices require the administration to the patients of specific contrast agents. These probes are paramagnetic or super-paramagnetic compounds able to reduce the longitudinal (T_1) and/or the transversal relaxation time (T_2) of the water protons in tissues, thus increasing the sensitivity of the analysis, reducing possible artefacts, and improving the quality of the collected images [1–3]. Currently, the T_1-MRI probes used in clinics are mainly based on linear and cyclic Gd^{3+}-chelates, because they are characterized by good thermodynamic stability and kinetic inertness and often well tolerated by the patients [1]. Nevertheless, the amount of contrast agents required for the analysis is very high (from mM to M) and recently some studies reported on the accumulation of these probes in the cerebral membrane. To overcome these problems, three different strategies may be followed: (i) the design of novel paramagnetic probes based on less toxic metals (e.g., Mn^{2+}, Fe^{3+}, ...), (ii) the optimization of the Gd(III)-chelates to enhance the relaxivity performances; and (iii) the development of nanoparticles containing a large amount of paramagnetic sites, with high relaxivity at magnetic fields used in clinic (1.5–3 Tesla). This last strategy is particularly interesting because it favors a lowering of the detection limit of the MRI technique and a reduction of the contrast agent amount to administrate [4].

The paramagnetic nanoparticles proposed in the literature are composed using a diamagnetic inorganic support functionalized with Gd(III)-chelates, opportunely modified to promote their chemical attachment on the surface [5–10]; or they are completely inorganic [11,12]. In the last case, the particles contain directly in the framework paramagnetic Gd^{3+} ions, often combined with other lanthanides with specific properties [11]. Recently, GdF$_3$, NaGdF$_4$, and Gd$_2$O$_3$ systems were synthesized and proposed for diagnostic applications [11]. These nanoparticles typically require a careful design, aiming to reduce the particles size and to increase the surface hydrophilicity. For instance, Van Veggel et al, recently demonstrated that the best relaxometric performances could be achieved by decreasing the particles size to below 5 nm, with a consequent increase of the surface to volume ratio [13]. The selection of the capping agents to confine on the particles surface is another important topic. Particles with different anions and polymer groups were synthesized in order to increase the water suspendibility [14–16].

Gd$_2$O$_3$ nanoparticles doped with luminescent lanthanides (Eu^{3+}, Tb^{3+}, Tm^{3+}, and Er^{3+}) [17–21] have been extensively studied as potential optical imaging and MRI probes. Furthermore, because of the largest atomic number of Yb^{3+} among the lanthanides, mixed oxide nanoparticles containing Yb^{3+} and Gd^{3+} ions were proposed as a potential dual CT (computed tomography) and MRI probes, obtaining interesting results [22].

On the base of these considerations, we adopted in this work a low cost one-pot procedure for the preparation of Gd$_2$O$_3$ nanoparticles, in order to reduce the reactions time and to eliminate the calcination steps that typically require high temperatures. Their surface was functionalized with citrate molecules able to stabilize the particles when dispersed in aqueous solution [23]. The idea to select citrate as chelating agent is motivated by the interesting results obtained for GdF$_3$ nanoparticles [23]. It was demonstrated that citrate molecules confer both high hydrophilicity to the particles surface, thus improving the interaction of the metal ions exposed on the surface with the water molecules and high negative charge density, favoring the stability of the final aqueous suspensions over the time. In parallel, Gd$_2$O$_3$ NPs were also doped with two different Yb^{3+} loading (5 and 10 mol %). The co-presence of Gd^{3+} and Yb^{3+} ions in the same particle can open the way to their possible use as dual MRI and CT contrast agents, as previously demonstrated for parent samples [22]. Specific attention was devoted to the investigation of their relaxometric properties as a function of the magnetic field and of their chemical stability in different conditions. These two aspects unfortunately are often missed in the literature and they require a deep comprehension to opportunely design very efficient MRI probes.

2. Results and Discussion

Gd$_2$O$_3$ nanoparticles and the derivative materials containing different Yb^{3+} loading were prepared by adapting a synthetic precipitation procedure reported in the literature [24]. In detail, Gd^{3+} and Yb^{3+} precursors were dissolved in a few mL of triethylene glycol (TEG) that contained sodium hydroxide. TEG molecules work as chelating agent for the metal ions, thus limiting the particle growth [25]. The reaction was carried at 210 °C for 1.30 h in the presence of citric acid in low molar amount, in respect to Gd^{3+} salt (more details are reported in the experimental section). As clearly demonstrated in the literature, the citric acid plays two specific roles: (i) it limits the particles size by replacing the TEG molecules on the particles surface and (ii) it improves the hydrophilic character of the particles surface and the aqueous dispersion [23]. Nanoparticles with low (hereafter named Gd$_2$O$_3$:Yb LL) and high Yb^{3+} loading (Gd$_2$O$_3$:Yb HL) were also prepared by introducing a solution of 5 and 10 mol % of Yb^{3+} salt along with the Gd^{3+} precursor in the reaction. A schematic view of the Gd$_2$O$_3$:Yb nanoparticles functionalized with citrate groups are reported in Figure 1A.

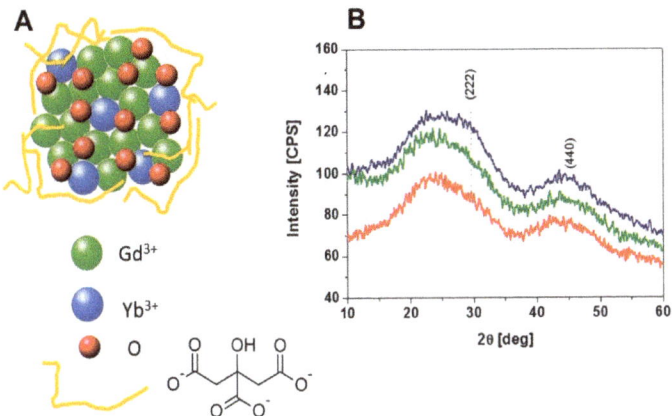

Figure 1. (**A**) Schematic view of Gd$_2$O$_3$:Yb nanoparticles covered on the surface with citrate; (**B**) X-ray profiles of Gd$_2$O$_3$ (red), Gd$_2$O$_3$:Yb (LL) (green), and Gd$_2$O$_3$:Yb (HL) (blue).

The structural properties of the samples were evaluated by an X-ray diffraction technique. The diffractogram of Gd$_2$O$_3$ NPs shows two wide bands centered at 29° and 44° 2θ (θ indicates the diffraction angle), attributed to the crystallographic planes 222 e 440, respectively (Figure 1B). This is typical of the body centered cubic structure, which has already been defined for these materials [26]. An additional component at low angles (ca. 22° 2θ) may be attributed to the organic fraction on the particles surface, which is in agreement with the literature data collected for parent samples. The X-ray profiles of the samples containing Yb^{3+} ions are completely comparable to that of Gd$_2$O$_3$, thus indicating that the incorporation in the structure of Yb^{3+} ions did not modify the structural features of these nanoparticles (Figure 1B). Furthermore, it is important to note that the presence of a wide band in the diffractogram of crystalline materials is typically associated to the nanometer nature of such samples. In light of these considerations, the average particles diameter (d) of all the samples was evaluated by applying the Debye–Scherrer equation (1) to the (222) reflection peak.

$$d = \frac{k\lambda}{B_d \cos \theta} \tag{1}$$

In the Debye–Scherrer equation, B_d is the full width at half the maximum intensity (FWHM) of the deconvoluted peak, λ is the X-ray wavelength, θ represents the diffraction angle and k is the Scherrer's constant that is 0.89 for spherical particles. The particle size estimated by X-ray diffraction (XRD) analysis for the Gd$_2$O$_3$ and derivative materials is approximately below 2.0 nm.

The chemical composition of the nanoparticles was estimated using elemental analysis (ICP-MS). The samples prior to the analysis were mineralized with concentrated nitric acid at a high temperature. Considering a density of 28.24 Gd/nm^3 [27], the amount of Gd^{3+} ions into Gd$_2$O$_3$ was estimated to be around 400 per particle. This number decreases to 378 and 349 when increasing the Yb^{3+} loading in Gd$_2$O$_3$:Yb LL and Gd$_2$O$_3$:Yb HL, respectively. Further, 22 and 51 Yb^{3+} ions per particle that corresponded to 5.5 and 12.7 mol % were determined in the two samples at low and high Yb^{3+} loading. The amount of citrate molecules exposed on the nanoparticles surface was quantified by CHN analysis, resulting to 2.38 mmol/g for Gd$_2$O$_3$ sample and 2.25 and 2.08 mmol/g for Gd$_2$O$_3$:Yb LL and Gd$_2$O$_3$:Yb HL, respectively.

The presence of citrate on the nanoparticles surface was also confirmed through infrared spectroscopy (IR). IR spectra of Gd$_2$O$_3$ and the samples such as Yb^{3+} ions were collected at room temperature and appeared very similar (Figure 2).

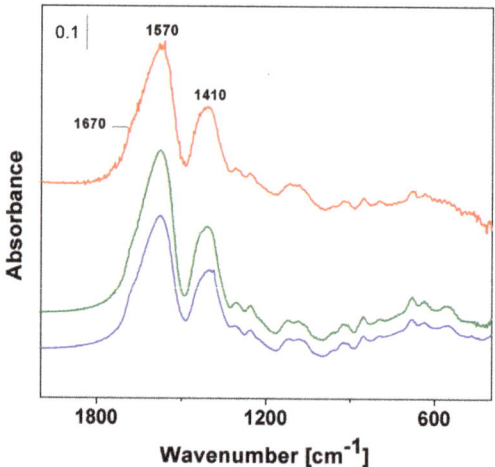

Figure 2. Fourier-transform infrared (FT-IR) spectra of Gd_2O_3 (red), Gd_2O_3:Yb (LL) (green) and Gd_2O_3:Yb (HL) (blue).

The two main peaks at 1570 cm^{-1} and 1410 cm^{-1} are clearly detectable and assigned to the asymmetric and symmetric stretching modes of the deprotonated COO$^-$ groups of citrate units. However, we must consider the band at 1570 cm^{-1} as a shoulder less intense at high wavenumbers, visible in all spectra that may be ascribed to the stretching vibrational mode of protonated COOH groups. The presence of this band suggests that a faction of pH-dependent protonated citric acid is also present on the particles surface.

The aqueous suspensions of the nanoparticles obtained directly by the synthesis procedure, without further modifications, were monitored by dynamic light scattering (DLS) analysis. The suspensions were visibly homogenous, as indicated in the digital photographs reported in Figure 3A. They appeared stable for ca, in which 1h without particles sedimentation did not require the use of surfactant or stabilizing agents to improve the particles suspension. Gd_2O_3 suspension shows hydrodynamic diameter of ca. 40 nm because of a partial particles aggregation. The aggregation state was more pronounced for Gd_2O_3:Yb LL and Gd_2O_3:Yb HL suspension, with hydrodynamic radius of ca. 120 nm. A possible explanation of these differences in the aggregates size may be related to the different charge density exposed on the surface.

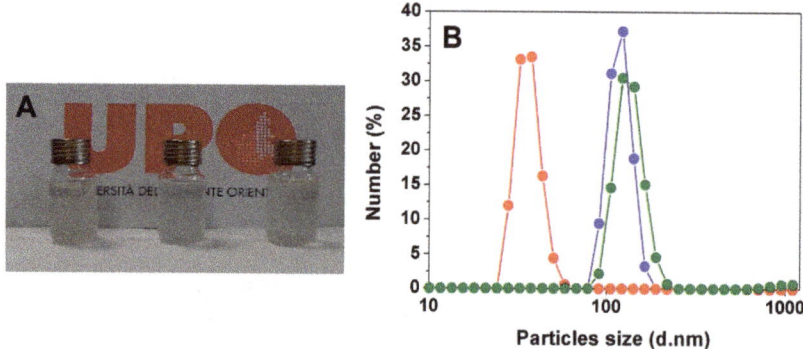

Figure 3. (**A**) Digital photograph of aqueous suspensions of Gd_2O_3, Gd_2O_3:Yb (LL) and Gd_2O_3:Yb (HL) (from left to right); (**B**) dynamic light scattering (DLS) analysis of Gd_2O_3 (red), Gd_2O_3:Yb (LL) (green) and Gd_2O_3:Yb (HL) (blue) suspensions.

To have more insights about this aspect, Z-potential analysis was performed on all samples in aqueous solutions, using the same experimental conditions as the DLS experiments. All nanoparticles showed a negative charge density with values of −14.0 mV for Gd_2O_3 and −7.0 and −10.7 mV for Gd_2O_3:Yb LL and Gd_2O_3:Yb HL suspensions. These values are likely to be associated with the presence of the citrate carboxylate groups that are not involved in the coordination with Gd^{3+} and/or Yb^{3+} ions, which agrees with IR data. Differences in the amount of protonated and deprotonated citrate groups can be responsible for the different charge density on the surface of the samples and their aggregation behavior.

Finally, a preliminary relaxometric study was carried out in aqueous solution as a function of the magnetic field applied, in order to evaluate the mechanisms responsible of the magnetic interaction with water molecules. The longitudinal (r_1) and transverse (r_2) proton relaxivities defined the efficacy of 1 mM concentration of paramagnetic center to enhance the relaxation rates of the water protons. The relaxivity values for all the samples were measured at 40 MHz and 60 MHz at 310 K and neutral pH, along with the derivative r_2/r_1 ratio (Table 1). The Gd^{3+} concentration in aqueous solutions was estimated by ICP-MS analysis.

Table 1. The r_1 and r_2/r_1 ratio values per gadolinium ion measured at 40 and 60 MHz (310 K).

Samples	r_1^{40} (mM^{-1} s^{-1})	r_1^{60} (mM^{-1} s^{-1})	r_2/r_1 (40 MHz)	r_2/r_1 (60 MHz)
Gd_2O_3	14.7	14.2	1.35	1.60
Gd_2O_3:Yb (LL)	14.7	14.7	1.39	1.92
Gd_2O_3:Yb (HL)	13.1	13.2	1.43	1.64

For all samples, the r_2/r_1 values do not change dramatically in the 20–60 MHz frequency range; they result to be below 2. This behavior suggests that these nanoparticles can be considered as positive MRI contrast agents [28,29]. The longitudinal relaxivity values calculated at high magnetic fields are in the 13–14 mM^{-1} s^{-1} range at 40 and 60 MHz, which is also the case for particles containing different Yb^{3+} loading. Moreover, these values are in line with those calculated for parent Gd_2O_3 samples with comparable size [30] and much higher than those observed for commercial Gd^{3+}-chelates contrast agents [28]. The enhancement of the relaxivity value is governed by the dipolar coupling occurring between the paramagnetic ions and the protons of water molecules. This process involves two mechanisms: a short-range interaction with the water molecules directly coordinated to the metal ions (inner sphere contribution (IS)) or involved in H-bond with polar groups in close proximity to the paramagnetic ions (second sphere mechanism (SS)) and a long-range interaction with the bulk water molecules in proximity to the particles surface (outer sphere process (OS)) [31]. The contribution of each mechanism interaction is related to the position of the paramagnetic ions in the final nanoparticle. Metal ions distributed inside the particles core contributes to the relaxivity enhancement trough OS mechanism, whereas the paramagnetic centers confined in the surface are accessible to the water molecules and then they contribute to SS and/or IS mechanisms. It is known that when the inner sphere contribution is active, its effect is markedly high and it determines the final relaxivity value. Considering this in terms of Gd_2O_3 nanoparticles, the Gd^{3+} ions are statistically distributed both inside and on the surface of the nanoparticles where the calculated relaxivity value is an average of conditions in which different mechanisms (IS, SS, and OS) are involved, which was additionally observed for GdF_3 and its parent materials [32].

To analyze in detail the role of the different mechanisms of interaction between the paramagnetic particles and the protons of water molecules, ^{1}H relaxivity value of the samples dispensed in aqueous solution are measured as a function of the proton Larmor frequency (Nuclear Magnetic Resonance Dispersion (NMRD)) [33] at 310 K from 0.01 to 500 MHz (Figure 4).

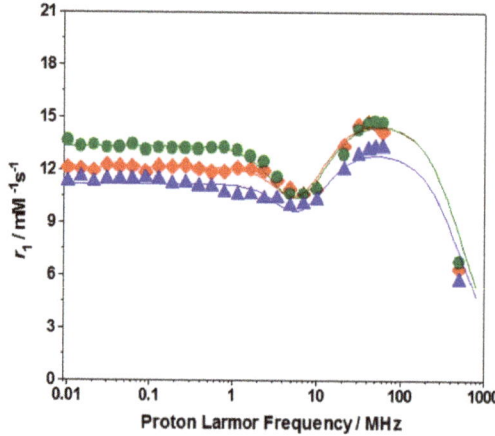

Figure 4. $1/T_1$ Nuclear Magnetic Resonance Dispersion (NMRD) profiles for Gd_2O_3 (red), Gd_2O_3:Yb (LL) (green), and Gd_2O_3:Yb (HL) (blue) at 310 K. The best-fit curves (solid lines) are calculated using the parameters of Table 2.

The NMRD profiles are typical of macromolecular systems with slow mobility, with a plateau at low filed, followed by a dispersion from 1 to 10 MHz, and finally a hump at high magnetic fields with a maximum close to 60 MHz (1.5 T), as previously observed for Gd_2O_3 and GdF_3 nanoparticles with comparable size (Figure 4) [30]. It is important to note that the relaxivity values calculated at the different magnetic fields are an average result considering that all the paramagnetic ions in the particle contribute in the same way to the relaxivity value. The best fit for the NMRD profile was obtained by applying the Solomon–Bloembergen–Morgan (SBM) [34] and Freed's [35] equations for the inner- and outer-sphere proton relaxation mechanisms and considering also the second sphere contribution. Some parameters were maintained during the fitting: the distance of closest approach for the outer-sphere contribution a_{GdH} was fixed to 4.0 Å; the distance between Gd^{3+} ion and the inner sphere water protons (r_{GdH}) to 3.1 Å; and the diffusion coefficient (D^{310}), attributed to the outer sphere water molecules diffusing close to the Gd^{3+} ions exposed on the surface, was fixed to 3.1×10^{-5} cm^2 s^{-1}. The more convincing results were obtained by considering one inner sphere water molecule (q = 1) coordinated to the Gd^{3+} ions with τ_R of 0.35 ns for all samples. The differences in the aggregates size for the samples are limited and we did not observe a clear and direct relationship between the particles size and the r_1 values during the best-fitting procedure. Furthermore, the good overlapping of the ^{1}H-NMRD profiles of Gd_2O_3 and Gd_2O_3:Yb LL testified these considerations Nevertheless, the SS mechanism was present and it corresponded to the presence of one water molecule for Gd_2O_3 and Gd_2O_3:Yb (LL) samples and 0.75 waters for Gd_2O_3:Yb (HL), with a distance of 3.5 Å and τ_R' of 0.20 ns for the first two systems and 0.17 ns for the particles with high Yb^{3+} loading. The electronic contributions correlated to the square of the zero-field splitting tensor, Δ^2. The correlation time describing the modulation of the zero-field splitting (τ_V) are comparable for all the samples and the relative values are reported in the Table 2.

Table 2. Selected best-fit parameters obtained from the analysis of the $1/T_1$ NMRD profiles (310 K) of all the nanoparticles.

Samples	Δ^2 (10^{19} s^{-2})	τ_V (ps)	q	τ_R (ns)	q'	τ_R' (ns)
Gd_2O_3	2.83 ± 0.17	37.0 ± 2.3	1	0.35 + 0.01	1	0.20 + 0.01
Gd_2O_3:Yb (LL)	2.77 ± 0.13	32.6 ± 1.6	1	0.35 + 0.01	1	0.20 + 0.02
Gd_2O_3:Yb (HL)	2.78 ± 0.21	32.8 ± 2.5	1	0.35 + 0.01	0.75	0.17 + 0.02

Finally, the stability of Gd_2O_3 nanoparticles was also monitored in reconstructed human serum (Seronorm) by measuring the longitudinal relaxation rate (R_1) values at 310 K and 40 MHz over the time. Despite a limited decrease of the relaxation rate after few minutes of *ca*, 10%, mainly ascribed to a possible alteration of the surface properties of the nanoparticles (i.e., replacement of the citrate ions by other anions present in the serenorm matrix), the values remained constant for 24 h. This test suggested that no detectable leaching of paramagnetic ions occurred in the matrix (Figure 5A).

In a second more severe test, the particles were treated with increased amount of a chelating agent, ethylendiamine tetracetic acid (EDTA). The behavior was different for the three samples. For Gd_2O_3, we assisted to a progressive decrease of r_1 value determined at 298 K and 40 MHz by increasing the EDTA amount. The EDTA/Gd^{3+} molar ratio of 0.5 was enough to promote the complete erosion of the nanoparticles with formation of the Gd(III)–EDTA chelate (Figure 5B). The particles bearing in the structure Yb^{3+} ions showed a completely different behavior. In the presence of limited amount of EDTA (EDTA/Gd^{3+} ratio of 0.3–0.5), the r_1 value increases and this was mainly evident for the sample with high Yb^{3+} loading. A further increase of the EDTA/Gd^{3+} molar ratio promoted a partial degradation of the nanoparticles with consequent decrease of the relaxivity values (Figure 5B). The partial increase of the relaxivity at the beginning was tentatively attributed to the complexation of the Yb^{3+} ions (not directly responsible of the relaxivity values observed for these NPs) on the surface and to the relative exposition of other internal Gd^{3+} ions, becoming more accessible to the water molecules.

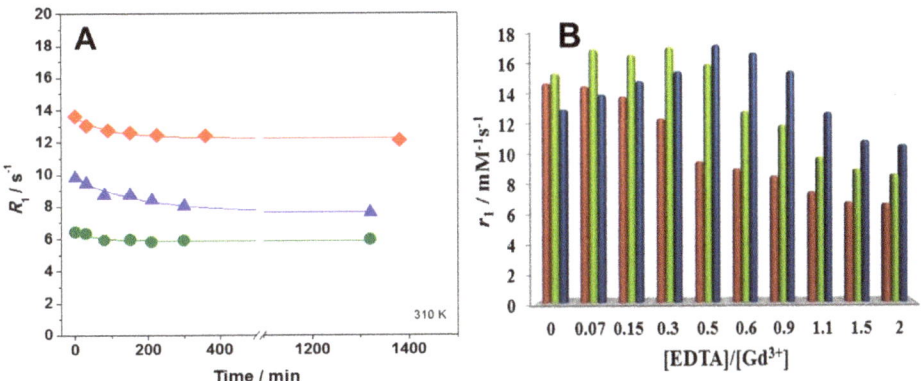

Figure 5. (**A**) Stability over the time of Gd_2O_3 (red), Gd_2O_3:Yb (LL) (green) and Gd_2O_3:Yb (HL) (blue) in Seronorm suspension at 40 MHz and 310 K; (**B**) r_1 values (40 MHz, 298 K) for Gd_2O_3 (red), Gd_2O_3:Yb (LL) (green) and Gd_2O_3:Yb (blue) as a function of [ethylendiammine tetracetic acid (EDTA)]/[Gd^{3+}] molar ratio.

3. Materials and Methods

Gd_2O_3 NPs: 2 mmol of $Gd(NO_3)_3 \cdot 6H_2O$ were dissolved in 10 mL of triethylene glycol (TEG). In parallel, 6 mmol of NaOH were added to other 10 mL of TEG. A third solution was prepared by dissolving 1.6 mmol of citric acid in 5 mL of TEG. All the solutions were stirred for 1 h at room temperature. In the second step, the first two solutions containing $Gd(NO_3)_3$ and NaOH were jointed together and maintained at 210 °C for 30 min. Then, the citric solution previously prepared was added and the final matrix was stirred at 210 °C for 1 h. A brown suspension was obtained. The solid phase containing Gd_2O_3 NPs was separated by centrifugation, and it was washed two times with 30 mL of ultrapure water. The particles were maintained in aqueous suspension in order to prevent particles aggregation.

Gd_2O_3:Yb NPs: A same procedure previously described for Gd_2O_3 was adopted for the preparation of mixed oxides containing low and high Ytterbium loading. The only difference is related to the molar amount of Gd^{3+} and Yb^{3+} precursor used in the reaction batch. In detail, for particles with lower Yb^{3+} amount (Gd_2O_3:Yb LL), 1.9 mmol of $Gd(NO_3)_3 \cdot 6H_2O$, and 0.1 mmol of Yb(NO$_3$)$_3$·5H$_2$O were used. On the contrary, 1.8 mmol and 0.2 mmol of Gd(NO$_3$)$_3$·6H$_2$O and Yb(NO$_3$)$_3$·5H$_2$O, respectively, were dissolved in TEG solvent for the synthesis of the particles with higher Yb^{3+} loading (Gd$_2$O$_3$:Yb HL).

Characterisation Methods

The amount of Gd^{3+} and Yb^{3+} in the final materials was estimated using ICP-MS Thermo Scientific X5 Series (Waltham, MA, USA). The samples were mineralized via dissolution in HNO$_3$ (5 mL) at 120 °C for 24 h. The amount of citrate in the nanoparticles was quantified by using Euro EA CHNSO Analyzer of HEKAtech GmbH Company (Wegberg, Germany). XRD patterns were obtained on a ARL XTRA48 diffractometer (Portland, OR, USA) by using Cu Kα radiation (λ = 1.54062 Å). IR spectra were collected in air at 298 K in the range (4000–400 cm^{-1}) and with a resolution of 4 cm^{-1} by using a Bruker Equinox 55 spectrometer (Milano, Italy). The samples were mixed with KBr (10 wt %). DLS and Z-potential experiments were performed on a suspension of the particles in aqueous medium by using a Zetasizer NanoZS instrument (Malvern, UK) operating in the particle size range from 0.6 nm to 6 mm and equipped with a He–Ne laser (λ = 633 nm).

The water proton longitudinal relaxation rates were measured with a Stelar Spinmaster spectrometer (Pavia, Italy) operating from 20 to 70 MHz at 310 K. The standard inversion–recovery method was employed (16 experiments, 2 scans) with a typical 90° pulse width of 3.5 μs. The temperature was controlled with a Stelar VTC-91 airflow heater equipped with a copper–constantan thermocouple. The proton $1/T_1$ NMRD profiles were measured on a fast field-cycling Stelar SmarTracer relaxometer over a continuum of magnetic field strengths from 0.00024 to 0.25 T (corresponding to 0.01–10 MHz proton Larmor frequencies). Additional data points in the range 20–70 MHz and at 500 MHz were obtained using a conventional spectrometer using WP 80 magnet interfaced with a STELAR console and Bruker NMR spectrometers, respectively.

4. Conclusions

In conclusion, Gd$_2$O$_3$ nanoparticles doped in the framework with different Yb^{3+} loading were successfully prepared through a fast and easily reproducible synthetic approach and they were decorated with hydrophilic citrate molecules. Nanoparticles that did not incorporate Yb^{3+} were also prepared as a reference material. In all cases, we obtained appreciable aqueous suspensions, which were stable and homogeneous without sedimentation. The samples showed relaxivity values at high magnetic fields, which were improved in respect to the clinically approved Gd(III)-chelates and in agreement with the results observed for comparable Gd$_2$O$_3$ and GdF$_3$ nanoparticles reported in the literature. ^{1}H NMR relaxometry data indicated that the relaxivity values were mainly determined by the inner sphere contribution, but also second sphere water molecules H-bonded to the polar groups of citrate contribute to the final relaxivity. The nanoparticles were chemically stable in physiological medium (human serum). Furthermore, Gd$_2$O$_3$ samples doped with Yb^{3+} ions and with particles size distribution in aqueous solution centered at ca. Moreover, 120 nm (comparable to the samples here discussed) were tested both in vitro and in vivo as dual contrast agents and a toxicological study demonstrated the good biocompatibility and safety of these nanoparticles [22]. These features along with the co-presence of Gd^{3+} and Yb^{3+} ions in the same inorganic structure make these nanoparticles suitable for dual MRI-CT diagnostic analyses.

Author Contributions: Conceptualization, Investigation, Writing-Original Draft Preparation, Supervision, Project Administration: F.C.; Methodology, Formal Analysis, Resources, Data Curation, Writing-Review & Editing, Visualization, Funding Acquisition: F.C. and G.G.

Funding: This research was supported by the Università del Piemonte Orientale (Ricerca locale 2016).

Acknowledgments: This study was realized under the auspices of COST Action European Network on NMR Relaxometry CA15209 and the Consorzio Interuniversitario di Ricerca in Chimica dei Metalli nei Sistemi Biologici (CIRCMSB). The authors appreciated the experimental support of Luigi Canobbio.

Conflicts of Interest: The authors declare no conflict of interest.

References

1. Doan, B.-T.; Meme, S.; Beloeil, J.-C. *The Chemistry of Contrast Agents in Medical Magnetic Resonance Imaging*; Merbach, A., Helm, L., Toth, E., Eds.; John Wiley & Sons: Chichester, UK, 2013; pp. 1–23.
2. Wahsner, J.; Gale, E.M.; Rodriguez-Rodriguez, A.; Caravan, P. Chemistry of MRI Contrast Agents: Current Challenges and New Frontiers. *Chem. Rev.* **2019**. [CrossRef] [PubMed]
3. Faulkner, S.; Blackburn, O.A. *The Chemistry of Molecular Imaging*; Long, N., Wong, W.-T., Eds.; John Wiley & Sons: Hoboken, NJ, USA, 2015; pp. 179–197.
4. Hanaoka, K.; Lubag, A.J.; Castillo-Muzquiz, A.; Kodadek, T.; Sherry, A.D. The detection limit of a Gd^{3+}-based T_1 agent is substantially reduced when targeted to a protein microdomain. *Magn. Reson. Imaging* **2008**, *26*, 608–617. [CrossRef] [PubMed]
5. Carniato, F.; Tei, T.; Dastrú, W.; Marchese, L.; Botta, M. Relaxivity modulation in Gd-functionalised mesoporous silicas. *Chem. Commun.* **2009**, 1246–1248. [CrossRef] [PubMed]
6. Carniato, F.; Tei, L.; Cossi, M.; Marchese, L.; Botta, M. A Chemical Strategy for the Relaxivity Enhancement of Gd^{III} Chelates Anchored on Mesoporous Silica Nanoparticles. *Chem. Eur. J.* **2010**, *16*, 10727–10734. [CrossRef] [PubMed]
7. Carniato, F.; Tei, L.; Arrais, A.; Marchese, L.; Botta, M. Selective Anchoring of Gd^{III} Chelates on the External Surface of Organo-Modified Mesoporous Silica Nanoparticles: A New Chemical Strategy to Enhance Relaxivity. *Chem. Eur. J.* **2013**, *19*, 1421–1428. [CrossRef] [PubMed]
8. Carniato, F.; Tei, L.; Botta, M. Gd-Based Mesoporous Silica Nanoparticles as MRI Probes. *Eur. J. Inorg. Chem.* **2018**, 4936–4954. [CrossRef]
9. Huang, W.-Y.; Davies, G.-L.; Davis, J.J. High signal contrast gating with biomodified Gd doped mesoporous nanoparticles. *Chem. Commun.* **2013**, *49*, 60–62. [CrossRef] [PubMed]
10. Bouchoucha, M.; C.-Gaudreault, R.; Fortin, M.-A.; Kleitz, F. Mesoporous Silica Nanoparticles: Selective Surface Functionalization for Optimal Relaxometric and Drug Loading Performances. *Adv. Funct. Mater.* **2014**, *24*, 5911–5923. [CrossRef]
11. Dong, H.; Du, S.-R.; Zheng, X.-Y.; Lyu, G.-M.; Sun, L.-D.; Li, L.-D.; Zhang, P.-Z.; Zhang, C.; Yan, C.-H. Lanthanide Nanoparticles: From Design toward Bioimaging and Therapy. *Chem. Rev.* **2015**, *115*, 10725–10815. [CrossRef] [PubMed]
12. Guillet-Nicolas, R.; Jean, N.; Bridot, L.; Seo, Y.; Fortin, M.-A.; Kleitz, F. Enhanced Relaxometric Properties of MRI "Positive" Contrast Agents Confined in Three-Dimensional Cubic Mesoporous Silica Nanoparticles. *Adv. Funct. Mater.* **2011**, *21*, 4653–4662. [CrossRef]
13. Johnson, N.J.J.; Oakden, W.; Stanisz, G.J.; Prosser, R.S.; van Veggel, F.C.J.M. Size-Tunable, Ultrasmall $NaGdF_4$ Nanoparticles: Insights into Their T_1 MRI Contrast Enhancement. *Chem. Mater.* **2011**, *23*, 3714–3722. [CrossRef]
14. Ahmad, M.W.; Kim, C.R.; Baeck, J.S.; Chang, Y.; Kim, T.J.; Bae, J.E.; Chaed, K.S.; Lee, G.H. Bovine Serum Albumin (BSA) and Cleaved-BSA Conjugated Ultrasmall Gd_2O_3 Nanoparticles: Synthesis, Characterization, and Application to MRI Contrast Agents. *Colloids Surf. A* **2014**, *450*, 67–75. [CrossRef]
15. Ahren, M.; Selegaard, L.; Soederlind, F.; Linares, M.; Kauczor, J.; Norman, P.; Kaell, P.-O.; Uvdal, K. A Simple Polyol-free Synthesis Route to Gd_2O_3 Nanoparticles for MRI Applications: An Experimental and Theoretical Study. *J. Nanopart. Res.* **2012**, *14*, 1006–1022. [CrossRef]
16. Cho, M.; Sethi, R.; Ananta narayanan, J.S.; Lee, S.S.; Benoit, D.N.; Taheri, N.; Decuzzi, P.; Colvin, V.L. Gadolinium Oxide Nanoplates with High Longitudinal Relaxivity for Magnetic Resonance Imaging. *Nanoscale* **2014**, *6*, 13637–13645. [CrossRef] [PubMed]
17. Hu, X.; Wang, M.; Miao, F.; Ma, J.; Shen, H.; Jia, N. Regulation of Multifunctional Mesoporous Core–shell Nanoparticles with Luminescence and Magnetic Properties for Biomedical Applications. *J. Mater. Chem. B* **2014**, *2*, 2265–2275. [CrossRef]
18. Liu, J.; Tian, X.; Luo, N.; Yang, C.; Xiao, J.; Shao, Y.; Chen, X.; Yang, G.; Chen, D.; Li, L. Sub-10 nm Monoclinic Gd_2O_3:Eu^{3+} Nanoparticles as Dual-Modal Nanoprobes for Magnetic Resonance and Fluorescence Imaging. *Langmuir* **2014**, *30*, 13005–13013. [CrossRef] [PubMed]
19. Chen, F.; Chen, M.; Yang, C.; Liu, J.; Luo, N.Q.; Yang, G.W.; Chen, D.H.; Li, L. Terbium-doped Gadolinium Oxide Nanoparticles Prepared by Laser Ablation in Liquid for Use as a Fluorescence and Magnetic Resonance Imaging Dual-modal Contrast Agent. *Phys. Chem. Chem. Phys.* **2015**, *17*, 1189–1196. [CrossRef] [PubMed]

20. Petoral, R.M.; Soederlind, F.; Klasson, A.; Suska, A.; Fortin, M.A.; Abrikossova, N.; Selegaard, L.; Kaell, P.-O.; Engstroem, M.; Uvdal, K. Synthesis and Characterization of Tb^{3+}-Doped Gd_2O_3 Nanocrystals: A Bifunctional Material with Combined Fluorescent Labeling and MRI Contrast Agent Properties. *J. Phys. Chem. C* **2009**, *113*, 6913–6920. [CrossRef]
21. Luo, N.; Yang, C.; Tian, X.; Xiao, J.; Liu, J.; Chen, F.; Zhang, D.; Xu, D.; Zhang, Y.; Yang, G.; et al. A General top-down Approach to Synthesize Rare Earth Doped-Gd_2O_3 Nanocrystals as Dualmodal Contrast Agents. *J. Mater. Chem. B* **2014**, *2*, 5891–5897. [CrossRef]
22. Liu, Z.; Pu, F.; Liu, J.; Jiang, L.; Yuan, Q.; Li, Z.; Ren, J.; Qu, X. PEGylated Hybrid Ytterbia Nanoparticles as High-Performance Diagnostic Probes for in Vivo Magnetic resonance and X-Ray Computed Tomography Imaging with Low Systemic Toxicity. *Nanoscale* **2013**, *5*, 4252–4261. [CrossRef] [PubMed]
23. Carniato, F.; Thangavel, K.; Tei, L.; Botta, M. Structure and dynamics of the hydration shells of citrate-coated GdF_3 nanoparticles. *J. Mater. Chem. B* **2013**, *1*, 2442–2446. [CrossRef]
24. Söderlind, F.; Pedersen, H.; Petoral, R.M., Jr.; Käll, P.-O.; Uvdal, K. Synthesis and characterisation of Gd_2O_3 nanocrystals functionalised by organic acids. *J. Colloid Interface Sci.* **2005**, *288*, 140–148. [CrossRef] [PubMed]
25. Caruntu, D.; Remond, Y.; Chou, N.H.; Jun, M.-J.; Caruntu, G.; He, J.; Goloverda, G.; O'Connor, C.; Kolesnichenko, V. Reactivity of 3d Transition Metal Cations in Diethylene Glycol Solutions. Synthesis of Transition Metal Ferrites with the Structure of Discrete Nanoparticles Complexed with Long-Chain Carboxylate Anions. *Inorg. Chem.* **2002**, *41*, 6137–6146. [CrossRef] [PubMed]
26. Wang, F.; Peng, E.; Zheng, B.; Li, S.F.Y.; Xue, J.M. Synthesis of Water-Dispersible Gd_2O_3/GO Nanocomposites with Enhanced MRI T_1 Relaxivity. *J. Phys. Chem. C* **2015**, *119*, 23735–23742. [CrossRef]
27. Faucher, L.; Guay-Bégin, A.A.; Lagueux, J.; Côté, M.-F.; Petitclerc, E.; Fortin, M.-A. Ultra-small gadolinium oxide nanoparticles to image brain cancer cells in vivo with MRI. *Contrast Media Mol. Imaging* **2011**, *6*, 209–218. [CrossRef] [PubMed]
28. Caravan, P.; Ellison, J.J.; McMurry, T.J.; Lauffer, R.B. Gadolinium(III) Chelates as MRI Contrast Agents: Structure, Dynamics, and Applications. *Chem. Rev.* **1999**, *99*, 2293–2352. [CrossRef] [PubMed]
29. Aime, S.; Geninatti Crich, S.; Gianolio, E.; Giovenzana, G.B.; Tei, L.; Terreno, E. High sensitivity lanthanide(III) based probes for MR-medical imaging. *Coord. Chem. Rev.* **2006**, *250*, 1562–1579. [CrossRef]
30. Bridot, J.-L.; Faure, A.-C.; Laurent, S.; Rivière, C.; Billotey, C.; Hiba, B.; Janier, M.; Josserand, V.; Coll, J.-L.; Vander Elst, L.; et al. Hybrid Gadolinium Oxide Nanoparticles: Multimodal Contrast Agents for in Vivo Imaging. *J. Am. Chem. Soc.* **2007**, *129*, 5076–5084. [CrossRef] [PubMed]
31. Botta, M. Second Coordination Sphere Water Molecules and Relaxivity of Gadolinium(III) Complexes: Implications for MRI Contrast Agents. *Eur. J. Inorg. Chem.* **2000**, 399–407. [CrossRef]
32. Carniato, F.; Tei, L.; Phadngam, S.; Isidoro, C.; Botta, M. $NaGdF_4$ Nanoparticles Coated with Functionalised Ethylenediaminetetraacetic Acid as Versatile Probes for Dual Optical and Magnetic Resonance Imaging. *ChemPlusChem* **2015**, *80*, 503–510. [CrossRef]
33. Helm, L.; Morrow, J.R.; Bond, C.J.; Carniato, F.; Botta, M.; Braun, M.; Baranyai, Z.; Pujales-Paradela, R.; Regueiro-Figueroa, M.; Esteban-Gómez, D.; et al. *Contrast Agents for MRI: Experimental Methods*; Pierre, V.C., Allen, M.J., Eds.; RSC: London, UK, 2018; pp. 11–242.
34. Solomon, I.; Bloembergen, N. Nuclear Magnetic Interactions in the HF Molecule. *J. Chem. Phys.* **1956**, *25*, 261. [CrossRef]
35. Freed, J.H. Dynamic effects of pair correlation functions on spin relaxation by translational diffusion in liquids. II. Finite jumps and independent T_1 processes. *J. Chem. Phys.* **1978**, *68*, 4034. [CrossRef]

 © 2019 by the authors. Licensee MDPI, Basel, Switzerland. This article is an open access article distributed under the terms and conditions of the Creative Commons Attribution (CC BY) license (http://creativecommons.org/licenses/by/4.0/).

Review

Nanomaterials with Tailored Magnetic Properties as Adsorbents of Organic Pollutants from Wastewaters

Marcos E. Peralta [1], Santiago Ocampo [1], Israel G. Funes [2], Florencia Onaga Medina [2], María E. Parolo [2] and Luciano Carlos [1,*]

[1] Instituto de Investigación y Desarrollo en Ingeniería de Procesos, Biotecnología y Energías Alternativas, PROBIEN (CONICET-UNCo), Buenos Aires 1400, Neuquén 8300, Argentina; marcos.peralta@probien.gob.ar (M.E.P.); santiago.ocampo@probien.gob.ar (S.O.)

[2] Centro de Investigación en Toxicología Ambiental y Agrobiotecnología, CITAAC (CONICET-UNCo), Facultad de Ingeniería, Universidad Nacional del Comahue, Buenos Aires 1400, Neuquén 8300, Argentina; israel.funes@fain.uncoma.edu.ar (I.G.F.); f.onagamedina@comahue-conicet.gob.ar (F.O.M.); maria.parolo@fain.uncoma.edu.ar (M.E.P.)

* Correspondence: luciano.carlos@probien.gob.ar

Received: 12 March 2020; Accepted: 30 March 2020; Published: 31 March 2020

Abstract: Water quality has become one of the most critical issue of concern worldwide. The main challenge of the scientific community is to develop innovative and sustainable water treatment technologies with high efficiencies and low production costs. In recent years, the use of nanomaterials with magnetic properties used as adsorbents in the water decontamination process has received considerable attention since they can be easily separated and reused. This review focuses on the state-of-art of magnetic core–shell nanoparticles and nanocomposites developed for the adsorption of organic pollutants from water. Special attention is paid to magnetic nanoadsorbents based on silica, clay composites, carbonaceous materials, polymers and wastes. Furthermore, we compare different synthesis approaches and adsorption performance of every nanomaterials. The data gathered in this review will provide information for the further development of new efficient water treatment technologies.

Keywords: adsorption; magnetic separation; nanotechnology; water treatments

1. Introduction

The deterioration of water quality by organic pollutants has become a global issue of concern that requires an effective solution. According to the European Environmental Agency Report, only around 40% of surface waters (rivers, lakes, and transitional and coastal waters) are in a good ecological status, and 38% are in a good chemical status [1]. Another report, on the quality of 100 rivers from 27 European countries concluded that only about 10% of the river water samples analyzed could be classified as "very clean" in terms of chemical pollution [2]. Pollutants present in water are classified as organic and inorganic and can reach surface waters and groundwaters mainly through industrial effluents, agricultural runoff, sewage plants and other human activities. In general, organic pollutants including aromatic compounds, dyes, pesticides, and antibiotics are toxic and/or resistant to microbial degradation. Once these organic pollutants enter the water bodies, water is no longer safe for drinking purpose and sometimes the complete removal of these pollutants from the water is a very difficult process.

Dyes are released by various industries, such as food, cosmetics and especially textiles [3,4]. According to incomplete statistics, over 35,000 metric tons of dyes are released into the environment [5]. Emerging contaminants, as pharmaceutical and personal care products, pesticides, food additives, surfactants, etc., are present in different water resources in the range of ng L^{-1}–µg L^{-1} [6]. These compounds are generally non-biodegradable or poorly biodegradable, thus becoming stable and persistent compounds when released into surface and groundwaters [7]. On the other hand, aromatic compounds, as phenols, anilines and polycyclic aromatic compounds, are also environmentally relevant contaminants and can be widely found in the effluents from dyestuffs, pharmaceuticals, petrochemicals, and other industries [8]. In recent years, there has been growing concern about the presence of these compounds in the aquatic environment due to their wide distribution and potential adverse health effects even at low concentration [8,9].

Various water treatment processes have been developed to reduce organic compounds' levels in waters, such as photocatalysis, ozonolysis, electrolysis, membrane process and adsorption. Among these methods, adsorption is one of the most promising techniques in water remediation due to its low cost, easy operation, the possibility of regeneration and feasibility of application in field conditions [3,6,10,11]. The overall adsorption process consists of the following key steps: (a) pollutant adsorption (b) recovery of the adsorbent for further reuse (c) adsorbent regeneration, and (d) management of both the regeneration solution and the saturated adsorbent. Adsorbent regeneration is a crucial economic factor for industrial applications [5]. Part of the cost that affects the application of the adsorption process lies in the possibility of reusing the adsorbent material several times. Therefore, the adsorbent material must have a good affinity to remove contaminants, be stable (i.e., not lose some of its components by leaching or bleeding) and must be regenerated and used in various adsorption/desorption cycles. To guarantee the reuse of the adsorbent, the separation process of the adsorbents from the aqueous medium is a key step to achieve a good performance and cost-effective treatment. Filtration or centrifugation techniques are often used for the separation process, but both techniques are not cost-effective and are difficult to handle when large volumes of water are used [11]. Moreover, adsorbents may lead to blockage of the filters or loss of adsorbent. To overcome these problems, the use of magnetic nanoadsorbents has been proposed [12,13]. Magnetic separation is an environmentally friendly alternative for the separation and recovery of nanomaterials, since it minimizes the use of solvents and auxiliaries, reduces the operation time, and is a cost-effective method compared to conventional separation processes as filtration or centrifugation [14]. For the reasons outlined above, magnetic adsorbents have emerged as a new generation of materials for decontamination processes [15]. In particular, the most commonly employed magnetic carriers for environmental applications are magnetite (Fe_3O_4) and maghemite (γ-Fe_2O_3) nanoparticles, because they are easy to synthesize, environmentally friendly, and have high saturation magnetization. The magnetic properties of iron oxide nanoparticles, in particular of those commonly used in environmental applications, are well documented and reviewed elsewhere and are beyond the scope of this review [16,17]. Bare magnetic iron oxide nanoparticles undergo oxidation/dissolution, especially in acid solutions [15,18], and co-aggregation because of their high surface energy [10,18], and even in environmental conditions their chemical stability can be affected [19], which limits the large-scale application of magnetic iron oxide nanoparticles. Therefore, a large number of functionalized magnetic nanoparticles and magnetic nanocomposites with new structures and surface properties have been produced to overcome the deficiencies of magnetic nanoparticles, provide more active sites and improve their aqueous stability and versatility. Currently, many different materials have been used in combination with magnetic nanoparticles to manufacture magnetic nanoadsorbents, such as silica, activated carbon, carbon nanotubes, polymers, metal–organic frame works, and clays, showing varying extents of effectiveness in removing the organic pollutants from water. Each type of adsorbent nanomaterial has different surface properties, surface chemical groups, and specific areas and pore size. The selection of the appropriate material represents a critical factor to guarantee efficient removal of the organic pollutant.

This review was designed to provide an overview of the synthesis methods, surface properties and application in organic pollutant removal of different magnetic nanoadsorbents. We will present recent scientific progress on the preparation of magnetic core–shell nanoparticles and nanocomposites paying special emphasis on materials with high adsorption capacities. Although there are many review articles in the literature that cover the application of magnetic nanomaterials in the environmental field [13,14,17,20–24], we believe that more specific data gathering is still needed on the synthesis methods and properties of magnetic nanoadsorbents that have been developed so far, and their efficiencies and limitations in the water treatments. This information will contribute to the field of rational design of nanomaterials for water treatment and will help in designing more efficient water treatment technologies to guarantee adequate water quality.

2. Silica-Based Materials

Silica (chemically SiO_2) has a three-dimensional network structure that consists of SiO_4 and ends with oxygen through siloxane groups (Si–O–Si) or silanol groups (Si–OH) [25]. Silanol groups offer a rich surface chemistry useful in the adsorption process, since they can complex some molecules and metal cations. This interaction can be improved by modifying the pH, due to their acid-basic behavior. In consequence, bare silica can be used as an efficient adsorbent for pollutants, such as dyes, heavy metals and aromatic compounds [26]. Silica nanoparticles can be porous or non-porous. Mesoporous silica ($mSiO_2$) is a unique class of synthetic porous material, with a pore size between 2 and 50 nm. Because of its porosity and higher surface area, mesoporous silica is often preferred as an adsorbent rather than non-porous silica. The highly adjustable structure, versatile surface chemistry, low production cost and simple synthesis procedure are the primary benefits that encourage their application in adsorption. Moreover, silanol groups allow the introduction of a wide variety of functional groups to the silica surface. Therefore, many surface functionalization methods, post-synthesis or during synthesis, have been developed to improve the adsorption capacity and selectivity of mesoporous silica [26–31]. The incorporation of iron and iron oxides, such as magnetite and maghemite, to the silica structure is usually carried out by two approaches. One method consists in covering a core of iron oxide nano- or micro-particles with a shell of non-porous silica or mesoporous silica to produce hierarchical structures known as "core–shell" particles (Figure 1A,B). These structures have the additional advantage of protecting the iron-based core from leaching and oxidation in water [32] and, at the same time, of reducing the tendency of aggregation of the particles. In the other method, magnetic particles are dispersed onto mesoporous silica. Wang and co-workers [33] prepared magnetite by the hydrothermal method and they covered with a non-porous silica shell via the Stöber method. The obtained $Fe_3O_4@SiO_2$ nanoparticles were tested for the removal of anionic dye Congo Red. As expected, the adsorption results were strongly pH dependent. The silanol groups at low pH values are protonated allowing electrostatic interactions between $Fe_3O_4@SiO_2$ nanoparticles and Congo Red molecules, which leads to a higher dye removal performance. These core–shell type particles can be further modified in different ways. Recently, Cao's group [34] used the polymer polyvinylpyrrolidone (PVP) to provide affinity to hydrophobic substances to $Fe_3O_4@SiO_2$ particles. They prepared magnetic cores by the miniemulsion method, then covered them with non-porous silica shell and finally with PVP. The resulted adsorbent ($Fe_3O_4@SiO_2$-PVP) was tested for the hydrophobic compound phenanthrene. The adsorption reaches the equilibrium fast (10 min), and the maximum adsorption capacity was high compared to previous PAH adsorption reports [35].

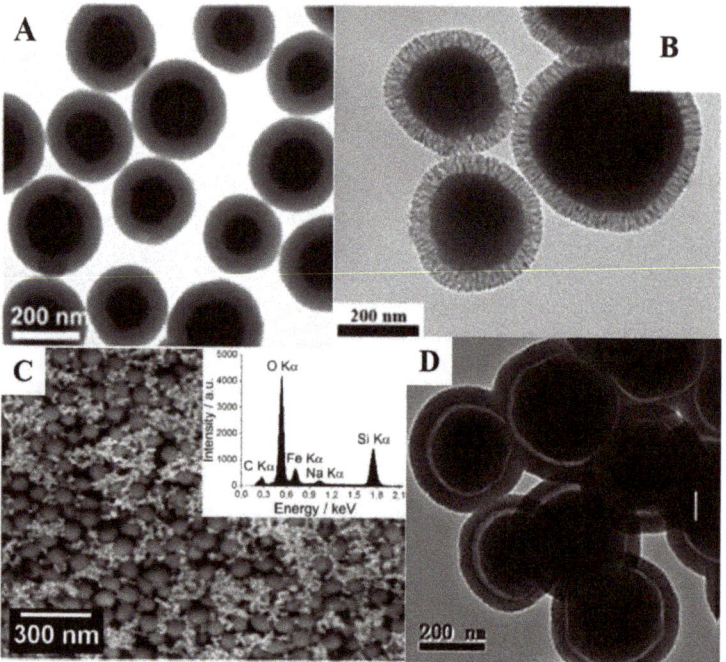

Figure 1. (**A**) TEM image of non-porous silica covering Fe$_3$O$_4$ nanoparticles (reproduced with permission from [36], published by Royal Society of Chemistry, 2015; (**B**) TEM image of mesoporous-silica covering Fe$_3$O$_4$ nanoparticles (reproduced from [37], published by Royal Society of Chemistry, 2018); (**C**) SEM images of Raspberry-like" supraparticle system (reproduced with permission from [38], published by ChemNanoMat, 2019); and (**D**) TEM micrographs of Fe$_3$O$_4$@SiO$_2$@h-mSiO$_2$ (reproduced with permission from [36], published by Royal Society of Chemistry, 2015).

The incorporation of silane groups is one of the most common procedures to provide affinity and versatility to an adsorbent for a specific target pollutant [39]. A silane is a group that consists in a central silicon atom bonded to alcoxy groups (usually methoxy and ethoxy) and at least a different moiety with a specific group, such as amino, thiol, hydroxy, vinyl, aliphatic chains, aromatic chains, etc. In a silanization reaction, the surface hydroxyl groups of the solid (e.g., Si–OH in SiO$_2$ or Fe–OH in Fe$_3$O$_4$) react with the alcoxy groups of the silane and form a covalent bond (Si–O–Si in SiO$_2$ and Fe–O–Si in Fe$_3$O$_4$). The work of Sasaki and Tanaka [40] is an example of a simple silanization procedure. They covered commercial magnetite with phenyltrimethoxysilane (PTMS) and studied the adsorption mechanism with several aromatic compounds. The new functionality incorporated by a silane allows further modifications. For instance, the incorporation of vinyl groups by grafting of 3-(methacryloxypropyl)trimethoxysilane (MPS) to Fe$_3$O$_4$@SiO$_2$ nanoparticles allows a radical polymerization reaction to obtain a polymer covering the nanoparticles [41,42]. Jiaqi and co-workers [43] lately proposed a dendritic-like structure to enlarge the magnetic silica nanoparticles area and thus improve their adsorption performance. For this, 3-chloropropyltriethoxysilane (CPTES) were grafted to Fe$_3$O$_4$@SiO$_2$ NPs; then a substitution reaction with ethylenediamine and finally a chemical reaction with maleic anhydride were performed. The obtained carboxylated ethylenediamine functionalized nanoparticles (Fe$_3$O$_4$@SiO$_2$-EDA-COOH) show remarkable magnetic saturation and high sorption capacity of cationic dye methylene blue and can be reused five cycles. However, the reported maximum sorption capacity of 43.15 mg g^{-1} corresponds to a relatively high pH of 10.

Wang's group [44] developed a novel MOF-based smart adsorbent named $Fe_3O_4@SiO_2@UiO-67$ for the simultaneous selective recognition, detection and removal of organophosphorus pesticide glyphosate for the first time. The prepared adsorbent contains Zr–OH groups with high affinity for phosphate groups, endowing it with an outstanding adsorption capacity for glyphosate. Furthermore, the adsorbent was able to be reused four times with no significant adsorption capacity decrease.

Other types of magnetic silica-based materials use nanoscale zero-valent iron (nZVI) instead of iron oxides as a core. nZVI is commonly applied as a catalyst because of its strong reduction capability. However, there are a few reports of their use to provide magnetic properties to silica-based materials. Li et al. [35] synthesized nZVI nanoparticles coated with silica and polydopamine using a two-step process. The obtained adsorbent ($Fe@SiO_2@PDA$) was applied in the removal of anthracene and phenanthrene from aqueous media, achieving maximum adsorption capacities of 0.185 and 0.367 mg g^{-1}, respectively. Interestingly, the adsorption efficiency of $Fe@SiO_2@PDA$ barely decreased after 10 cycles.

Recently, a novel particle system, so-called raspberry-like supraparticles, consisting in magnetite/maghemite and amorphous silica nanoparticles, was studied in the adsorption of methyl blue dye [38]. For the synthesis of these nanoparticles, the method of spray-drying was used: droplets from nanoparticle dispersions were generated in a hot chamber and the solvent was evaporated, forcing the remaining nanoparticles in the droplets together to form supraparticles. The supraparticle system consisted of nanoparticles of 10 nm diameter (Figure 1B) and showed a maximum adsorption capacity of 93 mg g^{-1}. Though this adsorption capacity is comparable or lower than previous reports, the adsorption kinetics results to be outstandingly fast, reaching equilibrium within 60 s. Further, the system can be regenerated either thermally or by acid treatment and can be reused consecutively with a slight loss in adsorption capacity.

Currently, it is of interest to have alternative silica sources rather than synthetic reactants, e.g., from industrial waste, to prepare silica-based materials and adsorbents. This approach could contribute to minimize waste and to apply more economic methods. For instance, silica were obtained from rice husk by calcination and acid leaching and with microwave assistance [45,46]. Additionally, solid waste coal gasification fine slag as the silica source successfully produced mesoporous glass microspheres with a specific surface area of 364 m^2 g^{-1} and an adsorption capacity of 140.57 mg g^{-1} in methylene blue removal [47]. Recently, silica obtained from rice husk has been used to incorporate an amorphous silica shell onto magnetite obtaining core–shell nanoparticles [45]. These particles were then amino-functionalized with (3-aminopropyl)trimethoxysilane (APTMS) and tested in the adsorption of methylene red dye.

The preparation of core–shell magnetic mesoporous silica is generally a multi-step procedure: synthesis of magnetic nanoparticles, coating with mesoporous silica, removal of template, and functionalization of the mesoporous silica shell. Among the many synthesis procedures for mesoporous silica [28], the hydrothermal method, based on the sol–gel Stöber method, is the most frequently reported for adsorbent preparation. This process consists in the polymerization of a silica precursor (e.g., TEOS) directed by micelles of a templating agent that lead to a mesoporous silica covering formerly prepared magnetic cores in a perpendicularly aligned pore arrange. Then, the templating agent remaining inside the pores is usually removed by calcination [37] or reflux in ethanol/acid water mix [48]. The pore size, particle size and morphology of mesoporous silica can be adjusted by selecting different template agents, by controlling pH and by using additives [29]. For instance, the synthesis of MCM-41 typically involves the cationic surfactant cetyltrimethylammonium bromide (CTAB); meanwhile, the neutral co-polymer with larger molecular weight Pluronic P-123 is used in preparation of SBA-15 [30]. Adding a co-solvent, like trimethylbenzene (TMB), to expand the micelle formed by the templating agent is another way to obtain wider pores [49]. Finally, the functionalization of mesoporous silica can be done post-synthesis or even during synthesis.

Zhang and co-workers [37] prepared these kinds of magnetic silica-based particles functionalized with cyclodextrins and they tested them as an adsorbent of the antibiotic doxycycline (DOX).

Cyclodextrins (CDs) are a class of macrocyclic molecules with a hydrophilic exterior and a lipophilic interior cavity, and they can form inclusion complexes with organic molecules, which can be trapped in the hydrophobic cavity through multiple interactions, including inclusion interactions, hydrophobic interactions, electrostatic attractions, and π–π interactions. First, magnetite was prepared via the solvothermal method and later these particles were covered with a non-porous silica layer through a modified Stöber method. The as-prepared Fe_3O_4@SiO_2 microspheres were put in a water solution of CTAB and triethanolamine; then, TEOS was added and a mesoporous silica layer covering the microspheres was formed. After calcination, Fe_3O_4@SiO_2@$mSiO_2$ microspheres were obtained (Figure 1C). To introduce covalently bound cyclodextrins, the Fe_3O_4@SiO_2@$mSiO_2$ microspheres were added to a basic water/ethanol solution containing APTMS and monochlorotriazinyl β-cyclodextrin (MCT-β-CD). The obtained Fe_3O_4@SiO_2@$mSiO_2$-CD microspheres showed a maximum DOX adsorption capacity of 78 mg g^{-1} at initial pH of 3.8. The pH of the DOX solution significantly affected the adsorption process, which is spontaneous at 298 K. The contact time required for antibiotic adsorption using this adsorbent is short compared to activated carbon.

In order to develop a simple synthesis procedure amenable to scale-up, some authors propose the functionalization of mesoporous silica during synthesis. One method consists in the co-condensation reaction adding a silane along with the primary silica precursor in the presence of a surfactant, thus obtaining structures known as periodic mesoporous organosilicas [50]. Cai's group reported the first one-pot synthesis of functionalized magnetic mesoporous silica composites for the adsorption removal of organic dyes from water [51]. The adsorption of methyl blue on the obtained adsorbent, here termed as Fe_3O_4@$mSiO_2$–C18, has a negligible influence of solution pH in the range of 4.5–9.5 with an optimal at pH 7.5 and a maximum adsorption capacity of 363 mg g^{-1} according to the Langmuir model. Furthermore, the adsorption equilibrium was achieved within a relatively short period of time (20 min), because of the high surface area and short adsorption path of mesoporous solid. Another approach is the work of Keller's group [52] in which the surfactant 3-(trimethoxysilyl)propyl-octadecyldimethyl-ammonium chloride (TPODAC) was used as a templating agent to prepare permanently confined micelle array core–shell nanoparticles. Maghemite cores were surrounded by TPODAC and TMB as a micelle-swelling agent, and then TEOS was added to covalently bind the surfactant onto the magnetic cores. Three different TMB:TPODAC weight ratio were tested. The authors studied the adsorption performance of the as-prepared nanoparticles (γ-Fe_2O_3@$mSiO_2$-TPODAC) on methyl orange, sulfamethoxazole, gemfibrozil, acenaphthene and phenanthrene. It was found that larger amounts of TMB (TMB:TPODAC 60%) results in a noteworthily higher sorption kinetic rate as well as slightly higher sorption capacity. In addition, the core–shell nanoparticles showed good adsorption capacity after five regeneration cycles. The nanoparticles show relatively low surface area and magnetic saturation; however, they proved to be a versatile adsorbent for dyes, ECs and PAHs.

Composites consisting of magnetic nanoparticles dispersed in a mesoporous silica matrix were also developed and applied as adsorbents. Ghanei [53] prepared iron oxide nanoparticles embedded in SBA-15 mesoporous silica functionalized with 3-methacryloxypropyltrimethoxysilane followed by a polymerization with acrylic acid monomer. The obtained adsorbent M-SBA-15/CPAA proved to have a noteworthy adsorption capacity towards Acid Blue 25 dye. However, the adsorption process proved to be strongly pH-dependent and the maximum corresponds to highly acid conditions (pH 2). Other composites functionalized via polymerization onto silica surface were used for removal of the dyes [54].

Hollow mesoporous silica spheres (Figure 1D), in contrast with conventional mesoporous silicas such as MCM-41 and SBA-15, have some additional features including low density, high adsorption capacity, high storage capacity and permeability, turning them into promising materials for applications in drug delivery, catalysis and adsorption [36]. Thus, the preparation of nanocomposites with magnetic cores inside the cavities of hollow mesoporous silica spheres, namely rattle-type or yolk-shell magnetic mesoporous silica nanocomposites, is of great interest for a variety of environmental applications. The study of Jin and co-workers [55] is the first report about the adsorption capacity of rattle-type nanoparticles. The synthesis consists in preparation of magnetic particles, covering with silica and a sacrificial template (resorcinol-formaldehyde resin polymer), a further layer of mesoporous silica and calcination to remove the organic matter. The prepared spheres, here named γ-Fe$_2$O$_3$@SiO$_2$@h-mSiO$_2$, present a similar fast removal rate of methyl blue to mesoporous MCM-41. The adsorption capacity increase with the pH, with a value of 41 mg g^{-1} at pH 7.2.

Table 1 summarizes the adsorption capacities and some physicochemical properties of the silica-based nanocomposites described in this section.

Table 1. Application of magnetic silica-based nanoadsorbents for organic pollutants removal from water.

Adsorbent	Surface Area (m^2 g^{-1})	Magnetic Saturation (emu g^{-1})	Organic Pollutant	Adsorption Capacity (mg g^{-1})	Ref.
Fe$_3$O$_4$@SiO$_2$	-	48.06	Congo red	50.54	[33]
Fe$_3$O$_4$@SiO$_2$-PVP	60.82	30.89	Phenanthrene	18.84	[34]
Fe$_3$O$_4$@SiO$_2$–VTEOS–DMDAAC	-	-	Methylene blue	109.89	[41]
Fe$_3$O$_4$@SiO$_2$-EDA-COOH	-	58.7	Methylene blue	43.15	[43]
Fe$_3$O$_4$@SiO$_2$@Zn–TDPAT	-	>20	Methylene blue Congo red	58.67 17.73	[56]
Fe$_3$O$_4$@SiO$_2$@UiO-67	-	20.9	Glyphosate	256.54	[44]
Fe@SiO$_2$@PDA	-	51.98	Anthracene Phenanthrene	0.484 0.184	[35]
Fe$_3$O$_4$/SiO$_2$ 10 nm SP	193	>25	Methylene blue	93	[38]
Fe$_3$O$_4$@SiO$_2$-NH$_2$	-	>40	Methylene red	81.39	[45]
Fe$_3$O$_4$@SiO$_2$@mSiO$_2$-CD	119	30.99	Doxycycline	78	[37]
Fe$_3$O$_4$@SiO$_2$-C18	303	22.62	Methylene blue	363.64	[51]
γ-Fe$_2$O$_3$@mSiO$_2$-TPODAC	1.63	7.09	Methyl orange Gemfibrozil Sulfamethoxazole Acenaphtene Phenanthrene	104 50 50 0.83 0.95	[52]
mMCM-41-*g*-p(GMA)-TAEA	185	19.6	Direct blue-6 Direct black-38	142.7 79.9	[54]
M-SBA-15/CPAA	159	2.68	Acid blue 25	909.09	[53]
γ-Fe$_2$O$_3$@SiO$_2$@h-mSiO$_2$	329	-	Methylene blue	41	[55]
MNCM-1	576	2.9	Methylene blue	248	[57]

3. Clay-Based Composites

Clays are naturally occurring adsorbents, although their hydrophilic character and natural negative charge limits the adsorption capacity of many hydrophobic organic pollutants and anionic compounds [4,58,59]. Despite this, their swelling capacity and ability to interleave and/or graft different substances makes them as a very attractive option to develop different composites. Several techniques, such as the incorporation of organic cations, polymers and metals using pillaring, have been used to modify the surface and structural properties of clays [60–63]. Adsorption studies using modified clay minerals as adsorbents have shown remarkable results [60,64,65]; however the small particle size and low density of these materials makes it difficult to separate them from the aqueous solution [66]. Therefore, conferring magnetic properties to clay minerals could open a wide range of possibilities for water treatment. The preparation of magnetic clay mineral composites has been the subject of various studies [3,4,6,10,11,67,68]. The interlayer space, channels, siloxane surfaces and edges of the clay minerals provide hosting sites to stabilize magnetic NPs, thus leading to the formation of magnetic nanoparticles/clay mineral nanocomposites [18]. Typical synthesis routes to prepare magnetic clay minerals nanocomposites are pillaring, coprecipitation and intercalation (Figure 2). Pillaring represents a simple synthesis route where two strategies are possible; the combination of magnetic NP and pillared clay mineral, where the pores of the pillared clay minerals host the magnetic NPs (route A1, Figure 2), or the use of magnetic NPs as pillars to expand the interlayer space of clay mineral in order to create magnetic pillared clay mineral nanocomposite (route A2, Figure 2). Co-precipitation is the other simple and widely used route of synthesis, based on the in-situ formation of the magnetic NPs in an aqueous dispersion of clay mineral (route B, Figure 2). The intercalation method is usually used in two ways; on the one hand, using surfactant-modified magnetic NP, which is stabilized into clay mineral (route C1, Figure 2), and on the other hand, the inclusion of magnetic NP into surfactant intercalated clay mineral, where the surfactants change the surface and expand the interlayer space of the clay minerals to match and facilitate the entrance of the magnetic NPs (route C2, Figure 2).

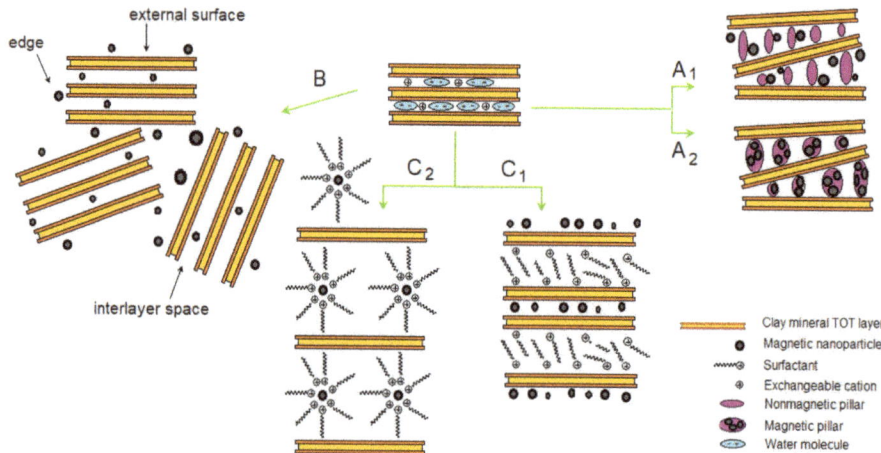

Figure 2. Scheme of typical synthesis routes to obtain magnetic nanoparticle/clay mineral 2:1 type (MNP/CM) nanocomposites: (A1) pillaring with nonmagnetic pillar; (A2) pillaring with magnetic pillar; (B) coprecipitation; (C1) intercalation of MNP into surfactant intercalated clay mineral; (C2) intercalation of surfactant-modified MNP into clay mineral.

Expansible and non-expansible clay minerals have been used to prepare magnetic nanocomposites. Magdy et al. [10] prepared magnetic nanocomposites of kaolin and magnetite (Fe_3O_4/kaolin) by the co-precipitation method in one step, and tested the prepared nanocomposites as adsorbents in the removal of the anionic Direct Red 23 dye. They achieved a complete removal of dye under the following set of operating conditions: initial dye concentration = 20 mg L^{-1}, adsorbent mass = 0.75 g, T = 25 °C and pH 7. Additionally, the adsorption followed the Langmuir isotherm with a maximum adsorbent capacity of 22.88 mg g^{-1}. On the other hand, Chang et al. [4] prepared magnetic nanocomposites from montmorillonite (Mt) and magnetite (Fe_3O_4/Mt) by a co-precipitation method. This study revealed that Fe_3O_4 nanoparticles are present on the surface of Mt. The magnetic nanocomposite showed a good adsorption efficacy (99.47%) in methylene blue removal.

Another possibility of very interesting modification results from the preparation of magnetic clay mineral nanocomposites with the incorporation of one or more extra components to improve some of the mentioned properties or incorporate new ones. Mu et al. [69] developed several magnetic nanocomposites combining in-situ intercalation, polymerization and coprecipitation techniques. They polymerized aniline molecule, during the synthesis of Fe_3O_4 in a suspension of Mt, obtaining polyaniline (PANI) and Fe_3O_4 supported on the surface of Mt (Mt/PANI/Fe_3O_4). They tested these materials in the adsorption of three dyes, methylene blue, brilliant green and congo red. One of the developed nanocomposites removes 99.6%, 96.2% and 98.1% of methylene blue, brilliant green and congo red, respectively, from 100 ppm dye solutions, with nanocomposite dosages of 1 g L^{-1}, at 25 °C.

They also studied the reuse of Mt/PANI/Fe$_3$O$_4$ using BM, and after five cycles the adsorption capacity did not decrease (in all cases it was close to 100%).

Arya and Philip [6] incorporated powdered-activated carbon, chitosan and sodium tripolyphosphate as a binding agent to magnetite nanoparticles supported in bentonite, with the aim of developing a new adsorbent compound capable of removing anionic, cationic, hydrophilic or hydrophobic contaminants. They studied the efficiency of this adsorbent to remove atenolol, ciprofloxacin and gemfibrozil from aqueous systems. The authors found a high removal of atenolol and ciprofloxacin (85% and 95% respectively, with an adsorbent dosage of 1.5 g L^{-1}), which can be attributed to the hydrophilic nature and the interchange capacity of clay–chitosan, since both pollutants are predominantly cationic at the working pH. On the other hand, the hydrophobic nature of activated carbon may be the main factor in the adsorption of genfibrozil (90% removal).

Diagboya and Dikio [11] developed magnetic adsorbent composites with feldspar clay (FLC), pericarp of oak fruits (PER) and magnetic nanoparticles. The nanocomposites were prepared by the co-precipitation method in the presence of FLC free of organic matter and sieved PER; then, the product obtained was pyrolyzed at 250 °C for 4 h. Magnetic adsorbents (BMF-0.5 and BMF-1) were used to study the removal of methylene blue. The authors performed three adsorption/desorption cycles of methylene blue; they observed that the adsorption capacity decreases in each cycle. The second cycle was lower than the first by approximately 7% for both nanocomposites and the third cycle decreased by a similar proportion.

On the other hand, Fizir et al. [68] developed magnetic halloysite nanotubes (MHNTs) by the method of co-precipitation followed by polymer grafting onto the nanocomposites, to adsorb norfloxacin. Combining the advantages of the high adsorption capacity and the magnetic properties of this biocompatible clay nanotube and the advantage of the polymer shell in improving the controlled and sustained release of the drug, they formulated a novel bioactive agent.

Beé et al. [3] used an extrusion method to obtain a magnetic adsorbent that was prepared by trapping maghemite nanoparticles (γ-Fe$_2$O$_3$) and montmorillonite (MMT) in cross-linked chitosan (CS) beads, in order to obtain an efficient adsorbent for cationic and anionic contaminants. They prepared pearls with different clay contents and performed adsorption experiments with methylene blue. They observed that the adsorption capacity of methylene blue increased as the amount of clay increased and the increase occurred over the entire pH range studied (3-12). The maximum uptake of methylene blue (82 mg g^{-1} at pH 9.9) was obtained with the material that had the highest proportion of clay; this indicates that the adsorption properties of the clay were not affected by encapsulation in the beads. Table 2 summarizes the adsorption capacities of the magnetic-clay-based nanocomposites described in this section.

Table 2. Application of magnetic clay-based nanoadsorbents for organic pollutants' removal from water.

Adsorbent	Surface Area ($m^2\ g^{-1}$)	Magnetic Saturation (emu g^{-1})	Organic Pollutant	Adsorption Capacity (mg g^{-1})	Ref.
Clay:chitosan:PAC:MNP	95	1.91	Atenolol	15.6	[6]
			Ciprofloxacin	39.1	
			Gemfibrozil	24.8	
Fe_3O_4/Mt	148	-	Methylene blue	106.4	[4]
Magnetic chitosan/clay beads	-	-	Methylene blue	82	[3]
MSEP	112	31.8	Atrazine	1.79	[67]
MHNTs	-	42.87	Norfloxacin	99.6	[68]
Fe_3O_4/kaolin	32	12.32	Direct red 23	22.88	[10]
Fe_3O_4-Sep	81	26.22	Bisphenol A	-	[66]
BMF-1	-	-	Methylene blue	14.93	[11]
BMF-0.5	-	-	Methylene blue	12.35	[11]
Mt/PANI/Fe_3O_4-2	-	36.52	Methylene blue	184.5	[69]

4. Carbon-Based Materials

Different carbon-based materials have been utilized for adsorption of pollutants, including activated carbon, graphitized carbon black, porous carbon, graphene oxide (GO) and carbon nanotubes (CNTs), due to the versatility of types of interactions that can accomplish with pollutants, such as electrostatic, hydrophobic and π–π interactions [70]. These materials differ in their structures and physicochemical characteristics, as pore size/shape, pore volume, surface area and surface functionality. For instance, CNTs can be considered as hollow graphitic nanomaterials comprising one (single-walled carbon nanotubes, SWNTs) or multiple (multiwalled carbon nanotubes, MWNTs) layers of graphene sheets, while GO is a derivative of graphene, which has abundant active functional groups such as carboxyl, hydroxyl and epoxy [71].

During the last decade, much effort has been devoted to developing efficient synthetic routes to shape-controlled, highly stable, and well-defined magnetic carbon hybrid nanocomposites. Several methods including the filling process, template-based synthesis, chemical vapor deposition, the hydrothermal/solvothermal method, the pyrolysis procedure, the sol–gel process and the self-assembly method can be used for the synthesis of high-quality magnetic carbon nanocomposites [72].

A list of magnetic carbon-based nanocomposites with their adsorption capacity for organic contaminants and some physicochemical properties has been summarized in Table 3. In some cases, the magnetization of the carbon-based materials enhances the adsorption properties when compared to the pristine carbon material. This is explained considering that magnetic nanoparticles can increase the porosity of carbon materials, which favor the diffusion of contaminants to more available adsorption sites.

Fe_3O_4 supported on reduced Grapheme Oxide (rGO) nanocomposite for the removal of harmful pesticides, namely simazine, simeton, atrazine, prometryn, and ametryn, was synthetized by Boruah et al. [73] using an eco-friendly in-situ solution chemistry approach, GO was synthesized from graphite powder and reduced to the rGO sheet using ascorbic acid as a reducing agent. Then, Fe_3O_4 nanoparticles were synthesized adopting the chemical co-precipitation method in the presence of rGO, yielding Fe_3O_4/rGO nanocomposite. Their study reveals that Fe_3O_4/rGO nanocomposite exhibits excellent adsorption performance towards the adsorption of the five pesticide molecules, compared to Fe_3O_4 nanoparticles and rGO sheets. The corresponding adsorption efficiencies for simazine, atrazine prometryn, ametryn, and simeton were found to be 88%, 75%, 91%, 93% and 81%, respectively.

In general, carbon-based nanomaterials have a unique π–π electronic structure that provides excellent properties to be used as adsorbents for the removal of aromatic compounds [74,75]. Yang et al. [76] compared the adsorption capacity of aromatic compounds (1-naphthylamine, 1-naphthol and naphthalene) between reduced graphene oxide/iron oxide composites (GO/Fe$_3$O$_4$) and multi-walled carbon nanotube/iron oxide composites (MWCNTs/Fe$_3$O$_4$). They found that electron–donor–acceptor (EDA) interaction was the primary adsorption mechanism and the higher polarity of adsorbates lead to higher adsorption capacity. In particular, GO/Fe$_3$O$_4$ showed better adsorption capacity than MWCNTs/Fe$_3$O$_4$, probably due to GO/Fe$_3$O$_4$ presenting more available and abundant adsorption sites.

Fan et al. [77] prepared magnetic β-cyclodextrin–chitosan/graphene oxide materials (MCCG) via a chemical route to remove methylene blue from aqueous solution. In this synthesis procedure, the carboxyl group of GO chemically reacts with the amine group of the magnetic β-cyclodextrin–chitosan composite to yield the magnetic carbon-based sorbent. MCCG showed excellent sorbent properties, benefiting from the high surface area of graphene oxide, hydrophobicity of β-cyclodextrin and the abundant amino and hydroxyl functional groups of chitosan. The maximum adsorption capacity of MCCG was 84.3 mg g^{-1}, which was higher than those reported for other adsorbents such as graphene/magnetite composite (43.8 mg g^{-1}), pyrophylite (4.2 mg g^{-1}), carbon nanotubes (46.2 mg g^{-1}), exfoliated graphene oxide (17.3 mg g^{-1}) and β-cyclodextrin–chitosan (50.12 mg g^{-1}) [78–80]. Among other magnetic carbonaceous materials with high sorption capacity to remove methylene blue, magnetic graphene sponge, (Fe$_3$O$_4$-GS) synthetized using a simple method, in which Fe$_3$O$_4$ nanoparticles are mixed with GO and subsequently lyophilized to obtain Fe$_3$O$_4$-GS, was reported by Yu et al. [81] This magnetic graphene sponge presents quite high adsorption capacity for MB (526 mg g^{-1}) and facile regeneration.

Alizadeh Fard and Barkdoll [82] prepared magnetic carbon nanotubes (MCN) to remove six micropollutants (Metolachlor, Bisphenol-A, Tonalide, Triclosan, Ketoprofen and Estriol) from water. In this work, to prevent the formation of other species of iron oxide, the synthesis of MCNs was performed separately: first, the magnetic nanoparticles were produced by a simple hydrothermal method. Then, the magnetic nanoparticles were mixed with the HNO$_3$-treated CNTs. MCN presented good adsorption capacity in which Bisphenol-A, Ketoprofen and Tonalide were the most effectively removed micropollutants, with 98%, 96% and 96% removal within 47 min, respectively.

Gong et al. [83] developed magnetic multi-wall carbon nanotube (MMWCNT) nanocomposite using a modified sol–gel process. The negatively charged MMWCNT surface over a wide pH range is one of the main characteristics that allows the removal of cationic pollutants. The adsorption properties of this adsorbent were examined using cationic dyes (methylene blue, neutral red and brilliant cresyl blue). They observed that dye adsorption capacity increased when pH increased from 3 to 7, while beyond 7, the adsorption capacity was not significantly altered. The comparison of the adsorption results with MWCNT and activated carbon indicated that MMWCNT nanocomposite showed the main advantage of the convenience of separation compared to the adsorption treatment in aqueous media. By using another synthesis strategy, Zhao et al. [84] prepared MWCNTs decorated with Fe$_3$O$_4$ nanoparticles modified with polyaniline (MWCNTs/Fe$_3$O$_4$/PANI) and used this material to test the adsorption of methyl orange and Congo red. The MWCNTs/Fe$_3$O$_4$/PANI showed high adsorption capacity toward the tested dyes (446.25 mg g^{-1} for methyl orange and 417.38 mg g^{-1} for congo red). The nitrogen containing functional groups of PANI, along with the synergistic effect between MWCNT π-conjugated bonds and PANI π-conjugated bonds, contribute to boosting the effective adsorption sites and thus increase the adsorption capacity toward MO and CR. Another important result obtained from the authors is that the presence of PANI prevented Fe$_3$O$_4$ nanoparticles from dissolving, and hence improved the stability of MWCNTs/Fe$_3$O$_4$/PANI in solution. In order to confer more carboxyl groups to magnetic carbon nanotube nanocomposites, Deng et al. [85] prepared carbon dot-modified magnetic carbon nanotubes (CMNTs) by three consecutive steps including the preparation of MWNTs, the synthesis of carbon dots and the surface modification of MWNTs using carbon dots. This adsorbent has a moderately high adsorption capacity of carbamazepine (65 mg g^{-1} at pH 7) and can be regenerated and reused up to six times with capacity loss less than 2.2%.

Liu et al. [86] assembled activated carbon and Fe$_3$O$_4$ nanoparticles (Fe$_3$O$_4$/AC) by a facile one-step thermal decomposition process and tested Fe$_3$O$_4$/AC for the adsorption of rhodamine B and methyl orange. Fe$_3$O$_4$/AC showed a high surface area (about 1200 m^2 g^{-1}), which was much higher than pure AC (about 750 m^2 g^{-1}), and wide pore size distribution. These properties benefit the adsorption and the rapid inter-diffusion of the dye molecules through interconnected channels. The obtained magnetic composite exhibited better adsorption of the dyes than its pure AC counterpart. Another approach to assemble Fe$_3$O$_4$ nanoparticles with activated carbon was reported by Shan et al. [87] through the ball milling method. They obtained two ultrafine magnetic activated carbon (Fe$_3$O$_4$/AC) and biochar (Fe$_3$O$_4$/BC) hybrid materials. The use of biochar drastically increased the surface area of the magnetic composite, while in case of AC the surface area decreased, probably because some pores in AC are blocked by the Fe$_3$O$_4$ nanoparticles produced in the milling process. However, in both cases, the authors obtained a higher adsorption capacity than pristine carbon materials using two pharmaceuticals as model contaminants (i.e., carbamazepine and tetracycline). On the other hand, Yu et al. [88] reported a higher sorption capacity of tetracycline (473 mg g^{-1}) using a magnetic graphene oxide sponge (Fe$_3$O$_4$-GOS) as sorbent. The material was prepared by lyophilizing the dispersion of Fe$_3$O$_4$ nanoparticles and graphene oxide (GO).

In the study performed by Lompe et al. [89] the adsorption capacity for nine micropollutants (Diclofenac, Fluoxetine, Estradiol, Norethindrone, Atrazine, Carbamazepine, Deethylatrazine, Sulfamethoxazole and Caffeine) on fresh and aged magnetic powdered activated carbon (MPAC) was investigated. They demonstrated that MPAC produced via co-precipitation can be customized with respect to its magnetic properties without compromising its adsorption capacity beyond the reduction expected for lower powdered activated carbon (PAC) contents. Maximum adsorption capacities of PAC and MPAC for all pollutants ranged between 1 and 80 mg g^{-1}.

Table 3. Application of magnetic carbon-based nanoadsorbents for organic pollutant removal from water.

Adsorbent	Surface Area ($m^2\ g^{-1}$)	Magnetic Saturation ($emu\ g^{-1}$)	Organic Pollutant	Adsorption Capacity ($mg\ g^{-1}$)	Ref.
GO/Fe_3O_4	272	-	1-Naphthylamine 1-Naphthol Naphthalene	2.85 2.70 2.63	[76]
$MWCNTs/Fe_3O_4$	77	-	1-Naphthylamine 1-Naphthol Naphthalene	1.45 1.13 1.05	[76]
Magnetic-cyclodextrin–chitosan/graphene oxide	402	55.1	Methylene blue	84.3	[77]
Fe_3O_4-GS	-	4.4	Methylene blue	526	[81]
MWCNT	61	-	Methylene blue Neutral red Brilliant cresyl blue	15.9 20.5 23.0	[83]
Fe_3O_4/AC	1200	16.5	Rhodamine B Methyl orange	182.4 150.3	[86]
$MWCNTs/Fe_3O_4$/PANI	-	42.9	Methyl orange Congo red	446.2 417.4	[84]
Fe_3O_4/BC	365	19.0	Carbamazepine Tetracycline	62.7 94.2	[87]
Fe_3O_4/AC	486	20.8	Carbamazepine Tetracycline	135.1 45.3	[87]
Fe_3O_4-GOS	-	1.1	Tetracycline	473	[88]
CMNTs	184	5.6	Carbamazepine	65	[85]
MPAC	430–780	5–30	Diclofenac, Fluoxetine, Estradiol, Norethindrone, Atrazine, Carbamazepine, Deethylatrazine, Sulfamethoxazole and Caffeine	1–80	[89]
MCN	-	-	Metolachlor, Bisphenol A, Tonalide, Triclosan, Ketoprofen and Estriol	18–28	[82]

5. Polymer-Based Materials

Polymers can be chemically anchored or physically adsorbed on magnetic nanoparticles to form a core–shell structure, acting as a protective layer and simultaneously providing active sites to adsorb pollutants. The core–shell structured can be synthesized by using the seed polymerization method and surface-modified Fe_3O_4 particles as the seeds [90]. This method allows the development of novel porous materials with high surface area and porosity to improve their adsorption performances [91].

Fe_3O_4@polyaniline, a typical Fe_3O_4-based magnetic core–shell material, was synthesized for the removal of pollutants such as humic acid, separation of organic dye, extraction and analysis of phenolic compounds and analysis of pyrethroids in tea drinks and polycyclic aromatic hydrocarbons [92]. As a magnetic core–shell material, polyaniline shows great promise, because it can effectively decrease the chance of aggregation, enhance the adsorption properties for organic pollutants via π–π and van der Waals interactions, and improve the stability of magnetic core–shell composites. However, the most appropriate composition of core–shell structure demands that the ratio between magnetic core and shell dimensions be optimized, since insufficient core magnetic oxide could determine a limited magnetic response and insufficient polymer component could result in poor adsorption capacity. Furthermore, to increase the maximum adsorption capacity and the versatility of the nanomaterials, several functional groups such as carboxylate, phosphate, sulfate, hydroxyl, amino and amide groups have been explored for the modification of conventional adsorbents. Hou et al. [93] prepared a core–shell nanoadsorbent based on Fe_3O_4 nanoparticles surface-modified with a copolymer, using 2,4-diaminophenol and formaldehyde for the adsorption of anionic dyes (amaranth, orange II and acid red 18) and obtained good adsorption capacities, fast adsorption processes and high saturation magnetization.

Liu et al. [94] reported the synthesis of magnetic nanospheres functionalized with β-cyclodextrin (β-CD) by the one-pot solvent thermal method using β-CD immobilized Fe_3O_4 magnetic nanoparticles with tetra-fluoroterephthalonitrile as the monomer. This material showed fast adsorption kinetics for methylene blue, high dispersibility in aqueous phase, a short equilibrium time (5 min), high recovery and good recyclability (keeping the adsorption efficiency above 86% after five uses). The maximum adsorption capacity of methylene blue was 305 mg g^{-1}.

During the last decade, porous organic polymers, including polymers of intrinsic microporosity [95], metal–organic frameworks (MOFs) [96], covalent organic frameworks (COFs) [97], porous aromatic frameworks [98] and hyper-cross-linked polymers (HCPs) [99] have attracted extensive interest because they have the advantages of low density, excellent chemical and physical stabilities, high surface area, easy control of pore size and functional modification [100]. MOFs are coordination polymers with intriguing structural motifs that can be self-assembled from organic ligands and metal ions or clusters of metal ions. In the last years, MOFs have been developing rapidly and have actually attracted extensive attention in the field of porous materials due to their high surface areas, with a range from 1000 to 10,000 m^2 g^{-1}, and a permanent porosity that is superior to those of porous materials such as activated carbon and zeolites [101]. While the use of MOFs as sorbents in gas phase is well-known, some MOFs show limitations in liquid phase adsorption due to their relatively low stability in water and their hydrophilic nature. However, in recent years, the development of water-stable MOFs has increased considerably [102]. Numerous different synthetic approaches, including slow diffusion, hydrothermal, electrochemical, mechanochemical, microwave-assisted heating and ultrasound can be applied to produce MOFs relying on the resulting structures and features [103]. Several magnetic adsorbents with MOF structure have been synthesized. Yang et al. [104] synthesized a new type of Fe_3O_4@MOF magnetic porous composite material with a core–shell structure of Fe_3O_4 nanoparticles coated with petal-like ZIF-67 crystals (Zeolitic imidazolate framework-67), in which Co^{2+} firstly combines with SO_3^{2-} provided by poly-styrenesulfonate sodium salt to form nucleation. This adsorbent showed high adsorption capacity for methyl orange (anionic dye) owing to the nature of Lewis base coordinated Co(II) and its high porosity. The equilibrium adsorption capacity was as high as 738 mg g^{-1}, which was significantly higher than other adsorbents like coconut shell activated carbon (368 mg g^{-1}) and clay (300 mg g^{-1}) [91]. Hamedi et al. [105] produced a magnetic MOF for the elimination of malachite green and methyl red from wastewaters. Metal–organic framework (MIL-101(Fe)) was prepared from $FeCl_3$ and terphthalic acid. Moreover, they used an extremely tinny film of 3, 4-dihydroxy-Lphenylalanine (PDopa) as an eco-friendly and effective binder between MIL-101(Fe) and Fe_3O_4 nanoparticles, in order to prevent detachment of the MOF from the magnetic material. In this regard, the authors were able to obtain the perfect capacities of adsorption (833 and 1250 mg g^{-1} for malachite green and methyl red, respectively). Moreover, the MIL-101(Fe)@PDopa@Fe_3O_4 adsorbent can be used almost four times to remove dyes. Wu et al. [106] prepared magnetic copper-based MOF (Fe_3O_4/HKUST-1) as an effective and recyclable adsorbent for the removal of ciprofloxacin and norfloxacin (two fluoroquinolone antibiotics) from aqueous solutions. The maximum adsorption capacities of the magnetic composites toward ciprofloxacin and norfloxacin reached 538 mg g^{-1} and 513 mg g^{-1}, respectively, noticeably higher than those values of most of the reported adsorbents for these two compounds.

Covalent organic frameworks (COFs), an emerging class of ordered crystalline porous polymers, are constructed from light elements and linked to organic monomers by strong covalent bonds that have an ordered π-structure with pore uniformity. In general, COFs are derived from MOF. They show comparable surface areas, ordered channel structures, well defined pore apertures, low densities, and thermal and chemical stability [91]. COFs with various functionalities and high crystallinity have been synthesized and several materials have been reported in the environmental field. Functionalized crystalline polyimide, polycationic and polyanionic COFs have been used for the removal of contaminants with varying physical and chemical properties. It is noteworthy that the size-sieving effect played a major role in the application of COFs for the removal of pollutants. Based on the literature, COFs' high surface areas (e.g., 3500 m^2 g^{-1}) give COFs more potential as adsorbents than zeolites or activated carbon [107]. Strategies to obtain water-stable materials with highly ordered structures and large surface areas are reviewed. By means of post-synthetic modification approaches, pore surfaces can be tuned to target specific contaminants based on size-dependent separation and charge-selective separation. Immobilization of COFs on solid substrates is a strategy to improve stability and dispersibility, while taking advantage of the properties of all materials present in the composite [108]. Yi et al. [109] prepared core–shell-structured magnetic COF nanocomposites (Fe_3O_4@COFs) for the adsorption of triclosan and triclocarban in aqueous solutions. Fe_3O_4@COFs was fabricated on the Fe_3O_4 nanoparticles using an in-situ growth strategy at room temperature via a condensation reaction of 1,3,5-tris(4-aminophenyl) benzene and terephthaldicarbox-aldehyde in the presence of dimethyl sulfoxide. The adsorption behaviors showed high adsorption capacity and fast adsorption of triclosan and triclocarban. Different magnetic COF composites were also reported for the extraction of bisphenols [110] and polycyclic aromatic hydrocarbons (PAHs) [111] from water.

Hyper-cross-linked polymers (HCPs) can be synthesized by using external crosslinker. The porosity can be improved by using various kinds of reactive monomers or different amounts of crosslinking. For instance, Hu et al. (2019) [100] have synthesized a magnetically hyper-cross-linked polymer using benzylamine and benzene as the reactive monomers, which combined the advantage of the amino-modified HCP (HCP-NH_2) and the magnetic Fe_3O_4 nanoparticles to adsorb organic pollutants such as tetracycline, ethyl orange, methylene blue, bisphenol A and 2,4-dichlorophenol in aqueous solution. The magnetic HCP presented high BET specific surface area (532.62 m^2 g^{-1}), well-developed mesoporous (0.3786 cm^3 g^{-1}) and good magnetic properties. The adsorption experiments indicated that the magnetic HCP-NH_2 showed highly efficient adsorption properties for organic contaminants. In particular, the maximum adsorption capacity of the magnetic HCP-NH_2 for tetracycline was 694 mg g^{-1} at 298 K, which was much higher than that of HCP-NH_2 (389 mg g^{-1}). Based on the adsorption performance, the authors propose that tetracycline removal was mainly driven by coordination interaction, cation exchange and hydrogen bonding interactions that involve polymer functional groups and magnetite surface groups.

Recently, various biomaterials based on natural polymers have been developed for improving adsorption capacities, increasing environmental compatibility and operating efficiently. Alginate-based composites have been extensively studied for applications in environmental sectors due to their biocompatible, nontoxic, and cost-effective properties [112]. Various alginate-based composites that enhanced adsorption performance have been reported for the removal of various pollutants including dyes, heavy metals, and antibiotics in water and wastewater. Environmental applications of alginate depend partly on the fact that the rich surface functional groups (e.g., carboxyl and hydroxyl) could capture metallic or cationic ions via ion exchange between the crosslinking cations and target pollutants. Alginate beads may serve as a stable matrix for other types of absorbents that are too fine in particle size and too difficult to separate from aqueous solution. Magnetic adsorbents can be developed by encapsulating magnetic functionalized nanoparticles in alginate beads along with different covalently cross-linked agents, such as molecules of different sizes and structures, including adipic dihydrazide, lysine, and poly-(ethylene glycol)-diamines [113,114]. Talbot et al. [115] reported the synthesis of alginate/maghemite nanoparticles for the adsorption of methylene blue. A co-precipitation method in alkaline medium followed by oxidation of magnetite into maghemite led to a stable colloidal dispersion, which was added to sodium alginate solution in order to obtain the magnetic nanocomposite. The nanocomposites showed high adsorption capacity in a wide pH range and a reuse performance higher than 98% even after ten adsorption/desorption cycles. On the other hand, Mohammadi et al. [116] fabricated superparamagnetic sodium alginate-coated Fe_3O_4

nanoparticles by a co-precipitation method, obtaining good removal efficiency of malachite green (48 mg g^{-1}).

Chitosan is another biopolymer that is widely used in environmental applications, owing to its low-cost source and environment-friendly nature. It has abundant reactive amino and hydroxyl groups, which can serve as binding sites turning it a potential adsorbent for organic pollutant removal [117,118]. Solubility of chitosan in an acidic medium limits its wide application in water treatment; thus, various cross-linking agents such as glutaraldehyde [119], sodium tripolyphosphate [92], and epichlorohydrin [120] offer an important pathway to improve its chemical stability and extend its potential applications. Zheng et al. [121] prepared poly([2-(methacryloxy)ethyl] trimethylammonium chloride)-grafted magnetic chitosan microparticles via free radical polymerization to obtain novel superior adsorbents with a huge electrostatic "force field" for capturing food dyes. The nanocomposite morphology was nearly spherical with approximately 125 µm average diameters, with a specific surface area of 150 m^2 g^{-1} and average pore diameters of 33.8 Å. Compared to unmodified adsorbents, the adsorption capacities toward Food Yellow 3 and Acid Yellow 23 were considerably enhanced after modification, indicating that the dyes were captured by electrostatic interaction and ion exchange. On the other hand, polymer-grafted magnetic microspheres (GMMs) were prepared by graft polymerization of 2-acrylamido-2-methylpropane sulfonic acid and acrylic acid onto the surface of chitosan/magnetite composite microspheres and were used as an adsorbent to remove methylene blue from aqueous solutions [122]. The authors reported a maximum adsorption capacity for methylene blue of 926 mg g^{-1}, which is notably higher than the reported values for other adsorbents. Table 4 shows the adsorption capacities of the polymer-based magnetic nanomaterials with higher organic pollutant uptake.

Table 4. Application of magnetic polymer-based nanoadsorbents for organic pollutants removal from water.

Adsorbent	Surface Area (m^2 g^{-1})	Magnetic Saturation (emu g^{-1})	Organic Pollutant	Adsorption Capacity (mg g^{-1})	Ref.
Bio-magnetic membrane capsules from PVA–alginate matrix	-	11.02	Malachite green	500	[123]
Magnetic nanocellulose from olive industry solid waste	-	21.4	Methylene blue	166.67	[124]
Fe$_3$O$_4$-amine-functionalized chitosan with p-Benzoquinone	-	17.5	Diclofenac sodium	469.48	[125]
Magnetic β-cyclodextrin porous polymer nanospheres	70.63	44.8	Methylene blue	305.8	[94]
Magnetic porphyrin-based porous organic polymer	310	45.9	Phenylurea herbicides	Metoxuron = 1.13, Mono-linuron = 0.95, Chlorotoluron = 0.86, Buturon = 1.10	[92]
Magnetic copper based metal–organic frameworks (MOF)	327.9	44	Fluoroquinolone antibiotics	Ciprofloxacin = 538 Norfloxacin = 513	[106]
Magnetic polyimide-Mg-Fe layered double hydroxides core–shell composite		26.38	Tetracycline 2,4-dichlorophenol Glyphosate	185.53 176.06 190.84	[126]
Magnetic mesoporous lignin from date palm pits	640	37.81	Diesel Gasoline	Diesel = 22370 Gasoil = 21010	[127]

6. Waste-Based Materials

Within the framework of the circular economy, several efforts have been made to add value to wastes, converting them into renewable raw material for the production of fine, bioplastic and auxiliary chemicals for technological and environmental applications. In particular, the modification of different industrial and household wastes into adsorbent materials has been the subject of much research, since it represents an economically sustainable method of waste valorization [128–130]. Among them, the most common are agricultural residues, for example the skin of different fruits and vegetables, remnants of branches and leaves, remnants of rice, corn, olive and many others. The reuse of waste from the paper industry, the metallurgical industry, cement and sludge from water treatment plants has also been studied [131].

For the synthesis of waste-derived magnetic materials, most studies use magnetite nanoparticles as the magnetic agent to modify the starting adsorbent sourced from wastes. Safarik and co-workers published two short studies in which they synthetized magnetic adsorbents by putting a stable suspension of magnetite in contact with powdered peanut husk and pine sawdust, respectively [132,133]. They tested these materials against a broad spectrum of organic dyes, obtaining, in particular, maximum adsorption capacities for crystal violet of 80.9 and 51.2 mg g^{-1}. Zuorro et al. [134] achieved remarkable adsorption performance for a waste-derived magnetic material obtained from a similar procedure, in which a suspension of coffee silverskin wastes was mixed with magnetite ferrofluid for 1 h at room temperature. This magnetic adsorbent was tested for methylene blue adsorption, achieved an adsorption capacity of 556 mg g^{-1} within 2 h at pH 6, and showed good reusability potential, with only a 14% decrease in adsorption after eight regeneration cycles with HCl. Minh et al. [135] also obtained a magnetic adsorbent from ground coffee wastes with an adsorption capacity toward methylene blue of 128 mg g^{-1}. On the other hand, Stan et al. [136] studied four different magnetite–starch materials prepared through an environmentally friendly synthetic route which consist of a co-precipitation method using water as the solvent and sodium bicarbonate as the precipitating agent. One of the four materials was made using bare magnetite and the other three used magnetite synthetized with fruit waste vegetable extract as surfactant. Removal efficiencies of optilan blue dye were between 72% and 89% for these materials. The authors also found that the use of fruit extract in the magnetite synthesis did not improve adsorption performance.

Another method to prepare waste-derived magnetic materials consists in the synthesis of magnetite nanoparticles in an aqueous system where the adsorbent is previously dissolved. Aydin and co-workers used a high-alkaline waste which is produced by the Bayer process in aluminum production called "red mud" and used it along with Fe(II) and Fe(III) salts to synthesize a magnetic-red mud material for adsorption of organophosphorus pesticides and antibiotics [137,138]. With an isoelectric point near eight, this material showed relatively low adsorption capacity towards pesticides, in the order of μg g^{-1}, and good performance against antibiotics with a maximum adsorption capacity of 200 mg g^{-1} for ciprofloxacin. Madrakian et al. [139] used the solid fraction of tea leaves previously washed and boiled, and tested the obtained magnetic adsorbent performance for seven common cationic and anionic organic dyes. The maximum adsorption capacities for the seven dyes were between 82 and 128 mg g^{-1}. Similar results were obtained by Madrakian et al. [140] using *Platanusorientalis* tree leaves as a raw material for the production of waste-derived magnetic materials. The adsorption capacities of five dyes were between 89 and 133 mg g^{-1}. The authors claim there is no significant loss in adsorption capacity after ten adsorption–desorption cycles for this material. Jodeh et al. [124] prepared magnetic cellulose-based materials from olive industry solid waste and tested the adsorption performance of these materials with methylene blue. In this study, the cellulose was extracted in a powder form by a multistep pulping and bleaching process. The extracted powder cellulose was converted to nanocrystalline cellulose (NCs) by acid hydrolysis and finally dispersed in an aqueous solution of Fe(II) and Fe(III) to obtain magnetic cellulose nanocrystalline through a co-precipitation method. The magnetic material showed good methylene blue adsorption capacity, ranging from 148 to 196 mg g^{-1}. Sun et al. [141] synthetized a magnetic material from wastes of *Vallisneria natans*, a widespread fast-growing aquatic plant, using a chemical co-precipitation method. This material achieved an outstanding methylene blue adsorption capacity (474 mg g^{-1}), almost constant in a wide range of pH values (from 5 to 9).

Yu et al. [142] prepared magnetic modified biomass from beer yeast and tested the adsorption potential by using methylene blue and basic violet as model dyes. The magnetic material was prepared by two steps that included the preparation of pyromellitic dianhydride (PMDA)-modified biomass in N,N-dimethylacetamide solution and the preparation of magnetic PMDA-modified biomass by a co-precipitation method under the assistance of ultrasound irradiation. The modified biomass achieved maximum adsorption capacities of 609 and 521 mg g^{-1} for methylene blue and basic violet, more than six times higher than the unmodified biomass adsorbent. Increasing pH showed a positive effect on adsorption capacities, showing values above 300 mg g^{-1} over a pH range between 3 and 9.

Core–shell magnetite nanoparticles, using soluble organic matter extracted from urban public park trimming and home gardening residues as a covering material, were used as sustainable and removable magnetic adsorbent materials [143–145]. The soluble organic matter (SBO) was obtained through 230 days of green residues composting followed by alkaline digestion and ultrafiltration of the soluble matter. Magnacca et al. [143] used the obtained Fe_3O_4-SBO nanoparticles for crystal violet removal from waters, and the results showed a good adsorption capacity (244 mg g^{-1}) at pH 7. On the other hand, Nistico et al. [144] studied the removal of polycyclic aromatic hydrocarbons (PAHs) by using Fe_3O_4-SBO nanoparticles. They found that a thermal treatment at 550 °C under nitrogen atmosphere of the Fe_3O_4-SBO nanoparticles enhance their adsorption performance towards PAHs.

Liu et al. [146] developed a synthesis procedure that uses iron mud. This procedure consist of a simple co-precipitation method where iron mud, previously dissolved with 2% HNO_3 overnight, acts as an Fe(III) source, and the addition of ascorbic acid (AA) reduces part of the Fe(III) to Fe(II) needed for the formation of magnetite at alkaline conditions. Increasing the AA/Fe^{3+} ratio during synthesis resulted in increased saturation magnetization values, crystallite size and reduced BET specific area. To further reduce the production cost, acid wastewater from a propylene plant was used to replace nitric acid in digestion of iron mud. Samples with higher crystallite size exhibited less BET specific surface area and less adsorption capacity. Results yield maximum methylene blue adsorption capacities of 87 mg g^{-1} for the material synthetized with a 0.1 AA/Fe^{3+} ratio. A list of magnetic waste-based nanomaterials with their adsorption capacity for organic contaminants is summarized in Table 5.

Table 5. Application of magnetic waste-based nanoadsorbents for organic pollutants removal from water.

Adsorbent	Surface Area (m g^{-1})	Magnetic Saturation (emu g^{-1})	Organic Pollutant	Adsorption Capacity (mg g^{-1})	Ref.
Magnetic Sawdust	-	-	Crystal violet	51.2	[132]
Magnetic Peanut husk	-	-	Acridine orange Bismark brown Crystal violet Safranin O	71.4 95.3 80.9 86.1	[133]
Fe$_3$O$_4$-Coffe Skin	-	-	Methylene blue	556	[134]
Fe$_3$O$_4$-Ground coffe waste	-	-	Methylene blue	128	[135]
Fe$_3$O$_4$-Starch Fe$_3$O$_4$-Av1-Starch Fe$_3$O$_4$-Wm-Starch	62 78 63	51 43 49	Optilan blue	119 111 63	[136]
Fe$_3$O$_4$-Red mud	84	12	Diazinon Malathion Parathiion Chlorpyrifos	1.9 1.7 2.9 3.9	[137]
Fe$_3$O$_4$-Red mud	84	12	Ciprofloxacin	200	[138]
Fe$_3$O$_4$-Tea Waste	-	-	Neutral red Reactive blue Congo red Janus green Methilene blue Crystal violet Thionine	127 88 83 130 119 114 128	[139]
Fe$_3$O$_4$-Tree leaves	-	-	Malachite green Neutral red Methylene blue Crystal violet Methyl violet	89 101 128 117 133	[140]
Fe$_3$O$_4$-Olive cristal cellulose	-	21	Methylene blue	166	[124]
Fe$_3$O$_4$-Acuatic Plant	7	3	Methylene blue	474	[141]
Fe$_3$O$_4$-Beer Yeast	-	-	Methylene blue Basic violet	609 521	[142]
Fe$_3$O$_4$-SBO	35	51	Crystal violet	244	[143]
Fe$_3$O$_4$ (MP-10)	70	9	Methylene blue	99.4	[144]
Fe$_3$O$_4$ (MP-3) Fe$_3$O$_4$ (MP-3w)	176 119	5 4	Methylene blue	87.3 56.7	[146]

7. Conclusions and Future Perspective

Magnetic core–shell nanoparticles and nanocomposites containing iron oxide nanoparticles are a very attractive option to be used in water treatment, due to their advantages in terms of improving some of the properties of unmodified materials for the adsorption of pollutants, namely: (i) a synergistic combination of adsorbent materials where a material is chemically anchored or physically adsorbed on magnetic nanoparticles; (ii) the use of low-cost materials; (iii) facilitation of the separation of the aqueous medium by a simple magnetic process, reducing the overall cost of the process; (iv) the possibility of being reused successively; (v) potential for change in the physicochemical properties of the nanoparticles and nanocomposites surface just by adjusting the experimental conditions, offering a high versatility to be applied to a wide range of contaminants (anionic, cationic, hydrophilic or hydrophobic). Therefore, nanoadsorbents with magnetic response has shown several advantages that could be used in the development water treatment technologies. In this review, the adsorption capacities of magnetic nanomaterials based on silica, clay, carbonaceous materials, polymers and waste were highlighted in order to provide valuable information related to their efficiency of various organic pollutants' removal.

Although magnetic core–shell nanoparticles and magnetic nanocomposites possess different structural characteristics and different removal mechanisms of organic molecules, their adsorption capacities for various organic pollutants are similar. It is worth mentioning that there are some cases where ultra-high adsorption uptake of organic contaminants was achieved. On the other hand, the preparation of nanoadsorbents by combining magnetic iron oxide nanoparticles with materials such as clays, carbon or polymers enable to control the aggregation of the magnetic nanoparticles and improve their performance in aqueous media.

Despite the growing interest in the synthesis of magnetic nanoadsorbents for water treatment in last years, nowadays there are few industrial applications and most of the research work is based on small-scale studies. We believe that further studies are needed to address large-scale applications, taking into account design aspects related to stability, adsorbent regeneration, and separation from the aqueous medium. This could help to identify the key operating conditions to achieve low-cost removal of the contaminant with minimal environmental impact. Along with these studies, the efficient production of magnetic nanomaterials at large scale must be investigated further, taking into consideration the environmental impact of manufacturing these materials.

Author Contributions: Conceptualization, L.C.; Investigation, M.E.P. (Marcos E. Peralta), S.O., I.G.F., F.O.M., M.E.P. (María E. Parolo) and L.C.; Writing—Original Draft Preparation, M.E.P. (Marcos E. Peralta), S.O., I.G.F., F.O.M., M.E.P. (María E. Parolo) and L.C.; Writing—Review & Editing, L.C.; Visualization, L.C.; Supervision, M.E.P. (María E. Parolo) and L.C.; Project Administration, M.E.P. (María E. Parolo) and L.C.; Funding Acquisition, M.E.P. (María E. Parolo) and L.C. All authors have read and agreed to the published version of the manuscript.

Funding: The authors acknowledge financial support from CONICET (PUE0067), ANPCyT (PICT 2017-1847) and the Universidad Nacional del Comahue (UNCo-04/I217). M.E.P., F.O.M. and S.O. thank CONICET for their research graduate grants. L.C. is research member of CONICET.

Conflicts of Interest: The authors declare no conflict of interest.

References

1. European Environment Agency. *European Waters Assessment of Status and Pressures 2018*; EEA Report; European Environment Agency: Luxembourg, 2018; ISBN 9789292139476.
2. Stuart, M.; Lapworth, D.; Crane, E.; Hart, A. Review of risk from potential emerging contaminants in UK groundwater. *Sci. Total Environ.* **2012**, *416*, 1–21. [CrossRef] [PubMed]
3. Bée, A.; Obeid, L.; Mbolantenaina, R.; Welschbillig, M.; Talbot, D. Magnetic chitosan/clay beads: A magsorbent for the removal of cationic dye from water. *J. Magn. Magn. Mater.* **2017**, *421*, 59–64. [CrossRef]
4. Chang, J.; Ma, J.; Ma, Q.; Zhang, D.; Qiao, N.; Hu, M.; Ma, H. Adsorption of methylene blue onto Fe_3O_4/activated montmorillonite nanocomposite. *Appl. Clay Sci.* **2016**, *119*, 132–140. [CrossRef]
5. Han, H.; Rafiq, M.K.; Zhou, T.; Xu, R.; Mašek, O.; Li, X. A critical review of clay-based composites with enhanced adsorption performance for metal and organic pollutants. *J. Hazard. Mater.* **2019**, *369*, 780–796. [CrossRef]
6. Arya, V.; Philip, L. Adsorption of pharmaceuticals in water using Fe_3O_4 coated polymer clay composite. *Microporous Mesoporous Mater.* **2016**, *232*, 273–280. [CrossRef]
7. Nagy, Z.M.; Molnár, M.; Fekete-Kertész, I.; Molnár-Perl, I.; Fenyvesi, É.; Gruiz, K. Removal of emerging micropollutants from water using cyclodextrin. *Sci. Total Environ.* **2014**, *485*, 711–719. [CrossRef]
8. Yang, K.; Wu, W.; Jing, Q.; Zhu, L. Aqueous adsorption of aniline, phenol, and their substitutes by multi-walled carbon nanotubes. *Environ. Sci. Technol.* **2008**, *42*, 7931–7936. [CrossRef]
9. Dailianis, S.; Tsarpali, V.; Melas, K.; Karapanagioti, H.K.; Manariotis, I.D. Aqueous phenanthrene toxicity after high-frequency ultrasound degradation. *Aquat. Toxicol.* **2014**, *147*, 32–40. [CrossRef]
10. Magdy, A.; Fouad, Y.O.; Abdel-Aziz, M.H.; Konsowa, A.H. Synthesis and characterization of Fe_3O_4/kaolin magnetic nanocomposite and its application in wastewater treatment. *J. Ind. Eng. Chem.* **2017**, *56*, 299–311. [CrossRef]

11. Diagboya, P.N.; Dikio, E.D. Scavenging of aqueous toxic organic and inorganic cations using novel facile magneto-carbon black-clay composite adsorbent. *J. Clean. Prod.* **2018**, *180*, 71–80. [CrossRef]
12. Reddy, D.H.K.; Yun, Y.S. Spinel ferrite magnetic adsorbents: Alternative future materials for water purification? *Coord. Chem. Rev.* **2016**, *315*, 90–111. [CrossRef]
13. Mehta, D.; Mazumdar, S.; Singh, S.K. Magnetic adsorbents for the treatment of water/wastewater—A review. *J. Water Process Eng.* **2015**, *7*, 244–265. [CrossRef]
14. Gómez-Pastora, J.; Bringas, E.; Ortiz, I. Recent progress and future challenges on the use of high performance magnetic nano-adsorbents in environmental applications. *Chem. Eng. J.* **2014**, *256*, 187–204. [CrossRef]
15. Larraza, I.; López-gónzalez, M.; Corrales, T.; Marcelo, G. Hybrid materials: Magnetite—Polyethylenimine—Montmorillonite, as magnetic adsorbents for Cr(VI) water treatment. *J. Colloid Interface Sci.* **2012**, *385*, 24–33. [CrossRef]
16. Nisticò, R.; Cesano, F.; Garello, F. Magnetic Materials and Systems: Domain structure visualization and other characterization techniques for the application in the materials science and biomedicine. *Inorganics* **2020**, *8*, 6. [CrossRef]
17. Nisticò, R. Magnetic materials and water treatments for a sustainable future. *Res. Chem. Intermed.* **2017**, *43*, 6911–6949. [CrossRef]
18. Chen, L.; Zhou, C.H.; Fiore, S.; Tong, D.S.; Zhang, H.; Li, C.S.; Ji, S.F.; Yu, W.H. Functional magnetic nanoparticle/clay mineral nanocomposites: Preparation, magnetism and versatile applications. *Appl. Clay Sci.* **2016**, *127*, 143–163. [CrossRef]
19. Yuan, P.; Fan, M.; Yang, D.; He, H.; Liu, D.; Yuan, A.; Zhu, J.X.; Chen, T.H. Montmorillonite-supported magnetite nanoparticles for the removal of hexavalent chromium [Cr(VI)] from aqueous solutions. *J. Hazard. Mater.* **2009**, *166*, 821–829. [CrossRef]
20. Mohammed, L.; Gomaa, H.G.; Ragab, D.; Zhu, J. Magnetic nanoparticles for environmental and biomedical applications: A review. *Particuology* **2017**, *30*, 1–14. [CrossRef]
21. Liu, J.; Qiao, S.Z.; Hu, Q.H.; Lu, G.Q. Magnetic nanocomposites with mesoporous structures: Synthesis and applications. *Small* **2011**, *7*, 425–443. [CrossRef]
22. Tang, S.C.N.; Lo, I.M.C. Magnetic nanoparticles: Essential factors for sustainable environmental applications. *Water Res.* **2013**, *47*, 2613–2632. [CrossRef] [PubMed]
23. Gao, F. An Overview of Surface-Functionalized Magnetic Nanoparticles: Preparation and Application for Wastewater Treatment. *ChemistrySelect* **2019**, *4*, 6805–6811. [CrossRef]
24. Gómez-Pastora, J.; Dominguez, S.; Bringas, E.; Rivero, M.J.; Ortiz, I.; Dionysiou, D.D. Review and perspectives on the use of magnetic nanophotocatalysts (MNPCs) in water treatment. *Chem. Eng. J.* **2017**, *310*, 407–427. [CrossRef]
25. Sharma, R.K.; Sharma, S.; Dutta, S.; Zboril, R.; Gawande, M.B. Silica-nanosphere-based organic-inorganic hybrid nanomaterials: Synthesis, functionalization and applications in catalysis. *Green Chem.* **2015**, *17*, 3207–3230. [CrossRef]
26. Cashin, V.B.; Eldridge, D.S.; Yu, A.; Zhao, D. Surface functionalization and manipulation of mesoporous silica adsorbents for improved removal of pollutants: A review. *Environ. Sci. Water Res. Technol.* **2018**, *4*, 110–128. [CrossRef]
27. Walcarius, A.; Mercier, L. Mesoporous organosilica adsorbents: Nanoengineered materials for removal of organic and inorganic pollutants. *J. Mater. Chem.* **2010**, *20*, 4478–4511. [CrossRef]
28. Wan, Y.; Zhao, D. On the Controllable Soft-Templating Approach to Mesoporous Silicates. *Chem. Rev.* **2007**, *107*, 2821–2860. [CrossRef]
29. Slowing, I.I.; Vivero-Escoto, J.L.; Trewyn, B.G.; Lin, V.S.Y. Mesoporous silica nanoparticles: Structural design and applications. *J. Mater. Chem.* **2010**, *20*, 7924–7937. [CrossRef]
30. Li, R.; Zhang, L.; Wang, P. Rational design of nanomaterials for water treatment. *Nanoscale* **2015**, *7*, 17167–17194. [CrossRef]
31. Diagboya, P.N.E.; Dikio, E.D. Silica-based mesoporous materials; emerging designer adsorbents for aqueous pollutants removal and water treatment. *Microporous Mesoporous Mater.* **2018**, *266*, 252–267. [CrossRef]
32. Cendrowski, K.; Sikora, P.; Zielinska, B.; Horszczaruk, E.; Mijowska, E. Chemical and thermal stability of core-shelled magnetite nanoparticles and solid silica. *Appl. Surf. Sci.* **2017**, *407*, 391–397. [CrossRef]

33. Wang, P.; Wang, X.; Yu, S.; Zou, Y.; Wang, J.; Chen, Z.; Alharbi, N.S.; Alsaedi, A.; Hayat, T.; Chen, Y.; et al. Silica coated Fe_3O_4 magnetic nanospheres for high removal of organic pollutants from wastewater. *Chem. Eng. J.* **2016**, *306*, 280–288. [CrossRef]
34. Wang, L.; Shen, C.; Cao, Y. PVP modified Fe_3O_4@SiO_2 nanoparticles as a new adsorbent for hydrophobic substances. *J. Phys. Chem. Solids* **2019**, *133*, 28–34. [CrossRef]
35. Li, J.; Zhou, Q.; Liu, Y.; Lei, M. Recyclable nanoscale zero-valent iron-based magnetic polydopamine coated nanomaterials for the adsorption and removal of phenanthrene and anthracene. *Sci. Technol. Adv. Mater.* **2017**, *18*, 3–16. [CrossRef] [PubMed]
36. Yue, Q.; Zhang, Y.; Wang, C.; Wang, X.; Sun, Z.; Hou, X.F.; Zhao, D.; Deng, Y. Magnetic yolk-shell mesoporous silica microspheres with supported Au nanoparticles as recyclable high-performance nanocatalysts. *J. Mater. Chem. A* **2015**, *3*, 4586–4594. [CrossRef]
37. Zhang, Y.; Jiang, F.; Huang, D.; Hou, S. A facile route to magnetic mesoporous core–shell structured silicas containing covalently bound cyclodextrins for the removal of the antibiotic doxycycline from water. *RSC Adv.* **2018**, *8*, 31348–31357. [CrossRef]
38. Oppmann, M.; Wozar, M.; Reichstein, J.; Mandel, K. Reusable Superparamagnetic Raspberry-Like Supraparticle Adsorbers as Instant Cleaning Agents for Ultrafast Dye Removal from Water. *ChemNanoMat* **2019**, *5*, 230–240. [CrossRef]
39. Hoffmann, F.; Cornelius, M.; Morell, J.; Fröba, M. Silica-based mesoporous organic–inorganic hybrid materials. *Angew. Chem. Int. Ed.* **2006**, *45*, 3216–3251. [CrossRef]
40. Sasaki, T.; Tanaka, S. Adsorption behavior of some aromatic compounds on hydrophobic magnetite for magnetic separation. *J. Hazard. Mater.* **2011**, *196*, 327–334. [CrossRef]
41. Chen, J.; Chen, H. Removal of anionic dyes from an aqueous solution by a magnetic cationic adsorbent modified with DMDAAC. *New J. Chem.* **2018**, *42*, 7262–7271. [CrossRef]
42. Peralta, M.E.; Jadhav, S.A.; Magnacca, G.; Scalarone, D.; Mártire, D.O.; Parolo, M.E.; Carlos, L. Synthesis and in vitro testing of thermoresponsive polymer-grafted core–shell magnetic mesoporous silica nanoparticles for efficient controlled and targeted drug delivery. *J. Colloid Interface Sci.* **2019**, *544*, 198–205. [CrossRef] [PubMed]
43. Jiaqi, Z.; Yimin, D.; Danyang, L.; Shengyun, W.; Liling, Z.; Yi, Z. Synthesis of carboxyl-functionalized magnetic nanoparticle for the removal of methylene blue. *Colloids Surf. A Physicochem. Eng. Asp.* **2019**, *572*, 58–66. [CrossRef]
44. Yang, Q.; Wang, J.; Chen, X.; Yang, W.; Pei, H.; Hu, N.; Li, Z.; Suo, Y.; Li, T.; Wang, J. The simultaneous detection and removal of organophosphorus pesticides by a novel Zr-MOF based smart adsorbent. *J. Mater. Chem. A* **2018**, *6*, 2184–2192. [CrossRef]
45. Ghorbani, F.; Kamari, S. Core–shell magnetic nanocomposite of Fe_3O_4@SiO_2@NH_2 as an efficient and highly recyclable adsorbent of methyl red dye from aqueous environments. *Environ. Technol. Innov.* **2019**, *14*, 100333. [CrossRef]
46. Peres, E.C.; Slaviero, J.C.; Cunha, A.M.; Hosseini-Bandegharaei, A.; Dotto, G.L. Microwave synthesis of silica nanoparticles and its application for methylene blue adsorption. *J. Environ. Chem. Eng.* **2018**, *6*, 649–659. [CrossRef]
47. Liu, S.; Chen, X.; Ai, W.; Wei, C. A new method to prepare mesoporous silica from coal gasification fine slag and its application in methylene blue adsorption. *J. Clean. Prod.* **2019**, *212*, 1062–1071. [CrossRef]
48. Brigante, M.; Parolo, M.E.; Schulz, P.C.; Avena, M. Synthesis, characterization of mesoporous silica powders and application to antibiotic remotion from aqueous solution. Effect of supported Fe-oxide on the SiO_2 adsorption properties. *Powder Technol.* **2014**, *253*, 178–186. [CrossRef]
49. Saikia, D.; Deka, J.R.; Wu, C.E.; Yang, Y.C.; Kao, H.M. pH responsive selective protein adsorption by carboxylic acid functionalized large pore mesoporous silica nanoparticles SBA-1. *Mater. Sci. Eng. C* **2019**, *94*, 344–356. [CrossRef]
50. Ganiyu, S.O.; Bispo, C.; Bion, N.; Ferreira, P.; Batonneau-Gener, I. Periodic Mesoporous Organosilicas as adsorbents for the organic pollutants removal in aqueous phase. *Microporous Mesoporous Mater.* **2014**, *200*, 117–123. [CrossRef]
51. Zhang, X.; Zeng, T.; Wang, S.; Niu, H.; Wang, X.; Cai, Y. One-pot synthesis of C18-functionalized core-shell magnetic mesoporous silica composite as efficient sorbent for organic dye. *J. Colloid Interface Sci.* **2015**, *448*, 189–196. [CrossRef]

52. Huang, Y.; Fulton, A.N.; Keller, A.A. Optimization of porous structure of superparamagnetic nanoparticle adsorbents for higher and faster removal of emerging organic contaminants and PAHs. *Environ. Sci. Water Res. Technol.* **2016**, *2*, 521–528. [CrossRef]
53. Ghanei, M.; Rashidi, A.; Tayebi, H.; Yazdanshenas, M.E. Removal of Acid Blue 25 from Aqueous Media by Magnetic-SBA-15/CPAA Super Adsorbent: Adsorption Isotherm, Kinetic, and Thermodynamic Studies. *J. Chem. Eng. Data* **2018**, *63*, 3592–3605. [CrossRef]
54. Arica, T.A.; Ayas, E.; Arica, M.Y. Magnetic MCM-41 silica particles grafted with poly(glycidylmethacrylate) brush: Modification and application for removal of direct dyes. *Microporous Mesoporous Mater.* **2017**, *243*, 164–175. [CrossRef]
55. Jin, C.; Wang, Y.; Tang, H.; Zhu, K.; Liu, X.; Wang, J. Versatile rattle-type magnetic mesoporous silica spheres, working as adsorbents and nanocatalyst containers. *J. Sol-Gel Sci. Technol.* **2016**, *77*, 279–287. [CrossRef]
56. Wo, R.; Li, Q.L.; Zhu, C.; Zhang, Y.; Qiao, G.F.; Lei, K.Y.; Du, P.; Jiang, W. Preparation and Characterization of Functionalized Metal–Organic Frameworks with Core/Shell Magnetic Particles (Fe_3O_4@SiO_2@MOFs) for Removal of Congo Red and Methylene Blue from Water Solution. *J. Chem. Eng. Data* **2019**, *64*, 2455–2463. [CrossRef]
57. Kittappa, S.; Pichiah, S.; Kim, J.R.; Yoon, Y.; Snyder, S.A.; Jang, M. Magnetised nanocomposite mesoporous silica and its application for effective removal of methylene blue from aqueous solution. *Sep. Purif. Technol.* **2015**, *153*, 67–75. [CrossRef]
58. Zhou, Y.; Lu, J.; Zhou, Y.; Liu, Y. Recent advances for dyes removal using novel adsorbents: A review. *Environ. Pollut.* **2019**, *252*, 352–365. [CrossRef]
59. Musso, T.B.; Parolo, M.E.; Pettinari, G.; Francisca, F.M. Cu(II) and Zn(II) adsorption capacity of three different clay liner materials. *J. Environ. Manag.* **2014**, *146*, 50–58. [CrossRef]
60. Betega de Paiva, L.; Rita, A.; Valenzuela, F.R. Organoclays: Properties, preparation and applications. *Appl. Clay Sci.* **2008**, *42*, 8–24. [CrossRef]
61. Parolo, M.E.; Pettinari, G.R.; Musso, T.B.; Sánchez-Izquierdo, M.P.; Fernández, L.G. Characterization of organo-modified bentonite sorbents: The effect of modification conditions on adsorption performance. *Appl. Surf. Sci.* **2014**, *320*, 356–363. [CrossRef]
62. Naranjo Pablo, M.; Sham Edgardo, L.; Rodriguez, C.E.; Torres Sánchez, R.M.; Monica, F. Identification and quantification of the interaction mechanisms between the cationic surfactant HDTMA-Br and montmorillonite. *Clays Clay Miner.* **2013**, *61*, 98–106. [CrossRef]
63. Roca Jalil, M.E.; Vieira, R.S.; Azevedo, D.; Baschini, M.; Sapag, K. Improvement in the adsorption of thiabendazole by using aluminum pillared clays. *Appl. Clay Sci.* **2013**, *71*, 55–63. [CrossRef]
64. Guégan, R. Organoclay applications and limits in the environment. *C. R. Chim.* **2018**, *22*, 132–141. [CrossRef]
65. Jaber, M.; Miehé-Brendlé, J. Organoclays. Preparation, Properties and Applications. In *Ordered Porous Solids*; Elsevier: Amsterdam, The Netherlands, 2009; pp. 31–49.
66. Xu, X.; Chen, W.; Zong, S.; Ren, X.; Liu, D. Magnetic clay as catalyst applied to organics degradation in a combined adsorption and Fenton-like process. *Chem. Eng. J.* **2019**, *373*, 140–149. [CrossRef]
67. Liu, H.; Chen, W.; Liu, C.; Liu, Y.; Dong, C. Magnetic mesoporous clay adsorbent: Preparation, characterization and adsorption capacity for atrazine. *Microporous Mesoporous Mater.* **2014**, *194*, 72–78. [CrossRef]
68. Fizir, M.; Dramou, P.; Zhang, K.; Sun, C.; Pham-Huy, C.; He, H. Polymer grafted-magnetic halloysite nanotube for controlled and sustained release of cationic drug. *J. Colloid Interface Sci.* **2017**, *505*, 476–488. [CrossRef]
69. Mu, B.; Tang, J.; Zhang, L.; Wang, A. Preparation, characterization and application on dye adsorption of a well-defined two-dimensional superparamagnetic clay/polyaniline/Fe_3O_4 nanocomposite. *Appl. Clay Sci.* **2016**, *132*, 7–16. [CrossRef]
70. Khajeh, M.; Laurent, S.; Dastafkan, K. Nanoadsorbents: Classification, preparation, and applications (with emphasis on aqueous media). *Chem. Rev.* **2013**, *113*, 7728–7768. [CrossRef]
71. Sherlala, A.I.A.; Raman, A.A.A.; Bello, M.M.; Asghar, A. A review of the applications of organo-functionalized magnetic graphene oxide nanocomposites for heavy metal adsorption. *Chemosphere* **2018**, *193*, 1004–1017. [CrossRef]
72. Zhu, M.; Diao, G. Review on the progress in synthesis and application of magnetic carbon nanocomposites. *Nanoscale* **2011**, *3*, 2748–2767. [CrossRef]

73. Boruah, P.K.; Sharma, B.; Hussain, N.; Das, M.R. Magnetically recoverable Fe_3O_4/graphene nanocomposite towards efficient removal of triazine pesticides from aqueous solution: Investigation of the adsorption phenomenon and specific ion effect. *Chemosphere* **2017**, *168*, 1058–1067. [CrossRef] [PubMed]
74. Hu, X.; Liu, J.; Mayer, P.; Jiang, G. Impacts of some environmentally relevant parameters on the sorption of polycyclic aromatic hydrocarbons to aqueous suspensions of fullerene. *Environ. Toxicol. Chem.* **2008**, *27*, 1868–1874. [CrossRef]
75. Yang, K.; Wang, X.; Zhu, L.; Xing, B. Competitive sorption of pyrene, phenanthrene, and naphthalene on multiwalled carbon nanotubes. *Environ. Sci. Technol.* **2006**, *40*, 5804–5810. [CrossRef] [PubMed]
76. Yang, X.; Li, J.; Wen, T.; Ren, X.; Huang, Y.; Wang, X. Colloids and Surfaces A: Physicochemical and Engineering Aspects Adsorption of naphthalene and its derivatives on magnetic graphene composites and the mechanism investigation. *Colloids Surf. A Physicochem. Eng. Asp.* **2013**, *422*, 118–125. [CrossRef]
77. Fan, L.; Luo, C.; Sun, M.; Qiu, H.; Li, X. Synthesis of magnetic β-cyclodextrin-chitosan/graphene oxide as nanoadsorbent and its application in dye adsorption and removal. *Colloids Surf. B Biointerfaces* **2013**, *103*, 601–607. [CrossRef]
78. Ai, L.; Zhang, C.; Chen, Z. Removal of methylene blue from aqueous solution by a solvothermal-synthesized graphene/magnetite composite. *J. Hazard. Mater.* **2011**, *192*, 1515–1524. [CrossRef]
79. Eom, D.; Prezzi, D.; Rim, K.T.; Zhou, H.; Lefenfeld, M.; Xiao, S.; Nuckolls, C.; Hybertsen, M.S.; Heinz, T.F.; Flynn, G.W. Structure and electronic properties of graphene nanoislands on CO(0001). *Nano Lett.* **2009**, *9*, 2844–2848. [CrossRef]
80. Ramesha, G.K.; Vijaya Kumara, A.; Muralidhara, H.B.; Sampath, S. Graphene and graphene oxide as effective adsorbents toward anionic and cationic dyes. *J. Colloid Interface Sci.* **2011**, *361*, 270–277. [CrossRef]
81. Yu, B.; Zhang, X.; Xie, J.; Wu, R.; Liu, X.; Li, H.; Chen, F.; Yang, H.; Ming, Z.; Yang, S.-T. Magnetic graphene sponge for the removal of methylene blue. *Appl. Surf. Sci.* **2015**, *351*, 765–771. [CrossRef]
82. Alizadeh Fard, M.; Barkdoll, B. Using recyclable magnetic carbon nanotube to remove micropollutants from aqueous solutions. *J. Mol. Liq.* **2018**, *249*, 193–202. [CrossRef]
83. Gong, J.L.; Wang, B.; Zeng, G.M.; Yang, C.P.; Niu, C.G.; Niu, Q.Y.; Zhou, W.J.; Liang, Y. Removal of cationic dyes from aqueous solution using magnetic multi-wall carbon nanotube nanocomposite as adsorbent. *J. Hazard. Mater.* **2009**, *164*, 1517–1522. [CrossRef] [PubMed]
84. Zhao, Y.; Chen, H.; Li, J.; Chen, C. Hierarchical MWCNTs/Fe_3O_4/PANI magnetic composite as adsorbent for methyl orange removal. *J. Colloid Interface Sci.* **2015**, *450*, 189–195. [CrossRef] [PubMed]
85. Deng, Y.; Ok, Y.S.; Mohan, D.; Pittman, C.U.; Dou, X. Carbamazepine removal from water by carbon dot-modified magnetic carbon nanotubes. *Environ. Res.* **2019**, *169*, 434–444. [CrossRef] [PubMed]
86. Liu, X.; Tian, J.; Li, Y.; Sun, N.; Mi, S.; Xie, Y.; Chen, Z. Enhanced dyes adsorption from wastewater via Fe_3O_4 nanoparticles functionalized activated carbon. *J. Hazard. Mater.* **2019**, *373*, 397–407. [CrossRef] [PubMed]
87. Shan, D.; Deng, S.; Zhao, T.; Wang, B.; Wang, Y.; Huang, J.; Yu, G.; Winglee, J.; Wiesner, M.R. Preparation of ultrafine magnetic biochar and activated carbon for pharmaceutical adsorption and subsequent degradation by ball milling. *J. Hazard. Mater.* **2016**, *305*, 156–163. [CrossRef] [PubMed]
88. Yu, B.; Bai, Y.; Ming, Z.; Yang, H.; Chen, L.; Hu, X.; Feng, S.; Yang, S.T. Adsorption behaviors of tetracycline on magnetic graphene oxide sponge. *Mater. Chem. Phys.* **2017**, *198*, 283–290. [CrossRef]
89. Lompe, K.M.; Vo Duy, S.; Peldszus, S.; Sauvé, S.; Barbeau, B. Removal of micropollutants by fresh and colonized magnetic powdered activated carbon. *J. Hazard. Mater.* **2018**, *360*, 349–355. [CrossRef]
90. Alaba, P.A.; Oladoja, N.A.; Sani, Y.M.; Ayodele, O.B.; Mohammed, I.Y.; Olupinla, S.F.; Daud, W.M.W. Insight into wastewater decontamination using polymeric adsorbents. *J. Environ. Chem. Eng.* **2018**, *6*, 1651–1672. [CrossRef]
91. Lv, S.W.; Liu, J.M.; Wang, Z.H.; Ma, H.; Li, C.Y.; Zhao, N.; Wang, S. Recent advances on porous organic frameworks for the adsorptive removal of hazardous materials. *J. Environ. Sci.* **2019**, *80*, 169–185. [CrossRef]
92. Zhou, Q.; Wang, Y.; Xiao, J.; Fan, H.; Chen, C. Preparation and characterization of magnetic nanomaterial and its application for removal of polycyclic aromatic hydrocarbons. *J. Hazard. Mater.* **2019**, *371*, 323–331. [CrossRef]
93. Huo, Y.; Wu, H.; Wang, Z.; Wang, F.; Liu, Y.; Feng, Y.; Zhao, Y. Preparation of core/shell nanocomposite adsorbents based on amine polymer-modified magnetic materials for the efficient adsorption of anionic dyes. *Colloids Surf. A Physicochem. Eng. Asp.* **2018**, *549*, 174–183. [CrossRef]

94. Liu, D.; Huang, Z.; Li, M.; Sun, P.; Yu, T.; Zhou, L. Novel porous magnetic nanospheres functionalized by β-cyclodextrin polymer and its application in organic pollutants from aqueous solution. *Environ. Pollut.* **2019**, *250*, 639–649. [CrossRef] [PubMed]
95. Budd, P.M.; Ghanem, B.S.; Makhseed, S.; McKeown, N.B.; Msayib, K.J.; Tattershall, C.E. Polymers of intrinsic microporosity (PIMs): Robust, solution-processable, organic nanoporous materials. *Chem. Commun.* **2004**, *4*, 230–231. [CrossRef] [PubMed]
96. Kumar, P.; Bansal, V.; Kim, K.H.; Kwon, E.E. Metal–organic frameworks (MOFs) as futuristic options for wastewater treatment. *J. Ind. Eng. Chem.* **2018**, *62*, 130–145. [CrossRef]
97. Han, S.S.; Furukawa, H.; Yaghi, O.M.; Iii, W.A.G. Covalent organic frameworks as exceptional hydrogen storage materials. *J. Am. Chem. Soc.* **2008**, *105*, 11580–11581. [CrossRef] [PubMed]
98. Ben, T.; Ren, H.; Shengqian, M.; Cao, D.; Lan, J.; Jing, X.; Wang, W.; Xu, J.; Deng, F.; Simmons, J.M.; et al. Targeted synthesis of a porous aromatic framework with high stability and exceptionally high surface area. *Angew. Chem. Int. Ed.* **2009**, *48*, 9457–9460. [CrossRef]
99. Germain, J.; Hradil, J.; Fréchet, J.M.J.; Svec, F. High surface area nanoporous polymers for reversible hydrogen storage. *Chem. Mater.* **2006**, *18*, 4430–4435. [CrossRef]
100. Hu, A.; Yang, X.; You, Q.; Liu, Y.; Wang, Q.; Liao, G.; Wang, D. Magnetically hyper-cross-linked polymers with well-developed mesoporous: A broad-spectrum and highly efficient adsorbent for water purification. *J. Mater. Sci.* **2019**, *54*, 2712–2728. [CrossRef]
101. Li, J.; Liu, Y.; Ai, Y.; Alsaedi, A.; Hayat, T.; Wang, X. Combined experimental and theoretical investigation on selective removal of mercury ions by metal organic frameworks modified with thiol groups. *Chem. Eng. J.* **2018**, *354*, 790–801. [CrossRef]
102. Wang, C.; Liu, X.; Keser Demir, N.; Chen, J.P.; Li, K. Applications of water stable metal-organic frameworks. *Chem. Soc. Rev.* **2016**, *45*, 5107–5134. [CrossRef]
103. Safaei, M.; Foroughi, M.M.; Ebrahimpoor, N.; Jahani, S.; Omidi, A.; Khatami, M. A review on metal–organic frameworks: Synthesis and applications. *TrAC Trends Anal. Chem.* **2019**, *118*, 401–425. [CrossRef]
104. Yang, Q.; Ren, S.S.; Zhao, Q.; Lu, R.; Hang, C.; Chen, Z.; Zheng, H. Selective separation of methyl orange from water using magnetic ZIF-67 composites. *Chem. Eng. J.* **2018**, *333*, 49–57. [CrossRef]
105. Hamedi, A.; Zarandi, M.B.; Nateghi, M.R. Highly efficient removal of dye pollutants by MIL-101(Fe) metal–organic framework loaded magnetic particles mediated by Poly L-Dopa. *J. Environ. Chem. Eng.* **2019**, *7*, 102882. [CrossRef]
106. Wu, G.; Ma, J.; Li, S.; Guan, J.; Jiang, B.; Wang, L.; Li, J.; Wang, X.; Chen, L. Magnetic copper-based metal organic framework as an effective and recyclable adsorbent for removal of two fluoroquinolone antibiotics from aqueous solutions. *J. Colloid Interface Sci.* **2018**, *528*, 360–371. [CrossRef]
107. Han, S.S.; Mendoza-Cortés, J.L.; Goddard, W.A. Recent advances on simulation and theory of hydrogen storage in metal-organic frameworks and covalent organic frameworks. *Chem. Soc. Rev.* **2009**, *38*, 1460–1476. [CrossRef]
108. Fernandes, S.P.S.; Romero, V.; Espiña, B.; Salonen, L.M. Tailoring Covalent Organic Frameworks to Capture Water Contaminants. *Chem. A Eur. J.* **2019**, *25*, 6461–6473. [CrossRef]
109. Li, Y.; Zhang, H.; Chen, Y.; Huang, L.; Lin, Z.; Cai, Z. Core-Shell Structured Magnetic Covalent Organic Framework Nanocomposites for Triclosan and Triclocarban Adsorption. *ACS Appl. Mater. Interfaces* **2019**, *11*, 22492–22500. [CrossRef]
110. Li, Y.; Yang, C.X.; Yan, X.P. Controllable preparation of core–shell magnetic covalent-organic framework nanospheres for efficient adsorption and removal of bisphenols in aqueous solution. *Chem. Commun.* **2017**, *53*, 2511–2514. [CrossRef]
111. He, S.; Zeng, T.; Wang, S.; Niu, H.; Cai, Y. Facile synthesis of magnetic covalent organic framework with three-dimensional bouquet-like structure for enhanced extraction of organic targets. *ACS Appl. Mater. Interfaces* **2017**, *9*, 2959–2965. [CrossRef]
112. Wang, B.; Wan, Y.; Zheng, Y.; Lee, X.; Liu, T.; Yu, Z.; Huang, J.; Ok, Y.S.; Chen, J.; Gao, B. Alginate-based composites for environmental applications: A critical review. *Crit. Rev. Environ. Sci. Technol.* **2019**, *49*, 318–356. [CrossRef]
113. Lee, K.Y.; Rowley, J.A.; Eiselt, P.; Moy, E.M.; Bouhadir, K.H.; Mooney, D.J. Controlling mechanical and swelling properties of alginate hydrogels independently by cross-linker type and cross-linking density. *Macromolecules* **2000**, *33*, 4291–4294. [CrossRef]

114. Urquiza, T.K.V.; Pérez, O.P.; Saldaña, M.G. Effect of the cross-linking with calcium ions on the structural and thermo-mechanical properties of alginate films. *Mater. Res. Soc. Symp. Proc.* **2011**, *1355*, 16–21.
115. Talbot, D.; Abramson, S.; Griffete, N.; Bée, A. pH-sensitive magnetic alginate/γ-Fe$_2$O$_3$ nanoparticles for adsorption/desorption of a cationic dye from water. *J. Water Process Eng.* **2018**, *25*, 301–308. [CrossRef]
116. Mohammadi, A.; Daemi, H.; Barikani, M. Fast removal of malachite green dye using novel superparamagnetic sodium alginate-coated Fe$_3$O$_4$ nanoparticles. *Int. J. Biol. Macromol.* **2014**, *69*, 447–455. [CrossRef] [PubMed]
117. Wen, Y.; Shen, C.; Ni, Y.; Tong, S.; Yu, F. Glow discharge plasma in water: A green approach to enhancing ability of chitosan for dye removal. *J. Hazard. Mater.* **2012**, *201*, 162–169. [CrossRef] [PubMed]
118. Li, T.T.; Liu, Y.G.; Peng, Q.Q.; Hu, X.J.; Liao, T.; Wang, H.; Lu, M. Removal of lead(II) from aqueous solution with ethylenediamine-modified yeast biomass coated with magnetic chitosan microparticles: Kinetic and equilibrium modeling. *Chem. Eng. J.* **2013**, *214*, 189–197. [CrossRef]
119. Park, S.I.; Kwak, I.S.; Won, S.W.; Yun, Y.S. Glutaraldehyde-crosslinked chitosan beads for sorptive separation of Au(III) and Pd(II): Opening a way to design reduction-coupled selectivity-tunable sorbents for separation of precious metals. *J. Hazard. Mater.* **2013**, *248*, 211–218. [CrossRef]
120. Tirtom, V.N.; Dinçer, A.; Becerik, S.; Aydemir, T.; Çelik, A. Comparative adsorption of Ni(II) and Cd(II) ions on epichlorohydrin crosslinked chitosan-clay composite beads in aqueous solution. *Chem. Eng. J.* **2012**, *197*, 379–386. [CrossRef]
121. Zheng, C.; Zheng, H.; Wang, Y.; Sun, Y.; An, Y.; Liu, H.; Liu, S. Modified magnetic chitosan microparticles as novel superior adsorbents with huge "force field" for capturing food dyes. *J. Hazard. Mater.* **2019**, *367*, 492–503. [CrossRef]
122. Xu, B.; Zheng, C.; Zheng, H.; Wang, Y.; Zhao, C.; Zhao, C.; Zhang, S. Polymer-grafted magnetic microspheres for enhanced removal of methylene blue from aqueous solutions. *RSC Adv.* **2017**, *7*, 47029–47037. [CrossRef]
123. Ali, I.; Peng, C.; Naz, I.; Lin, D.; Saroj, D.P.; Ali, M. Development and application of novel bio-magnetic membrane capsules for the removal of the cationic dye malachite green in wastewater treatment. *RSC Adv.* **2019**, *9*, 3625–3646. [CrossRef]
124. Jodeh, S.; Hamed, O.; Melhem, A.; Salghi, R.; Jodeh, D.; Azzaoui, K.; Benmassaoud, Y.; Murtada, K. Magnetic nanocellulose from olive industry solid waste for the effective removal of methylene blue from wastewater. *Environ. Sci. Pollut. Res.* **2018**, *25*, 22060–22074. [CrossRef] [PubMed]
125. Liang, X.X.; Omer, A.M.; Hu, Z.H.; Wang, Y.G.; Yu, D.; Ouyang, X.K. Efficient adsorption of diclofenac sodium from aqueous solutions using magnetic amine-functionalized chitosan. *Chemosphere* **2019**, *217*, 270–278. [CrossRef] [PubMed]
126. Wu, H.; Zhang, H.; Zhang, W.; Yang, X.; Zhou, H.; Pan, Z.; Wang, D. Preparation of magnetic polyimide@ Mg-Fe layered double hydroxides core-shell composite for effective removal of various organic contaminants from aqueous solution. *Chemosphere* **2019**, *219*, 66–75. [CrossRef] [PubMed]
127. Ahamad, T.; Naushad, M.; Alshehri, S.M. International Journal of Biological Macromolecules Ultra-fast spill oil recovery using a mesoporous lignin based nanocomposite prepared from date palm pits (Phoenix dactylifera L.). *Int. J. Biol. Macromol.* **2019**, *130*, 139–147. [CrossRef] [PubMed]
128. Silva, C.P.; Jaria, G.; Otero, M.; Esteves, V.I.; Calisto, V. Waste-based alternative adsorbents for the remediation of pharmaceutical contaminated waters: Has a step forward already been taken? *Bioresour. Technol.* **2018**, *250*, 888–901. [CrossRef] [PubMed]
129. Anastopoulos, I.; Bhatnagar, A.; Hameed, B.H.; Ok, Y.S.; Omirou, M. A review on waste-derived adsorbents from sugar industry for pollutant removal in water and wastewater. *J. Mol. Liq.* **2017**, *240*, 179–188. [CrossRef]
130. Sayehi, M.; Tounsi, H.; Garbarino, G.; Riani, P.; Busca, G. Reutilization of silicon- and aluminum- containing wastes in the perspective of the preparation of SiO$_2$-Al$_2$O$_3$ based porous materials for adsorbents and catalysts. *Waste Manag.* **2020**, *103*, 146–158. [CrossRef] [PubMed]
131. Gupta, V.K.; Carrott, P.J.M.; Ribeiro Carrott, M.M.L. Suhas Low-Cost adsorbents: Growing approach to wastewater treatmenta review. *Crit. Rev. Environ. Sci. Technol.* **2009**, *39*, 783–842. [CrossRef]
132. Safarik, I.; Lunackova, P.; Weyda, F.; Safarikova, M. Adsorption of water-soluble organic dyes on ferrofluid-modified sawdust. *Holzforschung* **2007**, *61*, 247–253. [CrossRef]
133. Safarik, I.; Safarikova, M. Magnetic fluid modified peanut husks as an adsorbent for organic dyes removal. *Phys. Procedia* **2010**, *9*, 274–278. [CrossRef]
134. Zuorro, A.; Di Battista, A.; Lavecchia, R. Magnetically modified coffee silverskin for the removal of xenobiotics from wastewater. *Chem. Eng. Trans.* **2013**, *35*, 1375–1380.

135. Minh, T.P.; Lebedeva, O.E. Adsorption Properties of a Magnetite Composite with Coffee Waste. *Russ. J. Phys. Chem. A* **2018**, *92*, 2044–2047. [CrossRef]
136. Stan, M.; Lung, I.; Soran, M.; Opris, O.; Leostean, C.; Popa, A.; Copaciu, F.; Diana, M.; Kacso, I.; Silipas, T.; et al. Starch-coated green synthesized magnetite nanoparticles for removal of textile dye Optilan Blue from aqueous media. *J. Taiwan Inst. Chem. Eng.* **2019**, *100*, 65–73. [CrossRef]
137. Aydin, S. Removal of Organophosphorus Pesticides from Aqueous Solution by Magnetic Fe_3O_4/Red Mud-Nanoparticles. *Water Environ. Res.* **2016**, *88*, 2275–2284. [CrossRef]
138. Aydin, S.; Aydin, M.E.; Beduk, F.; Ulvi, A. Removal of antibiotics from aqueous solution by using magnetic Fe_3O_4/red mud-nanoparticles. *Sci. Total Environ.* **2019**, *670*, 539–546. [CrossRef]
139. Madrakian, T.; Afkhami, A.; Ahmadi, M. Adsorption and kinetic studies of seven different organic dyes onto magnetite nanoparticles loaded tea waste and removal of them from wastewater samples. *Spectrochim. Acta Part A Mol. Biomol. Spectrosc.* **2012**, *99*, 102–109. [CrossRef]
140. Madrakian, E.; Ghaemi, E.; Ahmadi, M. Magnetic Solid Phase Extraction and Removal of Five Cationic Dyes from Aqueous Solution Using Magnetite Nanoparticle Loaded Platanusorientalis Waste Leaves. *Anal. Bioanal. Chem. Res.* **2016**, *3*, 279–286.
141. Sun, L.; Yuan, D.; Wan, S.; Yu, Z.; Dang, J. Adsorption Performance and Mechanisms of Methylene Blue Removal by Non-magnetic and Magnetic Particles Derived from the *Vallisneria natans* Waste. *J. Polym. Environ.* **2018**, *26*, 2992–3004. [CrossRef]
142. Yu, J.X.; Wang, L.Y.; Chi, R.A.; Zhang, Y.F.; Xu, Z.G.; Guo, J. A simple method to prepare magnetic modified beer yeast and its application for cationic dye adsorption. *Environ. Sci. Pollut. Res.* **2013**, *20*, 543–551. [CrossRef]
143. Magnacca, G.; Allera, A.; Montoneri, E.; Celi, L.; Benito, D.E.; Gagliardi, L.G.; Carlos, L. Novel magnetite nanoparticles coated with waste sourced bio- based substances as sustainable and renewable adsorbing materials. *ACS Sustain. Chem. Eng.* **2014**, *2*, 1518–1524. [CrossRef]
144. Nisticò, R.; Cesano, F.; Franzoso, F.; Magnacca, G.; Scarano, D.; Funes, I.G.; Carlos, L.; Parolo, M.E. From biowaste to magnet-responsive materials for water remediation from polycyclic aromatic hydrocarbons. *Chemosphere* **2018**, *202*, 686–693. [CrossRef]
145. Nisticò, R.; Celi, L.R.; Bianco Prevot, A.; Carlos, L.; Magnacca, G.; Zanzo, E.; Martin, M. Sustainable magnet-responsive nanomaterials for the removal of arsenic from contaminated water. *J. Hazard. Mater.* **2018**, *342*, 260–269. [CrossRef]
146. Liu, J.; Yu, Y.; Zhu, S.; Yang, J.; Song, J.; Fan, W.; Yu, H.; Bian, D.; Huo, M. Synthesis and characterization of a magnetic adsorbent from negatively-valued iron mud for methylene blue adsorption. *PLoS ONE* **2018**, *13*, e0191229. [CrossRef]

© 2020 by the authors. Licensee MDPI, Basel, Switzerland. This article is an open access article distributed under the terms and conditions of the Creative Commons Attribution (CC BY) license (http://creativecommons.org/licenses/by/4.0/).

Article

Efficient Separation of Heavy Metals by Magnetic Nanostructured Beads

Lisandra de Castro Alves, Susana Yáñez-Vilar *, Yolanda Piñeiro-Redondo and José Rivas

Applied Physic Department, NANOMAG Laboratory, Research Technological Institute, Universidade de Santiago de Compostela (USC), 15782 Santiago de Compostela, Spain; lisandracristina.decastro@usc.es (L.d.C.A.); yolanda.fayoly@gmail.com (Y.P.-R.); jose.rivas@usc.es (J.R.)
* Correspondence: susana.yanez@usc.es

Received: 9 April 2020; Accepted: 24 June 2020; Published: 26 June 2020

Abstract: This study reports the ability of magnetic alginate activated carbon (MAAC) beads to remove Cd(II), Hg(II), and Ni(II) from water in a mono-metal and ternary system. The adsorption capacity of the MAAC beads was highest in the mono-metal system. The removal efficiency of such metal ions falls in the range of 20–80% and it followed the order Cd(II) > Ni(II) > Hg(II). The model that best fitted in the ternary system was the Freundlich isotherm, while in the mono-system it was the Langmuir isotherm. The maximum Cd(II), Hg(II), and Ni(II) adsorption capacities calculated from the Freundlich isotherm in the mono-metal system were 7.09, 5.08, and 4.82 (mg/g) (mg/L)$^{1/n}$, respectively. Lower adsorption capacity was observed in the ternary system due to the competition of metal ions for available adsorption sites. Desorption and reusability experiments demonstrated the MAAC beads could be used for at least five consecutive adsorption/desorption cycles. These findings suggest the practical use of the MAAC beads as efficient adsorbent for the removal of heavy metals from wastewater.

Keywords: heavy metals; magnetite nanoparticles; adsorption; nanocomposite; hybrid; multi-metal; water

1. Introduction

Water pollution by heavy metals has become a serious problem due to the adverse effects on ecosystems and human health. More specifically, cadmium (Cd), mercury (Hg), lead (Pb), or nickel (Ni) are known to be highly carcinogenic and mutagenic at low concentrations, and may produce acute toxicity or even dead in living organisms, when present slightly above their allowed limits [1–3]. Although diverse technologies have been developed for the removal of heavy metals from water sources [4], there is still an urgent need for facile cleaning procedures that ensure high efficiency in the low concentration ranges. Chemical precipitation, ion precipitation, ion exchange, and adsorption are some of the most used techniques due to their potential for scaling up. Among them, the use of natural biopolymers, such as alginate, agarose, chitin and pectin [5–7] in metal biosorption from wastewaters has gained much attention in recent years. Alginate is a polysaccharide derived from brown algae and in the majority of the studies it has been used in the form of calcium alginate beads, due to its practical handling [8]. Nevertheless, the separation of the loaded biomaterial from the medium is often a problem. To overcome this problem, magnetite nanoparticles (Fe_3O_4-NPs) are being incorporated onto the biosorbent matrix [9–11] giving the possibility to magnetically manipulate and separate the hybrid materials from the water matrix. Magnetic alginate beads are a very attractive material with multiple properties such as high specific surface area, rapid recovery, cost-effectiveness, and chemical versatility, for which they are amenable to be combined with materials to increase their affinity for pollutants. Humic acid [12], Cyanex 302 [12] and microalgae [7] were incorporated into alginate beads for their affinity to metal ions. Different authors have shown the ability of several alginate beads composition to uptake metal ions from aqueous solutions. It was observed that magnetic alginate beads containing silica coated with iron carbide nanoparticles enhanced more the adsorption of copper ions than alginate beads alone [13]. In addition, it was studied that magnetic nanoparticles functionalized with citrate ions present an enhanced adsorption of Pb(II) metal ions from solution [14]. In one of our previous works [15], magnetic alginate beads tailored with commercial activated carbon revealed a high capacity for cadmium ions uptake. A 35% removal percentage of cadmium ions was achieved over 1 h with less than 15 mg of adsorbent used. These nanostructured beads revealed to be a great assessment for industrial use, for their high adsorption surface area, easy handling, and magnetic separation from any aqueous media. For these reasons, the same nanostructured beads were used in the present study and tested under more realistic conditions by studying their adsorption capacity when exposed to a mixture of heavy metals. With this aim, different adsorption tests were performed on mono-metal and ternary systems, comprising Cd(II), Hg(II) and Ni(II) metals ions, which are commonly found in waste water from industry and mining effluents. To gain insights in the adsorption mechanism other relevant aspects (e.g., metal-adsorbent mechanisms, metal distribution on beads surface and internal structure, metal desorption, reuse, porosity, among others) were studied in detail.

2. Results and Discussion

2.1. Morphology of Magnetic Beads

The morphology of the MAAC beads and commercial activated carbon was examined using scanning electron microscopy (SEM) and transmission electron microscopy (TEM). In Figure 1a, SEM micrograph shows that commercial activated carbon presents a disordered layer-like structure, which was further studied with TEM, reveals a wide variation in the layer size range between 40 µm to 320 µm (Figure 1b). SEM micrograph (Figure 1c) of a representative MAAC bead shows a spherical morphology with a porous and layer-like structure on the surface inherited from the precursor commercial activated carbon. The internal structure, as can be observed in Figure 1d, is a combination of activated carbon layers within the interconnected porous network of alginate in MAAC beads.

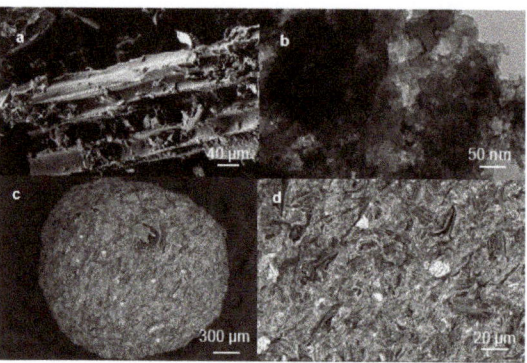

Figure 1. Commercial activated carbon SEM (**a**) and TEM (**b**) micrograph. SEM image of the MAAC bead surface (**c**) and internal structure (**d**).

2.2. Structural and Textural Characterization

The X-ray diffraction pattern of the MAAC beads in Figure 2a shows the presence of sharp diffraction peaks located at 2θ = 30.1, 35.5, 43.2, 53.5, 57.1, 62.7°. This is the characteristic diffraction pattern arising from the reflection of planes (022), (113), (004), (224), (115) and (044) corresponding to crystalline Fe_3O_4-NPs embedded in the porous beads. The MAAC bead's surface was also analyzed by Fourier transform infrared (FT-IR)spectroscopy. As shown in Figure 2b, the peaks at 3228 cm^{-1} and 1076 cm^{-1} are related to the –OH and –C–O stretching vibration bands of alginate, respectively, while peaks at 1585 and 1286 cm^{-1} are attributed to the asymmetric and symmetric stretching vibrations of the carboxyl groups of alginate, respectively [16]. Finally, the band at 558 cm^{-1} is due to collective vibrations of the magnetite lattice [17], which confirms the successful incorporation of Fe$_3$O$_4$-NPs into the alginate matrix.

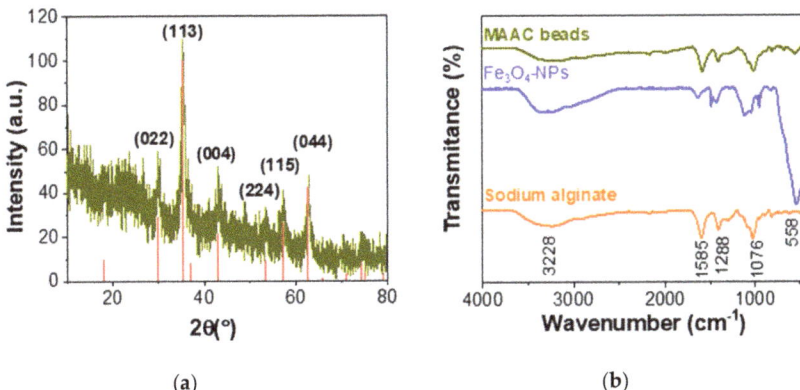

Figure 2. (a) X-ray diffraction and infrared spectra and (b) IR spectra of MAAC beads.

The mean particle size of the Fe$_3$O$_4$ nanoparticles employed for the synthesis of the MAAC beads were calculated from the X-ray diffraction (XRD) pattern (Figure S1) according to the linewidth of the (113) plane refraction peak using Scherrer equation:

$$L = K\lambda/\beta\cos\theta \qquad (1)$$

where L is the mean size of the ordered (crystalline) domains, λ is the X-ray wavelength, β is the width of the XRD peak at half height, K is a shape factor, about 0.9 for magnetite and θ is the Bragg angle The particle diameter calculated from the X-ray diffractogram, using the Scherrer equation was 9.58 nm. By TEM the image (Figure S1) we can observe nanoparticles with a size of 20 nm. The difference between TEM and XRD can be attributed to the fact that "crystallite size" is not synonymous with "particle size", while XRD is sensitive to the crystallite size inside the particles.

2.3. Specific Surface Are and Pore-Distribution

The textural properties of commercial activated carbon (commercial AC) and MAAC beads were analyzed with Brunauer–Emmett–Teller (BET) porosimetry and shown in Table 1. The surface area of commercial AC (849.32 m^2/g) is within the theoric range for activated carbons (500 to 3.000 m^2/g) [8]. The surface area of the MAAC beads is 107.13 m^2/g which is two orders of magnitude higher than the reported 6.25 m^2/g for calcium alginate beads [18], and can be ascribed to the presence of activated carbon. The pore volume of the MAAC beads (0.075 cm^3/g) was smaller than the one of commercial AC (0.28 cm^3/g). In addition, the pore diameter size of the MAAC beads (1.24 nm) is slightly smaller than the commercial AC (1.26 nm). This could be attributed to the deposition of the commercial AC on the surface of the MAAC beads (both surface and internally), leading to a complete filling of the smaller pores.

Table 1. Textural parameters of Commercial AC and MAAC beads.

Samples	BET Surface Area (S_{BET}) [1] (m^2/g)	Pore Volume (cm^3/g)	Pore Diameter [2] (nm)
Commercial AC	849.32	0.28	1.26
MAAC	107.13	0.075	1.24

[1] S_{BET} is the BET surface area evaluated at a relative pressure (p/p0) of 0.99. [2] Pore diameter calculated using the Barrett–Joyner–Halenda (BJH) method.

2.4. Magnetic Properties of MAAC Beads

Figure 3a shows the variation of magnetization, M, as a function of temperature of MAAC beads in the range 5 to 350 K in an external magnetic field of 100 Oe recorded in zero-field cooling (ZFC) and field cooling (FC). From the curves it is clearly observed the superimposition of the ZFC and FC curves take place at 275 K. The superimposition of ZFC and FC curves is one of the characteristic features of a superparamagnetic system. The magnetic content on the MAAC beads was calculated by thermogravimetric analyses (TGA), which was equal to 23%. Figure 3b illustrates the magnetization curves of bare Fe_3O_4-NPs and of the MAAC nanocomposite beads. The saturation of magnetization (Ms) for the synthesized Fe_3O_4-NPs (69.23 emu/g) was higher than the observed for the MAAC beads (48.62 emu/g). This may be attributed to the coating effect of alginate trapping the Fe_3O_4-NPs in the gel matrix. However, the MAAC beads have superparamagnetic behavior and are easily separated from solution with the help of an external magnetic force.

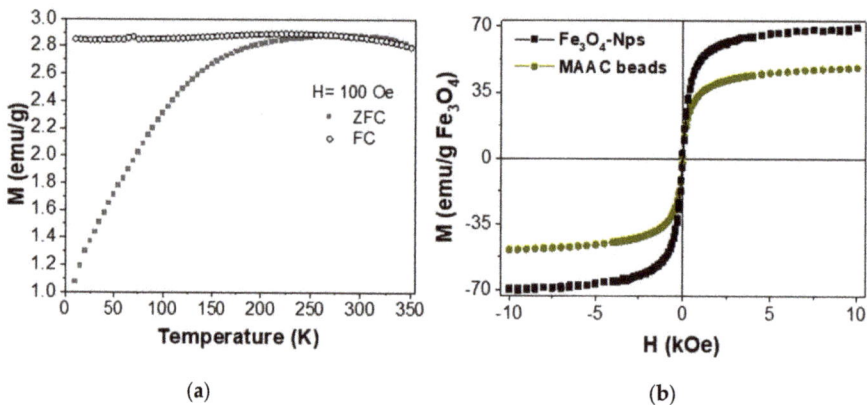

Figure 3. ZFC and FC curve recorded at 100 Oe (**a**) and magnetization curve of MAAC bead and Fe_3O_4-NPs at 25 °C (**b**).

3. Adsorption Study

3.1. Desorption and Reusability

The reusability of the MAAC bead was study by repetitive adsorption and desorption cycles of Cd(II) metal ions using 0.01 M HCl solution, as the desorption solution. The q (mg/g) desorption of cadmium was calculated directly from the amount of cadmium adsorbed and amount of cadmium desorbed using Equation (2):

$$q(desorption) = \frac{(M_{adsorb} - M_{desorb})V}{M} \quad (2)$$

where M_{adsorb} and M_{desorb} are the adsorbed and desorbed amount of metal ions (mg/g), respectively; V (L) is the volume of desorption solution and M (g) the mass of the MAAC beads used. The desorption percentage was calculated using the following Equation (3):

$$\% \text{ Desorption} = \frac{M_{desorb}}{M_{adsorb}} \times 100 \quad (3)$$

The desorption of cadmium from MAAC bead showed a fair desorption percentage over the five cycles, as can be observed in Figure 4. However, the adsorption capacity decreased with increasing regeneration cycle number, except for on cycle number two. The cadmium metal ions have entered inside of the beads structure after the first cycle (Figure S2) resulting in the maximum adsorption capacity on cycle two. Continuous adsorption resulted on internal pore saturation and consequently, on the decreased of adsorption capacity and increase amount of desorbed cadmium metal ions found in solution. Furthermore, the results suggest a reduction of the metal ions from the matrix in the first fourth cycles and more than 50% of metal was recovered. After the fifth cycle, the desorption percentage decreased to 20%, since cadmium metal ions remained inside of the beads structure after the third desorption cycle. Thus, it can be said that these beads have the potential to be reuse up to four cycles under the chosen conditions. For further studies, factors, such as time and desorbent concentration must be considered for maximum desorption capacity for more than 5 cycles. In previously studies [19], hydrochloric acid proved to be a good desorbent within 2 h of repetitive adsorption and desorption cycles.

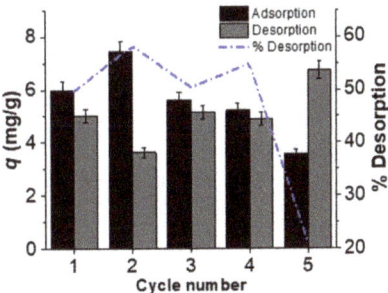

Figure 4. Adsorption and Desorption cycles of cadmium from MAAC beads (mean ± SE of 15 replicates).

3.2. Effect of pH

The influence of pH on Cd(II), Hg(II) and Ni(II) adsorption capacity by the MAAC beads was studied at pH range of (2.0 to 9.0). In Figure 5, the adsorbed metal ions per adsorbent mass q (mg/g) are presented for all tested pH values. As can be seen, the adsorption process for Cd(II) and Ni(II), was constant between the pH values of 4.5 to 7.0, followed by an increase at higher pH values. On contrary, Hg(II) adsorption depends in a non-predictable way on the pH, attaining a maximum adsorption capacity at pH 4.5, followed by a steep decrease at pH 6.5 and a large increment up to pH 9.0. The higher concentration of hydrolyzed ions such as H^+ allow an enhanced binding of Hg(II) metal ions to the sodium alginate surface [19]. Besides this binding mechanism being more enhanced at pH 2.0 for Hg(II) ions, at pH 4.5 all metal ions revealed to have a constant and high adsorption capacity at this pH, being the selected for the experiments at the mono and ternary system. The increasing adsorption at higher pH values for all metal ions may be attributed to the formation of hydroxyl ions [20].

At higher pH values, the decreased adsorption capacity by Cd(II) coincides with the decreasing concentration of Cd(II) and the precipitation of $Cd(OH)_2$ into solution. The $Cd(OH)_2$ ionic species are adsorbed by the MAAC beads occupying the available sites and preventing the further adsorption of Cd(II) ions [21]. For nickel an increasing trend at higher pH values was observed. This characteristic trend is attributed to the specific adsorption of cationic hydroxo-complexes, which is the pH range more favorable for the formation of these species [22]. The hydrolysis of mercury, on the other hand, begins at very low pH values (pH < 4.5) with the formation of Hg(OH)$^+$ and at pH values (pH > 4.5) the adsorption capacity suddenly increases because of the formation of mercury neutral species Hg(OH)$_2$ [23].

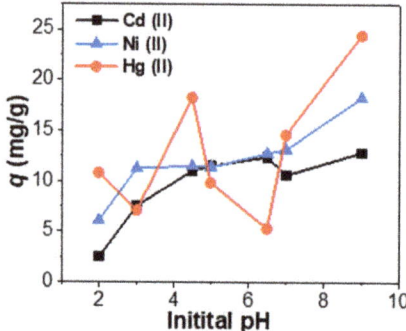

Figure 5. Effect of initial pH on Cd(II), Hg(II) and Ni(II) adsorption onto MAAC beads (mean standard deviation ± 0.3).

3.3. Adsorption of Mono and Ternary Systems

Batch adsorption experiments were performed using a defined amount of MAAC beads in mono and ternary systems containing Cd(II), Ni(II) and Hg(II) metal ions. The effect of the initial metal ion concentration on adsorption was studied for solutions prepared with a set of concentrations 10 to 250 mg/L, and applying previously optimized conditions (magnetic agitation at 300 rpm, using 14 mg of adsorbent during 6 h) at fixed pH = 4.5. The equilibrium adsorption capacity (q_e) was calculated according to the following equation [24]:

$$q_e = \frac{(C_0 - C_e)V}{M} \tag{4}$$

The removal efficiency ($R\%$) of Cd(II), Hg(II) and Ni(II) in the mono-component system was calculated using the following equation:

$$R\% = \frac{(C_0 - C_e)}{C_0} \times 100 \tag{5}$$

where q_e is the equilibrium adsorption capacity (mg/g); C_0 and C_e are the initial and equilibrium concentration (mg/L) of metal ions, respectively; V is the volume of working solution (L) and M is the weight (g) of adsorbent used. Alternatively, q, can be expressed in terms of molarity, q (mol/g), to gain insights into the number of moles (atoms) that adsorb on the cleaning beads (Figure S3).

In Figure 6, the adsorption capacity, q (mg/g), of Cd(II), Ni(II) and Hg(II) by the MAAC beads, for the mono and ternary metal adsorption batch tests are presented versus the initial concentration in the batch solution of each metal. From all tested metals under the current experimental conditions, it is evident that Cd(II) ions are preferentially adsorbed in both tests, being in the mono-metal case (Figure 6a) the adsorption capacity of Cd(II) ions twice the adsorption observed for Ni(II) and Hg(II).

In addition, in the mono-metal test, the adsorption of cadmium and nickel ions increases uniformly indicating ongoing adsorption process onto the MAAC beads, on contrary to mercury, which shows a decreased adsorption at the initial concentration of 150 mg/L. In the ternary metal system (Figure 6b), the adsorption capacity is generally smaller than in the mono-metal case, indicating a competition between the metal ions in the ternary system for available binding sites on the MAAC beads [25], with a striking exception at initial metal concentration of C = 150 mg/L, where all metal ions are adsorbed with more efficacy. Moreover, the similarities of the adsorption capacity curve of all ions in the ternary system with the mercury adsorption feature in the mono-metal system suggests that mercury has an important role modulating the adsorption mechanism, which will deserve future studies.

The removal efficiency decreases when the initial metal ions concentration is increased (Figure 7), a trend that was already observed in a previous study [15]. This behavior is mainly ascribed to the saturation of the available binding sites during the adsorption process, leading to a reduction of the adsorption capacity.

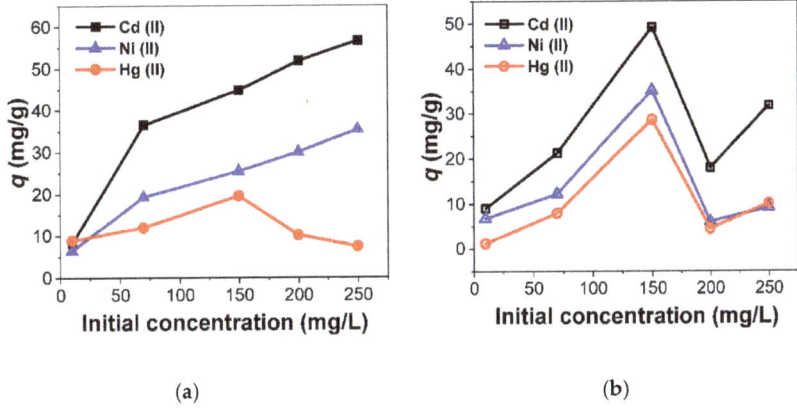

Figure 6. Effect of initial metal concentration on the adsorption capacity of MAAC beads for (**a**) mono-metal and (mean SD ± 1.45) (**b**) ternary experiment at pH 4.5 (mean SD ± 1.27).

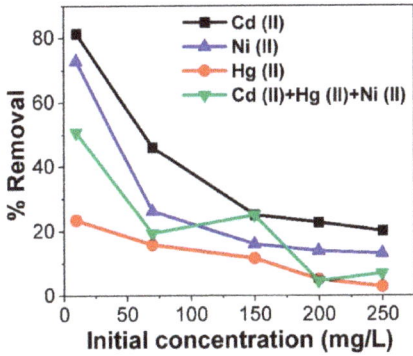

Figure 7. Removal efficiency or mono-metal and ternary experiments at pH 4.5.

3.4. Competitive Adsorption Evaluation

The interactive effect of Cd(II), Hg(II) and Ni(II) was investigated in the ternary system. For that, an evaluation ratio was introduced to assess the type of adsorption competition between each metal ion in the system and combined [26]. The evaluation ratio is expressed by the following Equation (6):

$$E = \frac{Q'_e}{Q_e} \tag{6}$$

where E, is the evaluation ratio; Q'_e (mg/g) is the amount of metal ions adsorbed in a ternary system and Q_e (mg/g) is the amount of metal ions adsorbed in the mono-metal system. If the evaluation ratio, $E > 1$, the presence of other metal ions have enhanced the adsorption of other metal ions in solution (synergism effect); when $E = 1$, this means that the presence of other metal ions would not influence the adsorption of another metal ion; and when $E < 1$, the presence of another metal ion would suppress the adsorption of one another (antagonism effect). The evaluation ratios of individual Cd(II), Hg(II) and Ni(II) in the ternary system and the whole mixture are listed in Table 2.

Table 2. Individual and sum of the evaluation ratios of Cd(II), Hg(II) and Ni(II) in the ternary system for MAAC beads.

Adsorbent	Initial Concentration (mg/L)	10	70	150	200	250
	Metal Ions			Evaluation Ratios		
MAAC	Cd(II)	1.08	0.58	1.10	0.35	0.56
	Hg(II)	0.14	0.66	1.46	0.44	1.36
	Ni(II)	1.05	0.63	1.38	0.20	0.26
	Cd(II) + Hg(II) + Ni(II)	0.68	0.71	1.48	0.32	0.48

Cadmium and nickel in both mono and ternary systems were most efficiently adsorbed than mercury (Figure 6a,b). On contrary, the study of the evaluation ratios in terms of competitive scenario reveals that mercury ions are the least suppressed of all three metal ions. The results compiled in Table 2, reveal an antagonistic effect ($E < 1$) almost in all concentrations in the individual metals and on the whole mixture of the ternary system, except at $C = 150$ (mg/L), where $E = 1.48$ ($E > 1$), indicating an average synergistic behavior between the metal ions. For high values of initial metal concentrations of $C = 200$ (mg/L) and $C = 250$ (mg/L), the degree of suppression is large for nickel and cadmium ions, while mercury adsorption is favored by a synergistic effect. Moreover, for $C \geq 70$ (mg/L) mercury is the least suppressed metal and benefits from an enhancement effect, in part due to the increase electrostatic repulsion among the cations that would limit the adsorption of the metal ion [27].

3.5. Adsorption Isotherms

To determine the adsorption capacity of the MAAC beads in a mono-metal and ternary system, adsorption studies with initial concentrations ranging from $C = 10$ (mg/L), to $C = 250$ (mg/L), were carried out. The adsorption equilibrium was studied fitting the experimental data to the linear equations of Langmuir and Freundlich isotherm models.

Langmuir isotherm model [28]:

$$\frac{C_e}{q_e} = \frac{1}{K_L q_m} + \frac{C_e}{q_m} \quad (7)$$

Freundlich isotherm model [29]:

$$\log q_e = \log K_F + \frac{1}{n} \log C_e \quad (8)$$

where q_e (mg/g) is the amount of metal ions adsorbed; C_e (mg/L) is the adsorbate concentration in solution, both at equilibrium; K_L (L/mg) is the Langmuir adsorption constant; and q_m (mg/g) is the maximum adsorption capacity for monolayer formation on the adsorbent. The value K_F can be defined as the adsorption or distribution coefficient and represents the quantity of metal ions adsorbed onto the beads. The value of $1/n$ indicates surface heterogeneity, which becomes more heterogeneous as its value gets closer to zero. A fundamental characteristic of the Langmuir isotherm is to predict the affinity between sorbate and sorbent using a dimensionless constant, known as separation factor R_L, which can be represented as:

$$R_L = \frac{1}{1 + K_L C_0} \quad (9)$$

where C_0 (mg/L) is the adsorbate initial concentration. The value of R_L stands between 0 and 1 for favorable adsorption, while $R_L > 1$ represents unfavorable adsorption, $R_L = 1$ represents linear adsorption and $R_L = 0$ for irreversible adsorption processes [28].

Table 3 lists the parameters of Langmuir and Freundlich isotherms models computed from the experimental tests using MAAC beads for mono-metal and ternary systems adsorption.

Table 3. Langmuir and Freundlich isotherm parameters.

Adsorption System	Metal Ions	Langmuir Parameters				Freundlich Parameters			
		q_m (mg/g)	K_L (L/mg)	R_L	R^2	n	$1/n$	K_F (mg$^{1-(1/n)}$L$^{1/n}$g^{-1})	R^2
Cd(II)	Cd(II)	59.17	0.048	0.103	0.981	2.52	0.397	7.09	0.983
Hg(II)	Hg(II)	25.00	0.040	0.118	0.922	3.94	0.254	5.08	0.865
Ni(II)	Ni(II)	37.04	0.031	0.144	0.945	2.79	0.358	4.82	0.995
Cd(II) + Hg(II) + Ni(II)	Cd(II)	56.18	0.029	0.339	0.758	2.49	0.401	6.11	0.889
	Hg(II)	181.82	0.007	0.164	0.681	1.50	0.668	3.35	0.901
	Ni(II)	172.71	0.001	0.456	0.561	1.14	0.878	0.35	0.953

The models of Langmuir and Freundlich equations in general described the data well, although the mono-metal system seems to be better described by Langmuir while the ternary system by the Freundlich isotherm, as suggested by the correlation coefficient values R^2. The Langmuir constant, K_L values were higher for the mono-metal system, revealing a higher adsorption capacity. Contrary to the ternary system, where the presence of other metal ions in the system decreased the adsorption capacity due to the competition of available adsorption sites onto the MAAC beads. The R_L parameter values stand below 1, indicating favorable and weakly reversible adsorption of studied metal ions onto MAAC beads. The values of n determined with the Freundlich equation were generally higher than 1.0, indicating heterogeneous adsorption process for all metal ions onto the beads. Among all metals, Cd(II) was the most highly adsorbed with a Freundlich K_F constant of 7.09 and 6.11 for both mono-metal and ternary system, respectively. In our study, the adsorption cannot be simply related to the physicochemical properties of metal cations, since cadmium with higher atomic weight and ionic radius than nickel was more intensely adsorbed [30]. Thus, adsorption in the ternary system was better described by the Freundlich adsorption isotherm, revealing heterogeneous surface with different affinity sites on MAAC beads.

4. Materials and Methods

4.1. Synthesis of MAAC Beads

The procedure of the MAAC beads synthesis is described in our previous study [15]. Briefly, Fe_3O_4-NPs were synthesize by reverse coprecipitation method [16]. In the synthesis, 15 mL of 1.0 M $FeCl_3·6H_2O$ (Alfa Aesar, Madrid, Spain) and 0.5 M $FeSO_4·7H_2O$ (Sigma, St. Louis, MO, USA) were mixed and added dropwise into a 3.5 M NH_4OH solution of 20 mL at 60 °C. The reaction proceeded for 30 min under mechanical agitation. The magnetic nanoparticles were then washed and re-dispersed in distilled water.

Beads were prepared in cross-linking solution using calcium chloride solutions as the cross-linking agent. Next, 2.0 g of sodium alginate (Sigma, St. Louis, MO, USA) was subsequently added to the previously prepared Fe_3O_4-NPs solution (35 mL). After obtaining a homogeneous solution, 3.0 g of commercial activated carbon was added and the solution was mechanically agitated for 4 h. The obtained suspension was added dropwise into a previously prepared bath of 0.13 M $CaCl_2$ (Sigma, St. Louis, MO, USA) and 450 μL Tween 20 (Fluka, Steinheim, Germany) under continuous magnetic speed of 450 rpm using a New Era NE-300 syringe pump (Biogen, Madrid, Spain). Beads were instantly formed and were left in the bath around 30 min for hardening. Afterwards, the beads were collected with a magnet and cleaned with distilled water. Finally, the magnetic beads were dried at 60 °C.

4.2. Effect of pH

The effect of pH on adsorption capacity of Cd(II), Hg(II) and Ni(II) metal ions was conducted individually by mixing (14 mg) of adsorbent with 20 mL of 10 mg/L^{-1} metal ions concentration. The adsorption capacity was studied at pH values of (2.0, 3.0, 4.5, 5.0, 6.5, 7.0, 9.0) under magnetic agitation (300 rpm) at ambient temperature for 6 h. The pH values were adjusted with 0.01 M HCl and 0.05 M NaOH using a Milwaukee pH51 waterproof (Aldo, Madrid, Spain). After that, the metal concentration present in the supernatant was determined.

4.3. Adsorption Study

Parameters such as adsorbent dosage, rotation speed and agitation type were already optimized in a previous study [16]. The effect of the initial metal concentration was studied with the following conditions (magnetic agitation of 300 rpm, 14 mg adsorbent dosage and contact time of 6 h). A set of solutions with a volume of 15 mL and varying metal concentration between 10 to 250 mg/L were prepared. The pH value was adjusted for each metal with 0.05 M HCL and 0.02 M NaCl using a pH meter Milwaukee pH51 waterproof (Aldo, Madrid, Spain). For the ternary adsorption system, solutions with equal concentrations (mg/L) of Cd(II), Ni(II) and Hg(II) were prepared with the same above experimental conditions used.

4.4. Desorption and Reusability

For the desorption experiments, hydrochloric acid as desorbent solution was used. First, the adsorption experiment was carried out by preparing 15 replicates of 20 mL of 10 mg/L solution of Cd(II), using (14 mg) of adsorbent. The solutions were magnetically agitated (300 rpm) for 1 h at ambient temperature. Before performing desorption, the beads were magnetically separated from solution and washed with distilled water. Subsequently, the beads were added to the eluting solution. The desorption was carried out by mixing the beads with 20 mL of 0.01 M HCl solution for 30 min under ultrasonic agitation. In the end of each adsorption and desorption cycles, the concentration of Cd(II) on the supernatant was measured. The beads surface and internal structure was analyzed by energy dispersive x-ray spectroscopy (EDX) mapping on each cycle for adsorption and desorption.

5. Characterization

The XRD measurements were performed using a Philips diffractometer (Panalytical, Callo End, UK) with Cu Kα radiation (λ = 1.5406 Å), with a step size of 0.02° and a counting time of 2 s per step from 10° to 80° (2θ). The Fourier transform infrared spectra (FT-IR) was recorded on a Varian FT-IR 670 (Varian, Palo Alto, CA, USA) spectrophotometer in the range 400–4000 cm^{-1}. The morphology of the activated carbon and hybrid magnetic beads were characterized by TEM using a JEOL JEM-1011 microscope (JEOL, Tokyo, Japan) operating at 100 kV and by SEM analysis using a ZEISS FE-SEM ULTRA Plus microscope operated at 30 kV (Zeiss, Oberkochen, Germany). The BET specific surface area and the pore size distribution of the samples were characterized under N$_2$ adsorption-desorption isotherms at 77 K using Micromeritics ASAP 2020 instrument (Micromeritics, Norcross, GA, USA). The TGA were studied using Perkin Elmer Pyris 7 (Perkin, Waltham, MA, USA) under an oxygen flow (20 mL/min) with heating rate of 50 °C up to 840 °C. A superconducting quantum interference device (SQUID) magnetometer (Quantum Design, Darmstadt, Germany) was used to analyze the magnetic properties of the MAAC beads. The FC and ZFC measurements were recorded at an applied field of 100 Oe by scanning between 5 and 350 K. Magnetic properties were assessed by measuring the magnetization curve using a vibrating sample magnetometer (VSM) (DMS, Massachusetts, MA, USA) with an applied field between −10 and 10 kOe at room temperature. The concentration of Cd(II), Hg(II) and Ni(II) ions was measured by inductively coupled plasma optical mission spectrophotometry (ICP-OES) using an emission spectrometer Perkin Elmer Model Optima 3300 DV (Perkin, Waltham, MA, USA).

6. Conclusions

Hybrid magnetic beads made by encapsulation of magnetite nanoparticles with sodium alginate and commercial activated carbon proved to be effective on the adsorption of Cd(II), Hg(II) and Ni(II) metal ions. The MAAC beads present a superparamagnetic behavior, although with a moderate saturation magnetization around 50 emu/g, their efficient magnetic separation from solution was allowed.

Their magnetic functionality is a crucial design parameter for industrial applications in which remote manipulation and extraction enhances the decontamination procedure. The quantitative studies on the equilibrium adsorption in the ternary metal system revealed a competitive adsorption process established between the different metal ions. The adsorption affinity, Cd(II) > Ni(II) > Hg(II), was clearly established in both mono and ternary system. The equilibrium for both systems was generally better described by the Freundlich model, indicating an heterogenous adsorption process. Cadmium metal ions were the most adsorbed by the MAAC beads, and consequently, it was the metal studied for the desorption experiment. The desorption rate was significant using hydrochloric acid and the beads tolerate well the desorption process without structure damage. These results support the potential application of the MAAC beads for the removal and recovery of metal ions from contaminated aqueous medium.

Supplementary Materials: The following are available online at http://www.mdpi.com/2304-6740/8/6/40/s1, Figure S1. X-ray diffractogram (a) and Transmission electron microscopy (TEM) (b) of Fe_3O_4 nanoparticles. Figure S2. Cadmium weight percentage obtained by EDX analysis of the surface and internal structure of the beads on the adsorption and desorption cycles. Figure S3. Adsorption capacity q (mol/g) at the mono-system (a) and ternary-system (b).

Author Contributions: Conceptualization, L.d.C.A. and S.Y.-V.; Methodology, L.d.C.A.; Investigation, L.d.C.A. and S.Y.-V.; Writing—original draft preparation, L.d.C.A.; Writing—review and editing, S.Y.-V., Y.P.-R., and J.R.; Supervision, Y.P.-R.; Project administration, J.R.; Funding acquisition, Y.P.-R. and J.R. All authors have read and agreed to the published version of the manuscript.

Funding: This research was supported by EP-INTERREG V A (POCTEP) Funds (project NANOEATERS/1378) and by the Consellería de Educación Program for Reference Research Groups project (GPC2017/015 and the Development of Strategic Grouping in Materials—AEMAT at the University of Santiago de Compostela under Grant No. ED431E2018/08, of the Xunta de Galicia.

Acknowledgments: The author acknowledges the technical support of the staff at the Ceramic Institute and the research group, from the Department of Analytical Chemistry, Nutrition and Bromatology (Faculty of Chemistry) from the University of Santiago de Compostela.

Conflicts of Interest: The authors declare no conflict of interest.

References

1. Chen, Q.Y.; Desmarais, T.; Costa, M. Metals and Mechanisms of Carcinogenesis. *Annu. Rev. Pharmacol. Toxicol.* **2019**, *59*, 537–554. [CrossRef]
2. Duruibe, J.O.; Ogwuegbu, M.O.C.; Egwurugwu, J.N. Heavy metal pollution and human biotoxic effects. *Int. J. Phys. Sci.* **2007**, *2*, 112–118.
3. Bryan, G.W. The effects of heavy metals (other than mercury) on marine and estuarine organisms. *Proc. R. Soc. Lond. Ser. B Biol. Sci.* **1971**, *177*, 389–410. [CrossRef]
4. Vilar, V.J.P.; Botelho, C.; Boaventura, R.A. Influence of pH, ionic strength and temperature on lead biosorption by Gelidium and agar extraction algal waste. *Process Biochem.* **2005**, *40*, 3267–3275. [CrossRef]
5. Lagoa, R.; Rodrigues, J.R. Evaluation of Dry Protonated Calcium Alginate Beads for Biosorption Applications and Studies of Lead Uptake. *Appl. Biochem. Biotechnol.* **2007**, *143*, 115–128. [CrossRef] [PubMed]
6. Arica, M.Y.; Bayramoğlu, G.; Yilmaz, M.; Bektaş, S.; Genç, Ö. Biosorption of Hg^{2+}, Cd^{2+}, and Zn^{2+} by Ca-alginate and immobilized wood-rotting fungus Funalia trogii. *J. Hazard. Mater.* **2004**, *109*, 191–199. [PubMed]
7. Bayramoğlu, G.; Tüzün, I.; Celik, G.; Yilmaz, M.; Arica, M.Y. Biosorption of mercury(II), cadmium(II) and lead(II) ions from aqueous system by microalgae Chlamydomonas reinhardtii immobilized in alginate beads. *Int. J. Miner. Process.* **2006**, *81*, 35–43. [CrossRef]
8. Silva, R.M.P.; Manso, J.P.H.; Rodrigues, J.R.; Lagoa, R. A comparative study of alginate beads and an ion-exchange resin for the removal of heavy metals from a metal plating effluent. *J. Environ. Sci. Health Part A* **2008**, *43*, 1311–1317. [CrossRef] [PubMed]
9. Venault, A.; Vachoud, L.; Pochat, C.; Bouyer, D.; Faur, C. Elaboration of chitosan/activated carbon composites for the removal of organic micropollutants from waters. *Environ. Technol.* **2008**, *29*, 1285–1296. [CrossRef] [PubMed]
10. Chang, Y.-C.; Chen, D.-H. Preparation and adsorption properties of monodisperse chitosan-bound Fe_3O_4 magnetic nanoparticles for removal of Cu(II) ions. *J. Colloid Interface Sci.* **2005**, *283*, 446–451. [CrossRef] [PubMed]

11. Escudero, C.; Fiol, N.; Villaescusa, I.; Bollinger, J.-C. Arsenic removal by a waste metal (hydr)oxide entrapped into calcium alginate beads. *J. Hazard. Mater.* **2009**, *164*, 533–541. [CrossRef] [PubMed]
12. Vreeker, R.; Li, L.; Fang, Y.; Appelqvist, I.; Mendes, E. Drying and Rehydration of Calcium Alginate Gels. *Food Biophys.* **2008**, *3*, 361–369. [CrossRef]
13. Ahmadpoor, F.; Shojaosadati, S.A.; Mousavi, S.Z. Magnetic silica coated iron carbide/alginate beads: Synthesis and application for adsorption of Cu(II) from aqueous solutions. *Int. J. Biol. Macromol.* **2019**, *128*, 941–947. [CrossRef] [PubMed]
14. Bée, A.; Talbot, D.; Abramson, S.; Dupuis, V. Magnetic alginate beads for Pb(II) ions removal from wastewater. *J. Colloid Interface Sci.* **2011**, *362*, 486–492. [CrossRef] [PubMed]
15. de Alves, L.C.; Yáñez-Vilar, S.; Piñeiro, Y.; Rivas, J. Novel Magnetic Nanostructured Beads for Cadmium(II) Removal. *Nanomaterials* **2019**, *9*, 356. [CrossRef] [PubMed]
16. Ben Hammouda, S.; Adhoum, N.; Monser, L. Synthesis of magnetic alginate beads based on magnetite nanoparticles for the removal of 3-methylindole from aqueous solution using Fenton process. *J. Hazard. Mater.* **2015**, *294*, 128–136. [CrossRef] [PubMed]
17. Xuan, S.; Hao, L.; Jiang, W.; Gong, X.; Hu, Y.; Chen, Z. Preparation of water-soluble magnetite nanocrystals through hydrothermal approach. *J. Magn. Magn. Mater.* **2007**, *308*, 210–213. [CrossRef]
18. Khoo, K.-M.; Ting, Y.-P. Biosorption of gold by immobilized fungal biomass. *Biochem. Eng. J.* **2001**, *8*, 51–59. [CrossRef]
19. Tripathi, A.; Meload, J.S.; D'Souza, S.F. Uranium (VI) recovery from aqueous medium using novel floating macroporous alginate-agarose-magnetite cryobeads. *J. Hazard. Mater.* **2013**, *246*, 87–95. [CrossRef]
20. Waller, P.A.; Pickering, W.F. Chemical Speciation & Bioavailability The effect of pH on the lability of lead and cadmium sorbed on humic acid particles The effect of pH on the lability of lead and cadmium sorbed on humic acid particles. *Chem. Speciat. Bioavailab.* **1993**, *5*, 11–22. [CrossRef]
21. Budimirovic, D.; Bajić, Z.; Velickovic, Z.; Milosevic, A.; Nikolic, J.B.; Drmanic, S.; Marinkovic, A. Removal of heavy metals from water using multistage functionalized multiwall carbon nanotubes. *J. Serbian Chem. Soc.* **2017**, *82*, 1175–1191. [CrossRef]
22. Zouboulis, A.; Kydros, K.; Zouboulis, A.I.; Kydros, K.A. Use of red mud for toxic metals removal: The case of nickel. *Artic. J. Chem. Technol. Biotechnol.* **2007**, *58*, 95–101. [CrossRef]
23. Cataldo, S.; Gianguzza, A.; Pettignano, A.; Villaescusa, I. Mercury(II) removal from aqueous solution by sorption onto alginate, pectate and polygalacturonate calcium gel beads. A kinetic and speciation based equilibrium study. *React. Funct. Polym.* **2013**, *73*, 207–217. [CrossRef]
24. Gong, J.L.; Wang, X.-Y.; Niu, C.; Chen, L.; Deng, J.-H.; Zhang, X.-R.; Niu, Q.-Y. Copper(II) removal by pectin–iron oxide magnetic nanocomposite adsorbent. *Chem. Eng. J.* **2012**, *185–186*, 100–107. [CrossRef]
25. Jeon, C.; Park, J.Y.; Yoo, Y.J. Novel immobilization of alginic acid for heavy metal removal. *Biochem. Eng. J.* **2002**, *11*, 159–166. [CrossRef]
26. Liu, X.; Xu, X.; Dong, X.; Park, J. Competitive Adsorption of Heavy Metal Ionsfrom Aqueous Solutions onto Activated Carbonand Agricultural Waste Materials. *Pol. J. Environ. Stud.* **2019**, *29*, 749–761. [CrossRef]
27. Mahamadi, C.; Nharingo, T. Competitive adsorption of Pb^{2+}, Cd^{2+} and Zn^{2+} ions onto Eichhornia crassipes in binary and ternary systems. *Bioresour. Technol.* **2010**, *101*, 859–864. [CrossRef]
28. Foo, K.Y.; Hameed, B. Insights into the modeling of adsorption isotherm systems. *Chem. Eng. J.* **2010**, *156*, 2–10. [CrossRef]
29. Febrianto, J.; Kosasih, A.N.; Sunarso, J.; Ju, Y.-H.; Indraswati, N.; Ismadji, S. Equilibrium and kinetic studies in adsorption of heavy metals using biosorbent: A summary of recent studies. *J. Hazard. Mater.* **2009**, *162*, 616–645. [CrossRef]
30. Bohli, T.; Ouederni, A.; Villaescusa, I. Simultaneous adsorption behavior of heavy metals onto microporous olive stones activated carbon: Analysis of metal interactions. *Euro-Mediterr. J. Environ. Integr.* **2017**, *2*, 19. [CrossRef]

 © 2020 by the authors. Licensee MDPI, Basel, Switzerland. This article is an open access article distributed under the terms and conditions of the Creative Commons Attribution (CC BY) license (http://creativecommons.org/licenses/by/4.0/).

MDPI
St. Alban-Anlage 66
4052 Basel
Switzerland
Tel. +41 61 683 77 34
Fax +41 61 302 89 18
www.mdpi.com

Inorganics Editorial Office
E-mail: inorganics@mdpi.com
www.mdpi.com/journal/inorganics